Practical Networking

Series Editor

Zhi-Li Zhang
Sci & Eng, 4-192 EE/CS Bldg
University of Minnesota, Dept of Comp
Minneapolis, MN, USA

Explosive growth in cloud and mobile computing coupled with new advances in systems and networking technologies as well as machine learning and artificial intelligence (AI) have revolutionized how networks and distributed systems are designed, developed, operated and managed. This is epitomized by data center networking where it has spurred a wholesale rethinking and re-designs from network architectures, to physical interconnects to routing, flow management and network application support. New networking paradigms and technologies such as software-defined networking, network function virtualization, smart NICs and software/hardware co-designs have emerged for better designing, operating, managing and evolving networks, and also enabled new visions such as "self-driving networks" and AIOps. The Practical Networking Series is centered on emerging topics in new networking paradigms, architectural designs, algorithms and mechanisms for primarily wired networks (from data center networks, enterprise networks to ISP networks), but also touches on "packet core networks" for emerging 5G and beyond cellular and wireless networks. Books in this series address these topics from both theoretical (e.g., new theoretical foundations, algorithms and performance analysis) and practical (e.g., new network mechanisms, protocols, APIs and standards, software frameworks) perspectives. Relatively short books on a timely and focused topic, research monographs, and textbooks are of interest. The Editor is seeking well written works by well-established researchers and practitioners in the networking field around the world, particularly Asia and North America.

Prospective Authors or Editors:

If you have an idea for a book, we would welcome the opportunity to review your proposal. Should you wish to discuss any potential project further or receive specific information regarding our book proposal requirements, please contact Zhi-Li Zhang or Susan Evans:

Zhi-Li Zhang
Department of Computer Science Department
University of Minnesota
4-192 Keller Hall, 200 Union Street SE
Minneapolis, MN 55455-0159
zhzhang@cs.umn.edu

Susan Evans
Senior Editor
Springer Nature
233 Springe Street
New York, NY 10013 USA
susan.evans@springernature.com

More information about this series at https://link.springer.com/bookseries/16325

Jorge Crichigno • Elie Kfoury • Elias Bou-Harb
Nasir Ghani

High-Speed Networks

A Tutorial

 Springer

Jorge Crichigno
College of Engineering and Computing
University of South Carolina
Columbia, SC, USA

Elie Kfoury
College of Engineering and Computing
University of South Carolina
Columbia, SC, USA

Elias Bou-Harb
Cyber Center for Security and Analytics
University of Texas at San Antonio
San Antonio, TX, USA

Nasir Ghani
Electrical Engineering
University of South Florida
Tampa, FL, USA

ISSN 2662-1703 ISSN 2662-1711 (electronic)
Practical Networking
ISBN 978-3-030-88843-5 ISBN 978-3-030-88841-1 (eBook)
https://doi.org/10.1007/978-3-030-88841-1

This Springer imprint is published by the registered company Springer Nature Switzerland AG
The registered company address is: Gewerbestrasse 11, 6330 Cham, Switzerland

Preface

Welcome to *High-Speed Networks*: A Tutorial. This book is the result of a journey of the authors who have been designing and deploying high-speed networks for several years, in particular Science Demilitarized Zones (Science DMZs). The Science DMZ is a high-speed network designed to facilitate the transfer of big science data.

As the popularity of high-speed networks and Science DMZs surged, the need for professionals with the skills to operate such infrastructures has increased. However, practitioners have been mostly trained to operate general-purpose networks, which have different characteristics from those of Science DMZs and other high-performance networks. At the time when the authors started designing and operating Science DMZs, the available material was limited to workshops organized by ESnet, the Scientific Networking Division at Lawrence Berkeley National Laboratory in the United States (U.S.).

This book tries to address the above gap. It provides practical knowledge and skills on Science DMZs and high-speed networks in general, which are reinforced with virtual laboratory experiments.

Audience

This book is for industry professionals and for students in computer science, information technology, and similar programs, who are interested in learning fundamental concepts related to high-speed networks and corresponding implementations. The book assumes minimal familiarity with networking, typically covered in an introductory networking course. It is appropriate for an upper-level undergraduate course, for a first-year graduate course, and for self-pace learning by industry professionals.

What is Unique About This Book?

The book delves into protocols and devices at different layers, from the physical infrastructure to application-layer tools and security appliances, that must be carefully considered for the optimal operation of Science DMZs and high-speed networks. In contrast to traditional books, the book is accompanied by hands-on virtual laboratory experiments that are conducted on a virtual platform.

The Virtual Platform and Virtual Laboratory Experiments

The virtual platform enables learners to immediately deploy virtual networks composed of an equipment pod (routers, switches, servers, firewalls, etc.) needed for mastering a topic. Experiments help learners to reinforce concepts and to learn how to optimally configure and manage network devices, based on real measurements and observations. Access to the platform is available for a fee and includes all material required to conduct the experiments. The URL of the virtual platform is: http://highspeednetworks.net/

Organization

The book follows a bottom-up approach. Chapter "Introduction to High-Speed Networks and Science DMZ" presents the motivation for Science DMZs and high-speed networks. Chapter "Network Cyberinfrastructure Aspects for Big Data Transfers" describes limitations of general-purpose networks when transferring large data sets across a Wide Area Network (WAN), and explores the cyberinfrastructure required to support such transfers. It also discusses different options a network may have to connect to other networks. Chapter "Data-Link and Network Layer Considerations for Large Data Transfers" describes attributes related to routers and switches, which have large impact on performance, including router's buffer size, maximum transmission unit (MTU), and others. Chapter "Impact of TCP on High-Speed Networks and Advances in Congestion Control Algorithms" discusses key features at the transport layer, such as TCP congestion control, pacing, and parallel connections. Chapter "Application and Security Aspects for Large Flows" presents application-layer tools used to support large data transfers. Chapter "Security Aspects" describes security challenges arising in Science DMZs and high-speed networks, and presents best practices. Chapter "Challenges and Open Research Issues" discusses challenges and open research issues.

Relevance of Networking Tools

The book provides a set of virtual laboratory experiments at the end of most chapters. All equipment pods are implemented with appliances running real protocol stacks. Examples include iPerf3, the Network Emulator (NETEM), traffic control (tc), and Zeek intrusion detection system. Recognizing the impact of Mininet on networking, the authors decided to use this network emulator to create topologies for the laboratory experiments. All tools are based on open-source software, which reflects industry trends. Specifically, over the years, the authors observed that open-source software has been increasingly used to design, build, test, and control networks. For example, since the original publication of the paper describing Mininet in 2010, Mininet has gained wide adoption in the industry and academia. Mininet's paper has received the ACM Test of Time Award.

Columbia, SC, USA   Jorge Crichigno
Columbia, SC, USA   Elie Kfoury
San Antonio, TX, USA   Elias Bou-Harb
Tampa, FL, USA   Nasir Ghani

Acknowledgement

The authors would like to express their gratitude to the U.S. National Science Foundation (NSF), Office of Advanced Cyberinfrastructure (OAC). This work would not be possible without NSF support. Part of the material was developed under the award numbers 1829698 and 1925484. The authors are also thankful to the Network Development Group (NDG) team who worked with Dr. Crichigno to deploy the virtual laboratory experiments. NDG's president, Richard Weeks, has constantly provided invaluable suggestions.

The first two authors would like to specially thank the members of the Cyberinfrastructure Laboratory at the University of South Carolina (USC) who helped create and test laboratory experiments, and the Department of Integrated Information Technology at USC for the conducive environment for writing this book.

Contents

Abbreviations

ACK	Acknowledgement
ACL	Access-Control List
AES	Advanced Encryption Standard
AMO	Atomic, Molecular, and Optical
BBR	Bottleneck Bandwidth and Round-Trip Time
BDP	Bandwidth-Delay Product
BGP	Border Gateway Protocol
BNL	Brookhaven National Laboratory (United States)
btlbw	Bottleneck Bandwidth
BWCTL	Bandwidth Test Controller
BYOD	Bring-Your-On-Device
CC*	Campus Cyberinfrastructure Program
CDF	Cumulative Distribution Function
CENIC	Corporation for Education Network Initiatives in California
CPI	Client Protocol Interpreter
CPU	Central Processing Unit
DMZ	Demilitarized Zone
DoS	Denial of Service
DTN	Data Transfer Node
DTP	Data Transfer Process
ESnet	Energy Science Network
FCC	Federal Communications Commission (United States)
FDT	Fast Data Transfer
FIB	Forwarding Information Base
FIC	File Integrity Check
FQ	Fair Queue
FT	Forwarding Table
FTP	File Transfer Protocol
GB	Gigabyte
Gbps	Gigabits Per Second
HOL	Head-Of-Line

HTCP	Hamilton Transmission Control Protocol
HTML	Hypertext Markup Language
HTTP	Hypertext Transfer Protocol
HTTPS	Hypertext Transfer Protocol Secure
IDS	Intrusion Detection System
IETF	Internet Engineering Task Force
IP	Internet Protocol
IPFIX	IP Flow Information Export
IPS	Intrusion Prevention System
IPsec	Internet Protocol Security
ISP	Internet Service Provider
KB	Kilobyte
Kbps	Kilobits Per Second
LAN	Local Area Network
LBNL	Lawrence Berkeley National Laboratory (United States)
LHC	Large Hadron Collider
MB	Megabyte
Mbps	Megabits Per Second
mdtmFTP	Multicore-Aware Data Transfer
Middleware	File Transfer Protocol
MPLS	Multi-Protocol Label Switching
MSS	Maximum Segment Size
MTU	Maximum Transmission Unit
NAT	Network Address Translator
netem	Network Emulator
NGIPS	Next Generation Intrusion Prevention System
NIC	Network Interface Card
NNMC	Northern New Mexico College
NP	Network Processor
NPAD	Network Path and Application Diagnostics
NREN	National Research and Education Network
NSF	National Science Foundation
NSFnet	National Science Foundation Network
NUMA	Non-Uniform Memory Access
OS	Operating System
OSCARS	On-Demand Secure Circuits and Reservation System
OSPF	Open Shortest Path First
OWAMP	One-Way Active Measurement Protocol
PB	Petabyte
POP	Point of Presence
RAM	Random Access Memory
RDMA	Remote Direct Memory Access
REN	Research and Education Network
RTT	Round-Trip Time
RTT_{min}	Minimum Round-Trip Time

SACK	Selective Acknowledgement
SCP	Secure Copy Protocol
SDMZ	Science Demilitarized Zone
SDN	Software Defined Networking
sFlow	Sampled Flow
SFTP	Secure File Transfer Protocol
SNMP	Simple Network Management Protocol
SPAN	Switched Port Analyzer
SPI	Server Protocol Interpreter
SQL	Structured Query Language
SSH	Secure Shell
TB	Terabyte
Tbps	Terabits Per Second
TCP	Transmission Control Protocol
UDP	User Datagram Protocol
UDT	UDP-based Data Transfer Protocol
UMA	Uniform Memory Access
UNM	University of New Mexico
uRPF	Unicast Reverse Path Forwarding
U.S.	United States
VLA	Very Large Array
VLAN	Virtual Local Area Network
VM	Virtual Machine
VOQ	Virtual Output Queueing
VPLS	Virtual Private LAN Service
VPN	Virtual Private Network
VXLAN	Virtual Extensible Local Area Network
WAN	Wide Area Network
WRN	Western Regional Network
XML	Extensible Markup Language

Introduction to High-Speed Networks and Science DMZ

This chapter provides a motivation for Science Demilitarized Zones (Science DMZs) and other high-speed network architectures designed for large data transfers. The chapter describes limitation of enterprise networks when used for large data transfers, current applications based on Science DMZs, and access to companion material and website.

1 Objective and Access to Accompanied Training Material

At present, there is an increasing need to deploy Science DMZs in support of big science data transfers. However, efforts to prepare researchers and other professionals with the right knowledge are limited to dispersed work by the academia and the industry. Despite the importance of Science DMZs, currently there is no structured material in the form of a book.

This book addresses this gap in the literature by presenting a comprehensive tutorial high-speed networks, while focusing on the Science DMZs architecture. Following a systematic approach through every layer of the protocol stack, this book integrates information and tools for a better understanding of the issues, key challenges, best practices, and future research directions related to Science DMZs.

The book is accompanied by hands-on virtual laboratory experiments conducted in a virtual platform, referred to as the Academic Cloud. Access to the Academic Cloud is available for a fee (six-month access) and includes all materials needed to conduct the experiments. The URL is:

http://highspeednetworks.net/

The virtual platform is accessible via a regular browser, as any website. The platform enables readers to complete hands-on laboratory exercises by creating

J. Crichigno et al., *High-Speed Networks*, Practical Networking,
https://doi.org/10.1007/978-3-030-88841-1_1

Fig. 1 Screen-capture of the website used as a companion platform for the book

virtual equipment pods. A pod is a set of virtual machines needed for the completion of a lab exercise. The pod can be as simple as a single isolated virtual machine, or as complex as autonomous systems with live traffic flowing to/from the Internet. Figure 1 shows a screen-capture of the website.

2   Motivation for Science DMZs

When the United States (U.S.) decided to build the interstate highway system in the 1950s, the country already had city streets and two-lane highways for daily-life transportation. While at first this system appeared to be redundant, the interstate highway system increased the ease of travel for Americans and the ability to transport goods from east to west, without stoplights [1].

Tracing similarities with the current cyberinfrastructure, today's general-purpose networks, also referred to as enterprise networks, are capable of efficiently transporting basic data. These networks support multiple missions, including organizations' operational services such as email, procurement systems, and web browsing. However, when transferring terabyte- and petabyte-scale science data, enterprise networks face many unsolved challenges [2]. Key issues preventing high throughput include slow processing by CPU-intensive security appliances, inability of routers and switches to absorb traffic bursts generated by large flows, end devices that are incapable of sending and receiving data at high rates, lack of data transfer applications that can exploit the available network bandwidth, and the absence of end-to-end path monitoring to detect failures.

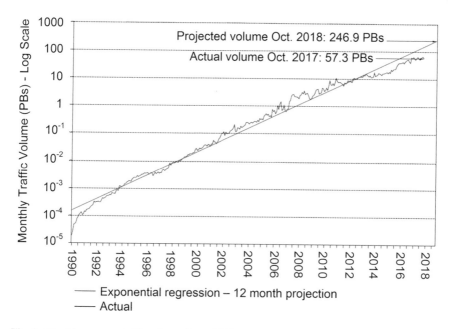

Fig. 2 Monthly average traffic volume through ESnet [3]

The need for a suitable cyberinfrastructure for large flows is illustrated in Fig. 2, which shows the monthly average traffic volume through the Energy Science network (ESnet) [3]. ESnet is a high-performance network that carries science traffic for the U.S. Department of Energy. As of 2018, this network is transporting tens of petabytes (PBs) per month, an increase of several orders of magnitude from some years ago.

In response to this challenge of transmitting big science data via a cyber-highway system without stoplights, ESnet developed the concept of Science Demilitarized Zone (Science DMZ or SDMZ) [4]. The Science DMZ is a network or a portion of a network designed to facilitate the transfer of big science data across wide area networks (WANs), typically at rates of 10 Gbps and above. In order to operate at such rates, this setup integrates the following key elements: (i) end devices, referred to as data transfer nodes (DTNs), that are built for sending/receiving data at a high rate over WANs; (ii) high-throughput paths connecting DTNs, instruments, storage devices, and computing systems. These paths are composed of highly capable routers and switches and have no devices that may induce packet losses. They are referred to as friction-free paths; (iii) performance measurement devices that monitor end-to-end paths over multiple domains; and (iv) security policies and enforcement mechanisms tailored for high-performance science environments.

3 Science DMZs Applications

The Science DMZ architecture is similar to building the interstate highway system, whereas stoplights are removed to permit the high-speed movement of large flows. The interconnection of Science DMZs is also analogous to the development of the National Science Foundation network (NSFnet) in 1985, one of the predecessors of today's Internet. NSF, the main government agency in the U.S. supporting research and education in science and engineering, established the NSFnet to link together five supercomputer centers that were then deployed across the U.S. [5]. With Science DMZs, institutions are similarly linked together and have access to a virtual co-location of data that may rest anywhere in the world through a high-speed data-sharing architecture. Along these lines, Fig. 3 highlights applications that currently exploit the Science DMZ architecture to transmit large flows from instruments to laboratories for data analysis. From very large to portable devices, these instruments generate a large amount of data in short periods of time.

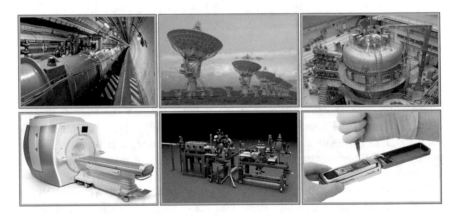

Fig. 3 Science DMZ data transfer applications. Top left: The Large Hadron Collider (LHC) produces approximately 30 PBs per year. Data is transmitted to multiple computing centers around the world. Photo courtesy of The European Organization for Nuclear Research [6]. Top center: The Very Large Array (VLA) is composed of 27 radio antennas of 25 m in diameter each. Daily data collection comprises several TBs, which are transmitted to research laboratories worldwide. Photo courtesy of the U.S. National Radio Astronomy Observatory [7]. Top right: Experimental Advanced Superconducting Tokamak. Data generated by the energy reactor is transmitted for analysis via a Science DMZ. Photo courtesy of ESnet [8]. Bottom left: magnetic resonance imaging scanner. Major brain imaging studies such as the Alzheimer's disease neuroimaging requires storage and transmission of multiple PBs of data [9]. Medical data can now be transported via medical Science DMZs [10, 11]. Photo courtesy of General Electric Healthcare [12]. Bottom center: Atomic, Molecular, and Optical (AMO) instrument. The instrument is used for a variety of experiments, such as illumination of single molecules. A single experiment can produce 150 to 200 TBs [13]. Photo courtesy of the U.S. SLAC National Accelerator Laboratory [14]. Bottom right: portable device for DNA and RNA sequencing, which generates tens of GBs of data per experiment [15]. Photo courtesy of Nanopore Technologies [16]

Table 1 Credentials to
access Client1 machine

Device	Account	Password
Client1	admin	password

Chapter 1—Lab 1: Introduction to Mininet

Overview
To conduct the experiment described in this section, please login into the Academic
Cloud at http://highspeednetworks.net/ and reserve a pod for Lab 1.

This lab provides an introduction to Mininet, a virtual testbed used for testing
network tools and protocols. It demonstrates how to invoke Mininet from the
command-line interface (CLI) utility and how to build and emulate topologies using
a graphical user interface (GUI) application.

Objectives
By the end of this lab, students should be able to:

1. Understand what Mininet is and why it is useful for testing network topologies.
2. Invoke Mininet from the CLI.
3. Construct network topologies using the GUI.
4. Save/load Mininet topologies using the GUI.

Lab Settings
The information in Table 1 provides the credentials of the machine containing
Mininet.

Lab Roadmap
This lab is organized as follows:

1. Section 4: Introduction to Mininet.
2. Section 5: Invoking Mininet using the CLI.
3. Section 6: Building and emulating a network in Mininet using the GUI.

4   Introduction to Mininet

Mininet is a virtual testbed enabling the development and testing of network tools
and protocols. With a single command, Mininet can create a realistic virtual network
on any type of machine (Virtual Machine (VM), cloud-hosted, or native) (Fig. 4).
Therefore, it provides an inexpensive solution and streamlined development running
in line with production networks. Mininet offers the following features:

- Fast prototyping for new networking protocols.
- Simplified testing for complex topologies without the need of buying expensive
  hardware.

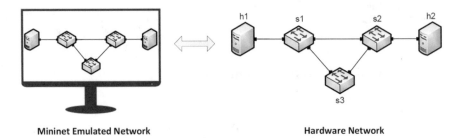

Fig. 4 Hardware network vs. Mininet emulated network

- Realistic execution as it runs real code on the Unix and Linux kernels.
- Open-source environment backed by a large community contributing extensive documentation.

Mininet is useful for development, teaching, and research as it is easy to customize and interact with it through the CLI or the GUI. Mininet was originally designed to experiment with *OpenFlow* and *Software-Defined Networking (SDN)*. This lab, however, only focuses on emulating a simple network environment without SDN-based devices.

Mininet's logical nodes can be connected into networks. These nodes are sometimes called containers, or more accurately, *network namespaces*. Containers consume sufficiently few resources that networks of over a thousand nodes have created, running on a single laptop. A Mininet container is a process (or group of processes) that no longer has access to all the host system's native network interfaces. Containers are then assigned virtual Ethernet interfaces, which are connected to other containers through a virtual switch. Mininet connects a host and a switch using a virtual Ethernet (veth) link. The veth link is analogous to a wire connecting two virtual interfaces, as illustrated below (Fig. 5).

Each container is an independent network namespace, a lightweight virtualization feature that provides individual processes with separate network interfaces, routing tables, and Address Resolution Protocol (ARP) tables.

Mininet provides network emulation opposed to simulation, allowing all network software at any layer to be simply run *as is*; i.e., nodes run the native network software of the physical machine. In a simulator environment on the other hand, applications and protocol implementations need to be ported to run within the simulator before they can be used.

5 Invoking Mininet Using the CLI

The first step to start Mininet using the CLI is to start a Linux terminal.

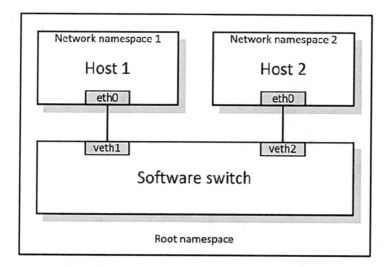

Fig. 5 Network namespaces and virtual Ethernet links

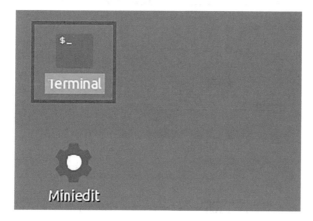

Fig. 6 Shortcut to open a Linux terminal

5.1 Invoking Mininet Using the Default Topology

Step 1 Launch a Linux terminal by holding the Ctrl+Alt+T keys or by clicking on the Linux terminal icon (Fig. 6).

The Linux terminal is a program that opens a window and permits you to interact with a command-line interface (CLI). A CLI is a program that takes commands from the keyboard and sends them to the operating system for execution.

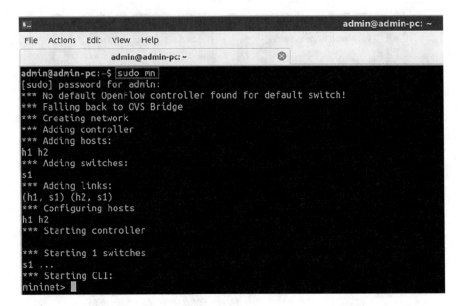

Fig. 7 Starting Mininet using the CLI

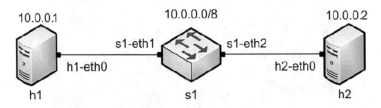

Fig. 8 Mininet's default minimal topology

Step 2 To start a minimal topology, enter the command sudo mn at the CLI. When prompted for a password, type password and hit enter (Fig. 7). Note that the password will not be visible as you type it.

The above command starts Mininet with a minimal topology, which consists of a switch connected to two hosts as shown below (Fig. 8).

When issuing the sudo mn command, Mininet initializes the topology and launches its command-line interface, which looks like this:

```
mininet>
```

Step 3 To display the list of Mininet CLI commands and examples on their usage, type the command help in the Mininet CLI (Fig. 9):

Step 4 To display the available nodes, type the command nodes : (Fig. 10).

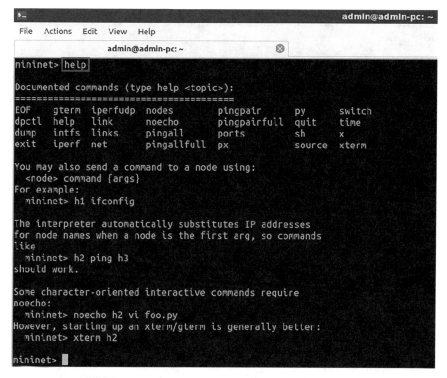

Fig. 9 Mininet's help command

Fig. 10 Mininet's nodes command

The output of this command shows that there are two hosts (host h1 and host h2) and a switch (s1).

Step 5 It is useful sometimes to display the links between the devices in Mininet to understand the topology. Issue the command net in the Mininet CLI to see the available links (Fig. 11).

The output of this command shows that:

1. Host h1 is connected using its network interface *h1-eth0* to the switch on interface *s1-eth1*

Fig. 11 Mininet's net command

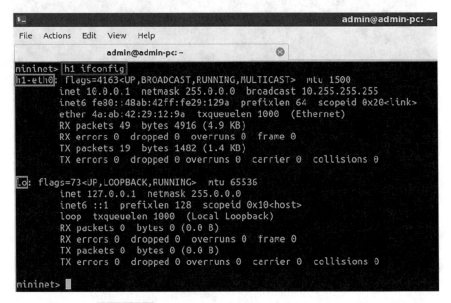

Fig. 12 Output of h1 ifconfig command

2. Host h2 is connected using its network interface *h2-eth0* to the switch on interface *s1-eth2*
3. Switch s1:

   (a) has a loopback interface *lo*
   (b) connects to *h1-eth0* through interface *s1-eth1*
   (c) connects to *h2-eth0* through interface *s1-eth2*

Mininet allows you to execute commands at a specific device. To issue a command for a specific node, you must specify the device first, followed by the command.

Step 6 Issue the command h1 ifconfig (Fig. 12).

This command executes the ⎸ifconfig⎸ Linux command on host h1. The command shows host h1's interfaces. The display indicates that host h1 has an interface *h1-eth0* configured with IP address 10.0.0.1, and another interface configured with IP address 127.0.0.1 (loopback interface).

5.2 Testing Connectivity

Mininet's default topology assigns the IP addresses 10.0.0.1/8 and 10.0.0.2/8 to host h1 and host h2, respectively. To test connectivity between them, you can use the command ⎸ping⎸. The ⎸ping⎸ command operates by sending Internet Control Message Protocol (ICMP) Echo Request messages to the remote computer and waiting for a response. Information available includes how many responses are returned and how long it takes for them to return.

Step 1 On the CLI, type ⎸h1 ping 10.0.0.2⎸. This command tests the connectivity between host h1 and host h2. To stop the test, press ⎸Ctrl+c⎸. The figure below shows a successful connectivity test. Host h1 (10.0.0.1) sends four packets to host h2 (10.0.0.2) and successfully received the expected responses (Fig. 13).

Step 2 Stop the emulation by typing ⎸exit⎸ (Fig. 14).

The command ⎸sudo mn -c⎸ is often used on the Linux terminal (not on the Mininet CLI) to clean a previous instance of Mininet (e.g., after a crash).

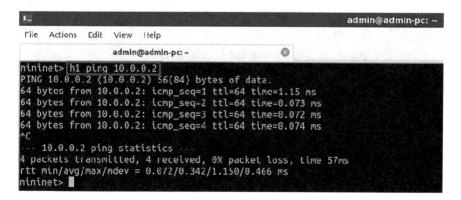

Fig. 13 Connectivity test between host h1 and host h2

Fig. 14 Stopping the emulation using exit

6 Building and Emulating a Network in Mininet Using the GUI

In this section, you will use the application MiniEdit to deploy the topology illustrated below (Fig. 15). MiniEdit is a simple GUI network editor for Mininet.

6.1 Building the Network Topology

Step 1 A shortcut to MiniEdit is located on the machine's Desktop. Start MiniEdit by clicking on MiniEdit's shortcut (Fig. 16). When prompted for a password, type password .

MiniEdit will start, as illustrated below (Fig. 17).

The main buttons are:

1. *Select*: allows selection/movement of the devices. Pressing *Del* on the keyboard after selecting the device removes it from the topology.
2. *Host*: allows addition of a new host to the topology. After clicking this button, click anywhere in the blank canvas to insert a new host.
3. *Switch*: allows addition of a new switch to the topology. After clicking this button, click anywhere in the blank canvas to insert the switch.
4. *Link*: connects devices in the topology (mainly switches and hosts). After clicking this button, click on a device and drag to the second device to which the link is to be established.
5. *Run*: starts the emulation. After designing and configuring the topology, click the run button.
6. *Stop*: stops the emulation.

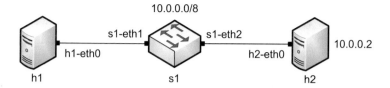

Fig. 15 Lab topology

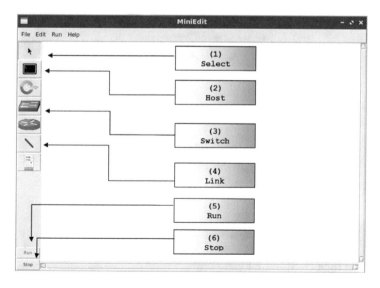

Fig. 16 MiniEdit Desktop shortcut

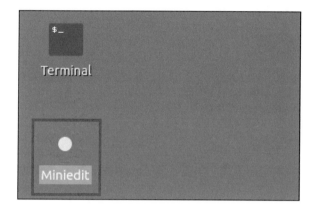

Fig. 17 MiniEdit Graphical User Interface (GUI)

Step 2 To build the topology of Fig. 15, two hosts and one switch must be deployed. Deploy these devices in MiniEdit, as shown below (Fig. 18).

Use the buttons described in the previous step to add and connect devices. The configuration of IP addresses is described in Step 3.

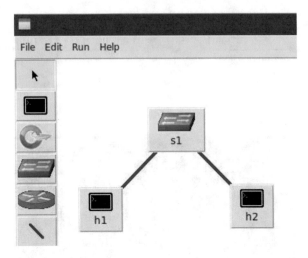

Fig. 18 MiniEdit's topology

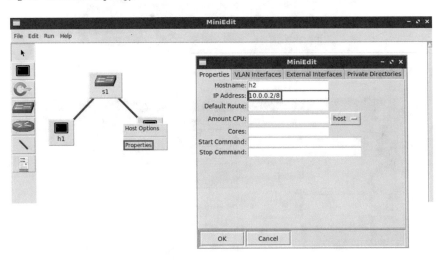

Fig. 19 Configuration of a host's properties

Step 3 Configure the IP addresses at host h1 and host h2. Host h1's IP address is 10.0.0.1/8 and host h2's IP address is 10.0.0.2/8. A host can be configured by holding the right-click and selecting properties on the device. For example, host h2 is assigned the IP address 10.0.0.2/8 in the figure below (Fig. 19).

6.2   Testing Connectivity

Before testing the connection between host h1 and host h2, the emulation must be started.

Step 1 Click on the *Run* button to start the emulation (Fig. 20). The emulation will start and the buttons of the MiniEdit panel will gray out, indicating that they are currently disabled.

Step 2 Open a terminal on host h1 by holding the right-click on host h1 and selecting *Terminal* (Fig. 21). This opens a terminal on host h1 and allows the execution of commands on the host h1. Repeat the procedure on host h2.
   The network and terminals at host h1 and host h2 will be available for testing (Fig. 22).

Step 3 On host h1's terminal, type the command │ ifconfig │ to display its assigned IP addresses (Fig. 23). The interface *h1-eth0* at host h1 should be configured with the IP address 10.0.0.1 and subnet mask 255.0.0.0.
   Repeat Step 3 on host h2. Its interface *h2-eth0* should be configured with IP address 10.0.0.2 and subnet mask 255.0.0.0.

Step 4 On host h1's terminal, type the command │ ping 10.0.0.2 │ (Fig. 24). This command tests the connectivity between host h1 and host h2. To stop the test, press │ Ctrl+c │. The figure below shows a successful connectivity test. Host h1 (10.0.0.1) sent six packets to host h2 (10.0.0.2) and successfully received the expected responses.

Step 5 Stop the emulation by clicking on the *Stop* button (Fig. 25).

Fig. 20 Starting the emulation

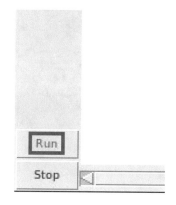

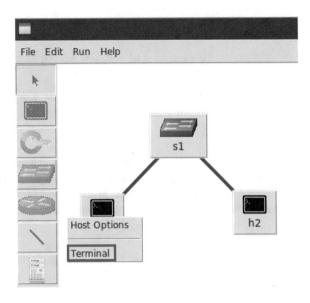

Fig. 21 Opening a terminal on host h1

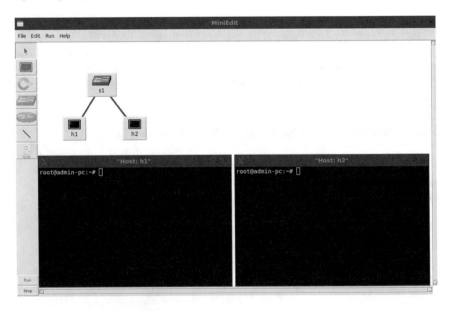

Fig. 22 Terminals at host h1 and host h2

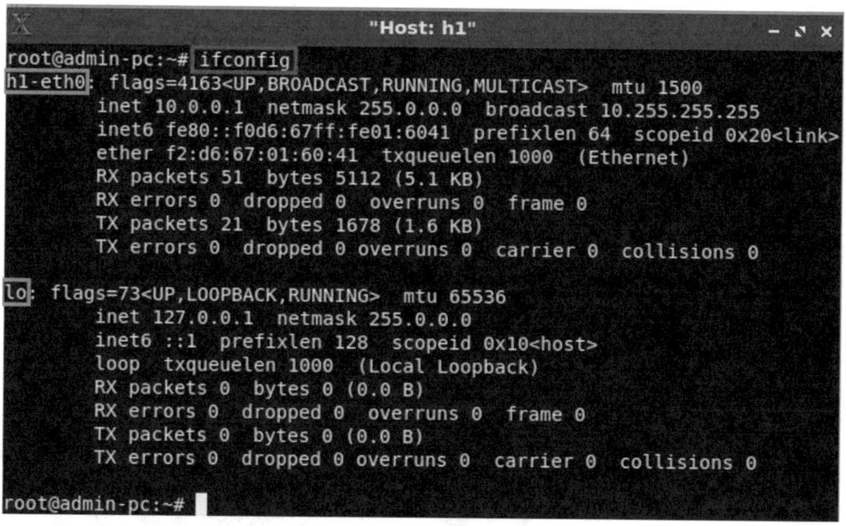

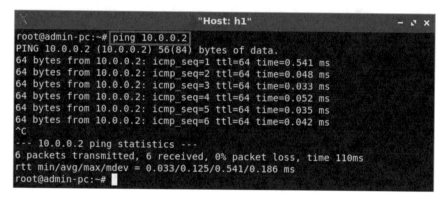

Fig. 23 Output of | ifconfig | command on host h1

Fig. 24 Connectivity test using | ping | command

6.3   Automatic Assignment of IP Addresses

In the previous section, you manually assigned IP addresses to host h1 and host h2. An alternative is to rely on Mininet for an automatic assignment of IP addresses (by default, Mininet uses automatic assignment), which is described in this section.

Step 1 Remove the manually assigned IP address from host h1. Hold right-click on host h1, *Properties* (Fig. 26). Delete the IP address, leaving it unassigned, and press the *OK* button as shown below. Repeat the procedure on host h2.

Fig. 25 Stopping the emulation

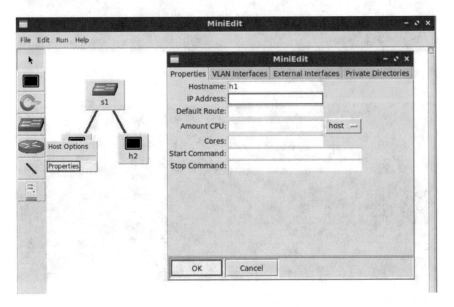

Fig. 26 Host h1 properties

Step 2 Click on *Edit*, *Preferences* button (Fig. 27). The default IP base is 10.0.0.0/8. Modify this value to 15.0.0.0/8, and then press the *OK* button.

Step 3 Run the emulation again by clicking on the *Run* button. The emulation will start and the buttons of the MiniEdit panel will be disabled.

Step 4 Open a terminal on host h1 by holding the right-click on host h1 and selecting Terminal (Fig. 28).

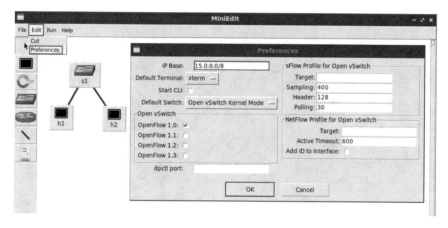

Fig. 27 Modification of the IP Base (network address and prefix length)

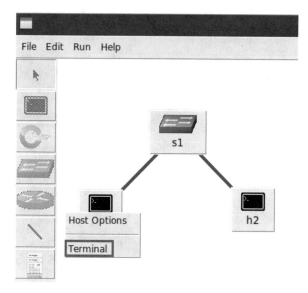

Fig. 28 Opening a terminal on host h1

Step 5 Type the command ifconfig to display the IP addresses assigned to host h1 (Fig. 29). The interface *h1-eth0* at host h1 now has the IP address 15.0.0.1 and subnet mask 255.0.0.0.

You can also verify the IP address assigned to host h2 by repeating Steps 4 and 5 on host h2's terminal. The corresponding interface *h2-eth0* at host h2 has now the IP address 15.0.0.2 and subnet mask 255.0.0.0.

Step 6 Stop the emulation by clicking on *Stop* button.

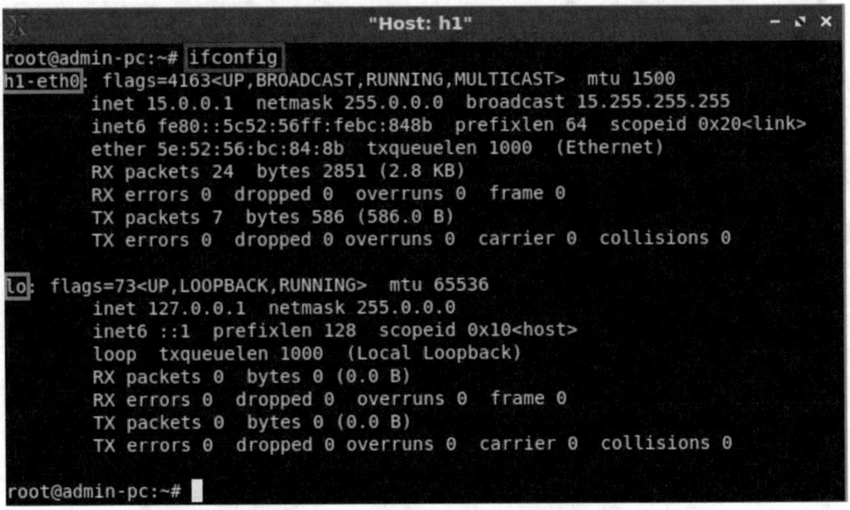

"Host: h1"
root@admin-pc:~# ifconfig
h1-eth0: flags=4163<UP,BROADCAST,RUNNING,MULTICAST> mtu 1500
inet 15.0.0.1 netmask 255.0.0.0 broadcast 15.255.255.255
inet6 fe80::5c52:56ff:febc:848b prefixlen 64 scopeid 0x20<link>
ether 5e:52:56:bc:84:8b txqueuelen 1000 (Ethernet)
RX packets 24 bytes 2851 (2.8 KB)
RX errors 0 dropped 0 overruns 0 frame 0
TX packets 7 bytes 586 (586.0 B)
TX errors 0 dropped 0 overruns 0 carrier 0 collisions 0
lo: flags=73<UP,LOOPBACK,RUNNING> mtu 65536
inet 127.0.0.1 netmask 255.0.0.0
inet6 ::1 prefixlen 128 scopeid 0x10<host>
loop txqueuelen 1000 (Local Loopback)
RX packets 0 bytes 0 (0.0 B)
RX errors 0 dropped 0 overruns 0 frame 0
TX packets 0 bytes 0 (0.0 B)
TX errors 0 dropped 0 overruns 0 carrier 0 collisions 0
root@admin-pc:~#

Fig. 29 Output of ifconfig command on host h1

MiniEdit
File Edit Run Help
New
Open
Save
Export Level 2 Script
Quit
s1
h1
h2
Save the topology as...
Directory: /home/admin/lab_topologies
Lab 2.mn Lab 8.mn Lab 14.mn
Lab 3.mn Lab 9.mn Lab 15.mn
Lab 4.mn Lab 10.mn Lab 16.mn
Lab 5.mn Lab 11.mn Lab 17.mn
Lab 6.mn Lab 12.mn Lab 18.mn
Lab 7.mn Lab 13.mn Lab 19.mn
File name: MyTopology
Save
Files of type: Mininet Topology (*.mn)
Cancel

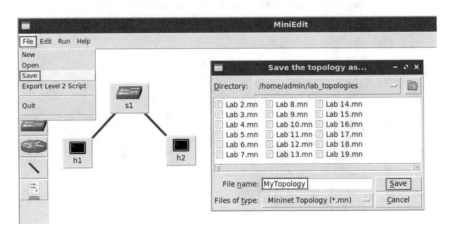

Fig. 30 Saving the topology

6.4 Saving and Loading a Mininet Topology

It is often useful to save the network topology, particularly when its complexity increases. MiniEdit enables you to save the topology to a file.

Step 1 To save your topology, click on *File* then *Save* (Fig. 30). Provide a name for the topology and save on your machine.

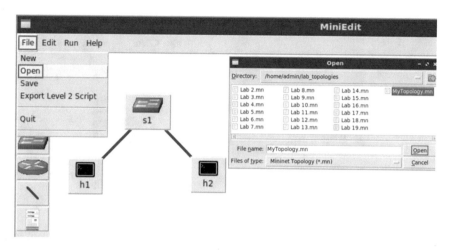

Fig. 31 Opening a topology

Step 2 To load the topology, click on *File* then *Open*. Locate the topology file and click on *Open* (Fig. 31). The topology will be loaded again to MiniEdit.

The upcoming labs' topologies are already built and stored in the folder */home/admin/lab_topologies* located in the Client's home directory. The *Open* dialog is used to avoid manually rebuilding each lab's topology.

Chapter 1—Lab 2: Introduction to iPerf3

Overview
To conduct the experiment described in this section, please login into the Academic Cloud at http://highspeednetworks.net/ and reserve a pod for Lab 2.

This lab briefly introduces iPerf3 and explains how it can be used to measure and test network throughput in a designed network topology. It demonstrates how to invoke both client-side and server-side options from the command-line utility.

Objectives
By the end of this lab, students should be able to:

1. Understand throughput and how it differs from bandwidth in network systems.
2. Create iPerf3 tests with various settings on a designed network topology.
3. Understand and analyze iPerf3's test output.
4. Visualize iPerf3's output using a custom plotting script.

Lab Settings
The information in Table 2 provides the credentials of the machine containing Mininet.

Table 2 Credentials to
access Client1 machine

Device	Account	Password
Client1	admin	password

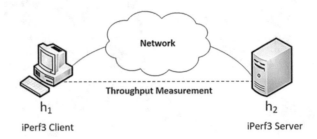

Fig. 32 Throughput measurement with iPerf3

Lab Roadmap

This lab is organized as follows:

1. Section 7: Introduction to iPerf3.
2. Section 8: Lab topology.
3. Section 9: Using iPerf3 (client and server commands).
4. Section 10: Plotting iPerf3's results.

7   Introduction to iPerf3

Bandwidth is a physical property of a transmission media that depends on factors such as the construction and length of wire or fiber. To network engineers, bandwidth is the maximum data rate of a channel, a quantity measured in bits per second (bps). Having a high-bandwidth link does not always guarantee high network performance. In fact, several factors may affect the performance such as latency, packet loss, jitter, and others.

In the context of a communication session between two end devices along a network path, *throughput* is the rate in bps at which the sending process can deliver bits to the receiving process. Because other sessions will be sharing the bandwidth along the network path, and because these other sessions will recur, the available throughput can fluctuate with time. Note, however, that sometimes the terms throughput and bandwidth are used interchangeably.

iPerf3 is a real-time network throughput measurement tool. It is an open-source, cross-platform client–server application that can be used to measure the throughput between the two end devices (Fig. 32). A typical iPerf3 output contains a timestamped report of the amount of data transferred and the throughput measured.

Measuring throughput is particularly useful when experiencing network bandwidth issues such as delay, packet loss, etc. iPerf3 can operate on Transmission Control Protocol (TCP), User Datagram Protocol (UDP), and Stream Control Transmission Protocol (SCTP).

In iPerf3, the user can set *client* and *server* configurations via options and parameters and can create data flows to measure the throughput between the two end hosts in a unidirectional or bidirectional way. iPerf3 outputs a timestamped report of the amount of data transferred and the throughput measured.

8 Lab Topology

Let us get started with creating a simple Mininet topology using MiniEdit. The topology uses 10.0.0.0/8, which is the default network assigned by Mininet (Fig. 33).

Step 1 A shortcut to MiniEdit is located on the machine's Desktop. Start MiniEdit by clicking on MiniEdit's shortcut (Fig. 34). When prompted for a password, type password .

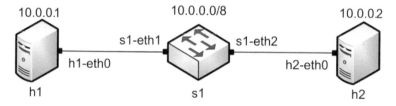

Fig. 33 Mininet's default minimal topology

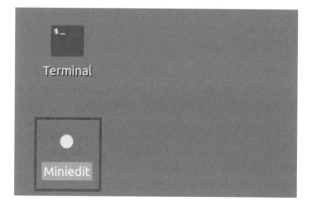

Fig. 34 MiniEdit shortcut

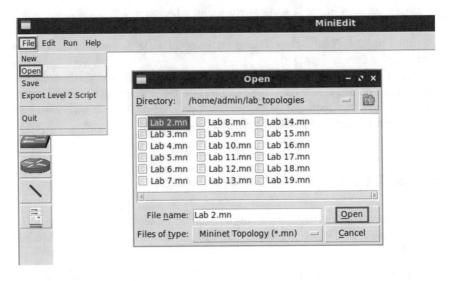

Fig. 35 MiniEdit's *Open* dialog

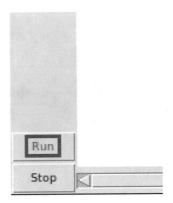

Fig. 36 Running the emulation

Step 2 On MiniEdit's menu bar, click on *File* then *Open* to load the lab's topology. Locate the *Lab 2.mn* topology file in the default directory, */home/admin/lab_topologies*, and click on *Open* (Fig. 35).

Step 3 Before starting the measurements between host h1 and host h2, the network must be started. Click on the *Run* button located at the bottom left of MiniEdit's window to start the emulation (Fig. 36).

The above topology uses 10.0.0.0/8, which is the default network assigned by Mininet.

8.1 Starting Host h1 and Host h2

Step 1 Hold the right-click on host h1 and select *Terminal* (Fig. 37). This opens the terminal of host h1 and allows the execution of commands on that host.

Step 2 Test connectivity between the end-hosts using the ⎡ping⎤ command. On host h1, type the command ⎡ping 10.0.0.2⎤ (Fig. 38). This command tests the connectivity between host h1 and host h2. To stop the test, press ⎡Ctrl+c⎤. The figure below shows a successful connectivity test.

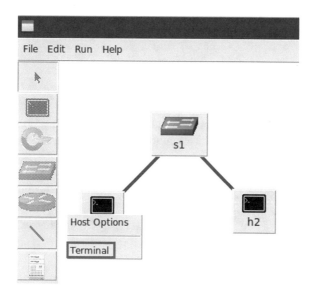

Fig. 37 Opening a terminal on host h1

```
                            "Host: h1"                          - ˅ x
root@admin-pc:~# ping 10.0.0.2
PING 10.0.0.2 (10.0.0.2) 56(84) bytes of data.
64 bytes from 10.0.0.2: icmp_seq=1 ttl=64 time=0.370 ms
64 bytes from 10.0.0.2: icmp_seq=2 ttl=64 time=0.078 ms
64 bytes from 10.0.0.2: icmp_seq=3 ttl=64 time=0.082 ms
64 bytes from 10.0.0.2: icmp_seq=4 ttl=64 time=0.080 ms
^C
--- 10.0.0.2 ping statistics ---
4 packets transmitted, 4 received, 0% packet loss, time 77ms
rtt min/avg/max/mdev = 0.078/0.152/0.370/0.126 ms
root@admin-pc:~# ▊
```

Fig. 38 Connectivity test using ⎡ping⎤ command

The figure above indicates that there is connectivity between host h1 and host h2. Thus, we are ready to start the throughput measurement process.

9   Using iPerf3 (Client and Server Commands)

Since the initial setup and configuration are done, it is time to start a simple through-put measurement. The user interacts with iPerf3 using the $\boxed{\text{iperf3}}$ command. The basic $\boxed{\text{iperf3}}$ syntax used on both the client and the server is as follows:

```
iperf3 [-s|-c] [ options ]
```

9.1   Starting Client and Server

Step 1 Hold the right-click on host h2 and select *Terminal* (Fig. 39). This opens the terminal of host h2 and allows the execution of commands on that host.

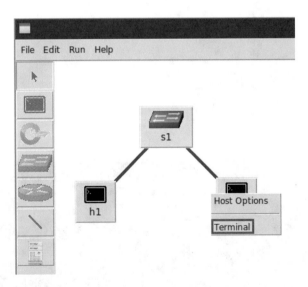

Fig. 39 Opening a terminal on host h2

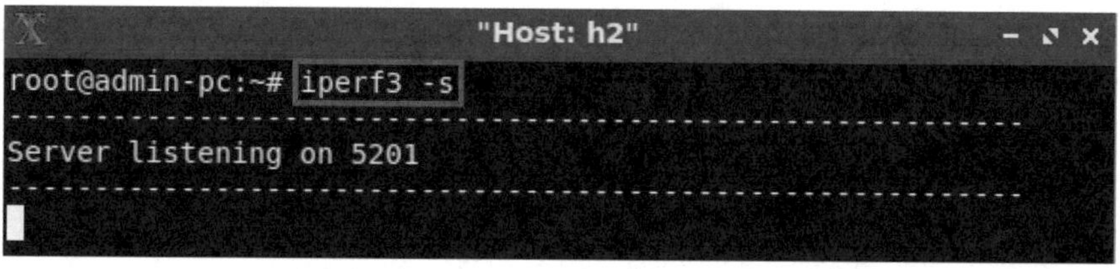

Fig. 40 Host h2 running iPerf3 server

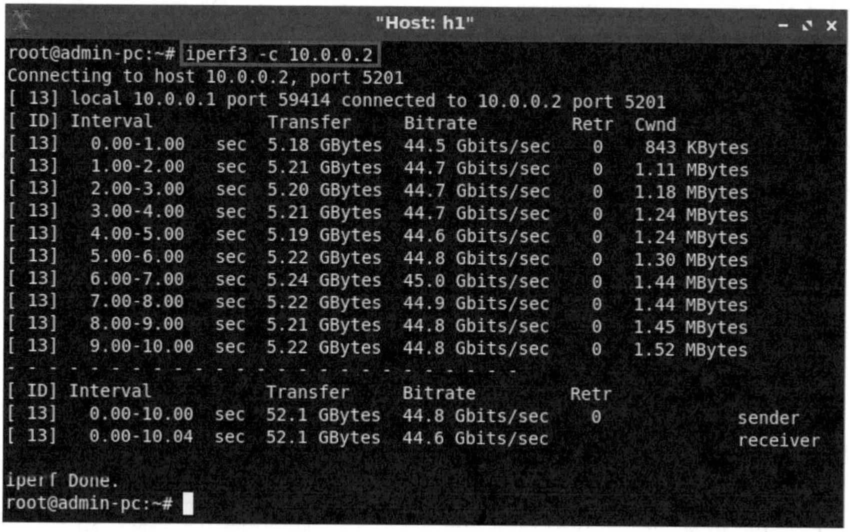

Fig. 41 Host h1 running iPerf3 as client

Step 2 To launch iPerf3 in server mode, run the command iperf3 -s in host h2's terminal as shown in the figure below (Fig. 40):

```
iperf3 -s
```

The parameter -s in the command above indicates that the host is configured as a server. Now, the server is listening on port 5201 waiting for incoming connections.

Step 3 Now to launch iPerf3 in client mode, run the command iperf3 -c 10.0.0.2 in host h1's terminal as shown in the figure below (Fig. 41):

```
iperf3 -c 10.0.0.2
```

The parameter -c in command above indicates that host h1 is configured as a client. The parameter 10.0.0.2 is the server's (host h2) IP address. Once the test is completed, a summary report on both the client and the server is displayed containing the following data:

- *ID*: identification number of the connection.
- *Interval*: time interval to periodically report throughput. By default, the time interval is 1 second.
- *Transfer:* how much data was transferred in each time interval.
- *Bitrate:* the measured throughput in each time interval.
- *Retr:* the number of TCP segments retransmitted in each time interval. This field increases when TCP segments are lost in the network due to congestion or corruption.
- *Cwnd:* indicates the congestion windows size in each time interval. TCP uses this variable to limit the amount of data the TCP client can send before receiving the acknowledgement of the sent data.

The summarized data, which starts after the last dashed line, shows the total amount of transferred data is 52.1 Gbyte and the throughput 44.8 Gbps.

Step 4 In order to stop the server, press $\boxed{\text{Ctrl+c}}$ in host h2's terminal. The user can see the throughput results in the server side too. The summarized data on the server is similar to that of the client side's and must be interpreted in the same way.

9.2   Setting Transmitting Time Period

Setting the transmission time period is configured solely on the client. To change the default transmission time, apply the following steps:

Step 1 Start the iPerf3 server on host h2 (Fig. 42).

```
iperf3 -s
```

Step 2 Start the iPerf3 client with the $\boxed{\text{-t}}$ option followed by the number of seconds (Fig. 43).

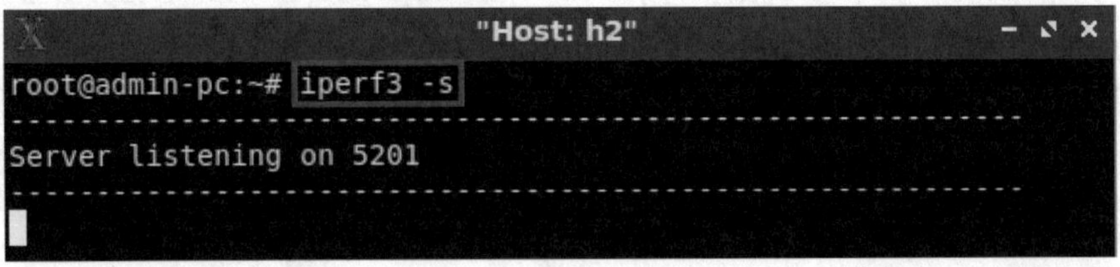

Fig. 42 Host h2 running iPerf3 as server

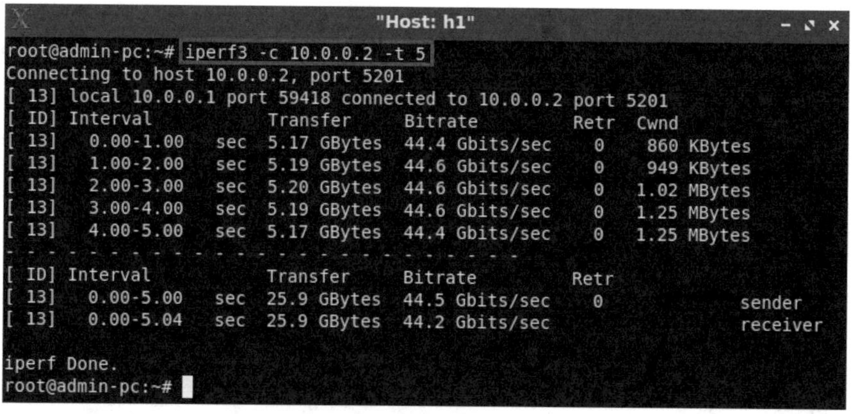

```
X                              "Host: h1"                        - ⤢ x
root@admin-pc:~# iperf3 -c 10.0.0.2 -t 5
Connecting to host 10.0.0.2, port 5201
[ 13] local 10.0.0.1 port 59418 connected to 10.0.0.2 port 5201
[ ID] Interval           Transfer     Bitrate         Retr  Cwnd
[ 13]   0.00-1.00   sec  5.17 GBytes  44.4 Gbits/sec    0    860 KBytes
[ 13]   1.00-2.00   sec  5.19 GBytes  44.6 Gbits/sec    0    949 KBytes
[ 13]   2.00-3.00   sec  5.20 GBytes  44.6 Gbits/sec    0   1.02 MBytes
[ 13]   3.00-4.00   sec  5.19 GBytes  44.6 Gbits/sec    0   1.25 MBytes
[ 13]   4.00-5.00   sec  5.17 GBytes  44.4 Gbits/sec    0   1.25 MBytes
- - - - - - - - - - - - - - - - - - - - - - - - -
[ ID] Interval           Transfer     Bitrate         Retr
[ 13]   0.00-5.00   sec  25.9 GBytes  44.5 Gbits/sec    0              sender
[ 13]   0.00-5.04   sec  25.9 GBytes  44.2 Gbits/sec                   receiver

iperf Done.
root@admin-pc:~# █
```

Fig. 43 Host h1 transmitting for 5 s

```
iperf3 -c 10.0.0.2 -t 5
```

The above command starts an iPerf3 client for a 5-s time period transmitting at an average rate of 44.5 Gbps.

Step 3 In order to stop the server, press Ctrl+c in host h2's terminal. The user can see the throughput results in the server side too.

9.3 Setting Time Interval

In this test, the user will configure the client to perform a throughput test with 2-seconds reporting time interval on both the client and the server. Note the default 1-second interval period in Fig. 43.

The -i option allows setting the reporting interval time in seconds. In this case the value should be set to 2 seconds on both the client and the server.

Step 1 Setting the interval value on the server (host h2's terminal) (Fig. 44):

```
iperf3 -s -i 2
```

Step 2 Setting the interval value on the client (host h1's terminal) (Fig. 45):

```
iperf3 -c 10.0.0.2 -i 2
```

Note that the -i option can be specified differently on the client and the server. For example, if the -i option is specified with the value 3 on the client only, then

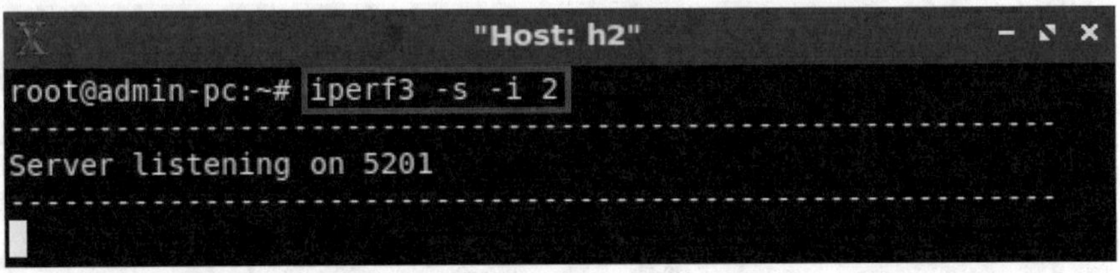

```
"Host: h2"                                   – ↘ ✕
root@admin-pc:~# iperf3 -s -i 2
- - - - - - - - - - - - - - - - - - - - - - - - - - - - - - - - - - - - - - - - -
Server listening on 5201
- - - - - - - - - - - - - - - - - - - - - - - - - - - - - - - - - - - - - - - - -
```

Fig. 44 Host h2 running iPerf3 as server

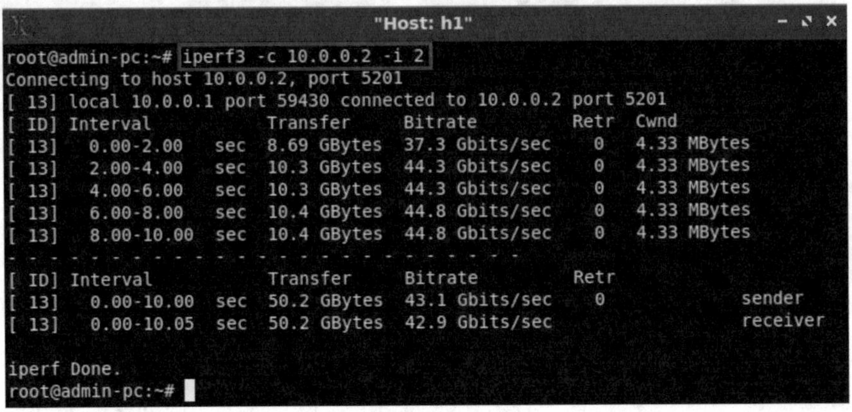

```
"Host: h1"                                   – ↘ ✕
root@admin-pc:~# iperf3 -c 10.0.0.2 -i 2
Connecting to host 10.0.0.2, port 5201
[ 13] local 10.0.0.1 port 59430 connected to 10.0.0.2 port 5201
[ ID] Interval           Transfer     Bitrate         Retr  Cwnd
[ 13]   0.00-2.00   sec  8.69 GBytes  37.3 Gbits/sec    0   4.33 MBytes
[ 13]   2.00-4.00   sec  10.3 GBytes  44.3 Gbits/sec    0   4.33 MBytes
[ 13]   4.00-6.00   sec  10.3 GBytes  44.3 Gbits/sec    0   4.33 MBytes
[ 13]   6.00-8.00   sec  10.4 GBytes  44.8 Gbits/sec    0   4.33 MBytes
[ 13]   8.00-10.00  sec  10.4 GBytes  44.8 Gbits/sec    0   4.33 MBytes
- - - - - - - - - - - - - - - - - - - - - - - - - - - -
[ ID] Interval           Transfer     Bitrate         Retr
[ 13]   0.00-10.00  sec  50.2 GBytes  43.1 Gbits/sec    0           sender
[ 13]   0.00-10.05  sec  50.2 GBytes  42.9 Gbits/sec                receiver

iperf Done.
root@admin-pc:~# ▉
```

Fig. 45 Host h1 and host h2 reporting every 2 s

the client will be reporting every 3 seconds while the server will be reporting every
second (the default ‾-i‾ value).

Step 3 In order to stop the server, press ‾Ctrl+c‾ in host h2's terminal. The user
can see the throughput results in the server side too.

9.4   Changing the Number of Bytes to Transmit

In this test, the client is configured to send a specific amount of data by setting
the number of bytes to transmit. By default, iPerf3 performs the throughput
measurement for 10 seconds. However, with this configuration, the client will keep
sending packets until all the bytes specified by the user were sent.

Step 1 Type the following command on host h2's terminal to start the iPerf3 server
(Fig. 46):

```
iperf3 -s
```

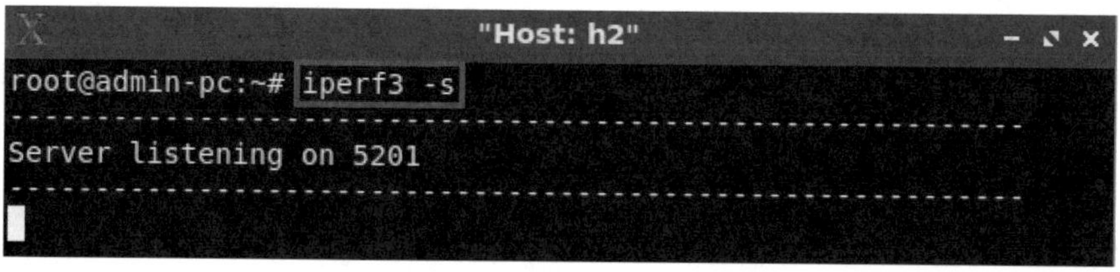

Fig. 46 Host h2 running iPerf3 as server

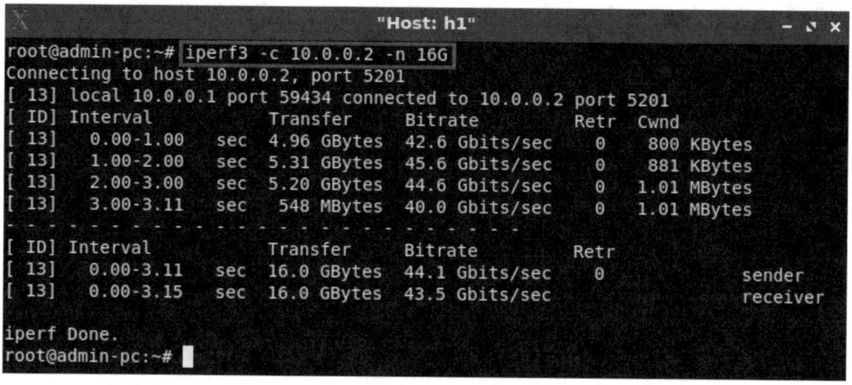

Fig. 47 Host h1 sending 16 Gbps of data

Step 2 This configuration is only set on the client (host h1's terminal) using the
-n option as follows:

```
iperf3 -c 10.0.0.2 -n 16G
```

The -n option in the above command indicates the amount of data to transmit:
16 Gbytes. The user can specify other scale values, for example, 16M is used to
send 16 Mbytes (Fig. 47).

Note the total time spent for sending the 16 Gbytes of data is 3.11 s and not the
default transmitting time used by iPerf3 (10 s).

Step 3 In order to stop the server, press Ctrl+c in host h2's terminal. The user
can see the throughput results in the server side too.

9.5   Specifying the Transport-Layer Protocol

So far, the throughput measurements were conducted on the TCP protocol, which is
the default configuration protocol. In order to change the protocol to UDP, the user
must invoke the option -u on the client side. Similarly, the option –sctp is used

```
X                                    "Host: h2"                    –  ↗  ✕
root@admin-pc:~# iperf3 -s
- - - - - - - - - - - - - - - - - - - - - - - - - - - - - - - - - - - - - - - - -
Server listening on 5201
- - - - - - - - - - - - - - - - - - - - - - - - - - - - - - - - - - - - - - - - -
▮
```

Fig. 48 Host h2 running iPerf3 as server

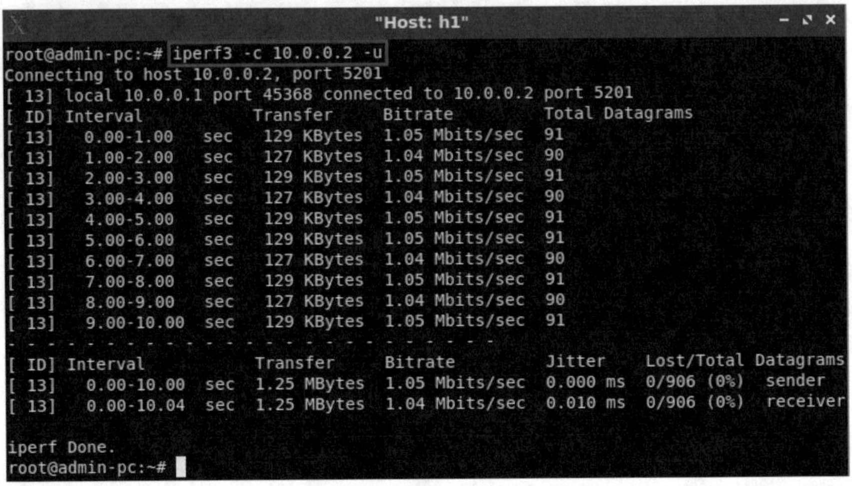

```
X                                      "Host: h1"                    –  ↗  ✕
root@admin-pc:~# iperf3 -c 10.0.0.2 -u
Connecting to host 10.0.0.2, port 5201
[ 13] local 10.0.0.1 port 45368 connected to 10.0.0.2 port 5201
[ ID] Interval           Transfer     Bitrate         Total Datagrams
[ 13]   0.00-1.00   sec   129 KBytes  1.05 Mbits/sec  91
[ 13]   1.00-2.00   sec   127 KBytes  1.04 Mbits/sec  90
[ 13]   2.00-3.00   sec   129 KBytes  1.05 Mbits/sec  91
[ 13]   3.00-4.00   sec   127 KBytes  1.04 Mbits/sec  90
[ 13]   4.00-5.00   sec   129 KBytes  1.05 Mbits/sec  91
[ 13]   5.00-6.00   sec   129 KBytes  1.05 Mbits/sec  91
[ 13]   6.00-7.00   sec   127 KBytes  1.04 Mbits/sec  90
[ 13]   7.00-8.00   sec   129 KBytes  1.05 Mbits/sec  91
[ 13]   8.00-9.00   sec   127 KBytes  1.04 Mbits/sec  90
[ 13]   9.00-10.00  sec   129 KBytes  1.05 Mbits/sec  91
- - - - - - - - - - - - - - - - - - - - - - - - - - - - - - - - - - - -
[ ID] Interval           Transfer     Bitrate         Jitter    Lost/Total Datagrams
[ 13]   0.00-10.00  sec  1.25 MBytes  1.05 Mbits/sec  0.000 ms  0/906 (0%)  sender
[ 13]   0.00-10.04  sec  1.25 MBytes  1.04 Mbits/sec  0.010 ms  0/906 (0%)  receiver

iperf Done.
root@admin-pc:~# ▮
```

Fig. 49 Host h1 sending UDP datagrams

for the SCTP protocol. iPerf3 automatically detects the transport-layer protocol on the server side.

Step 1 Start the iPerf3 server on host h2 (Fig. 48).

```
iperf3 -s
```

Step 2 Specify UDP as the transport-layer protocol using the -u option as follows (Fig. 49):

```
iperf3 -c 10.0.0.2 -u
```

Once the test is completed, it will show the following summarized data:

- *ID, Interval, Transfer, Bitrate*: same as TCP.
- *Jitter*: the difference in packet delay.
- *Lost/Total*: indicates the number of lost datagrams over the total number sent to the server (and percentage).

After the dashed lines, the summary is displayed, showing the total amount of transferred data (1.25 Mbytes) and the maximum achieved bandwidth (1.05 Mbps), over a time period of 10 s. The Jitter, which indicates in milliseconds (ms) the variance of time delay between data packets over a network, has a value of 0.010 ms. Finally, the lost datagrams value is 0 (zero) and the total datagram that the server has received was 906, and thus, the loss rate is 0%. These values are reported on the server as well.

Step 3 In order to stop the server, press Ctrl+c in host h2's terminal. The user can see the throughput results in the server side too.

9.6 Changing Port Number

If the user wishes to measure throughput on a specific port, the -p option is used to configure both the client and the server to send/receive packets or datagrams on the specified port.

Step 1 Start the iPerf3 server on host h2. Use the -p option to specify the listening port (Fig. 50).

```
iperf3 -s -p 3250
```

Step 2 Start the iPerf3 client on host h1. Use the -p option to specify the server's listening port (Fig. 51).

```
iperf3 -c 10.0.0.2 -p 3250
```

Step 3 In order to stop the server, press Ctrl+c in host h2's terminal. The user can see the throughput results in the server side too.

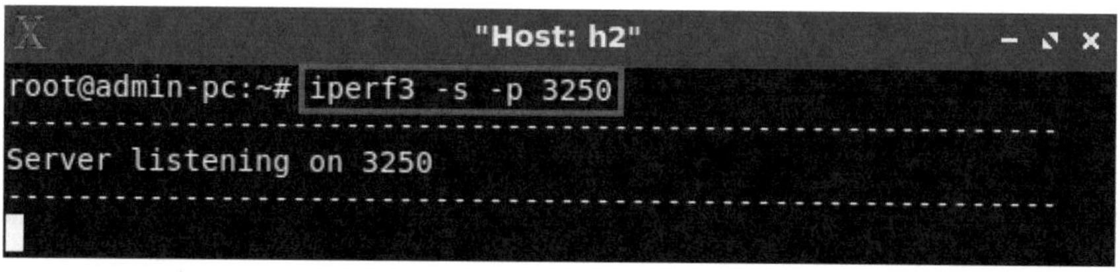

Fig. 50 Host h2 running iPerf3 as server on port 3250

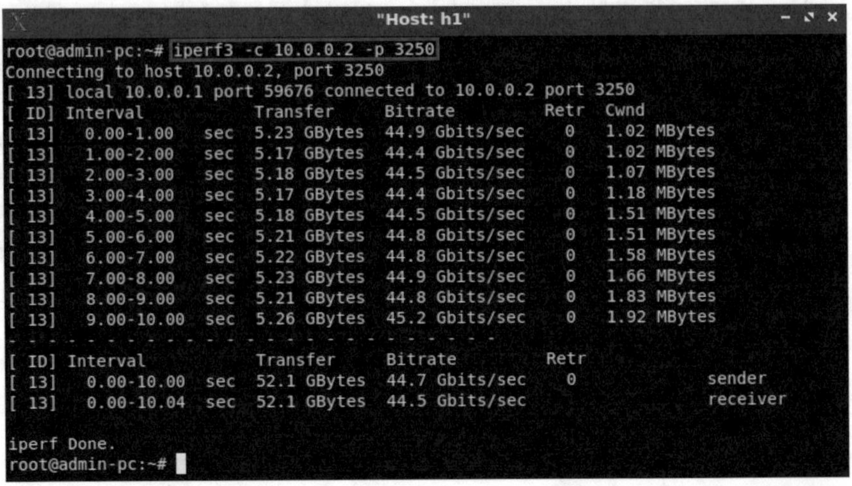

```
X                                   "Host: h1"                        - ⤢ x
root@admin-pc:~# iperf3 -c 10.0.0.2 -p 3250
Connecting to host 10.0.0.2, port 3250
[ 13] local 10.0.0.1 port 59676 connected to 10.0.0.2 port 3250
[ ID] Interval           Transfer     Bitrate         Retr  Cwnd
[ 13]   0.00-1.00   sec  5.23 GBytes  44.9 Gbits/sec    0   1.02 MBytes
[ 13]   1.00-2.00   sec  5.17 GBytes  44.4 Gbits/sec    0   1.02 MBytes
[ 13]   2.00-3.00   sec  5.18 GBytes  44.5 Gbits/sec    0   1.07 MBytes
[ 13]   3.00-4.00   sec  5.17 GBytes  44.4 Gbits/sec    0   1.18 MBytes
[ 13]   4.00-5.00   sec  5.18 GBytes  44.5 Gbits/sec    0   1.51 MBytes
[ 13]   5.00-6.00   sec  5.21 GBytes  44.8 Gbits/sec    0   1.51 MBytes
[ 13]   6.00-7.00   sec  5.22 GBytes  44.8 Gbits/sec    0   1.58 MBytes
[ 13]   7.00-8.00   sec  5.23 GBytes  44.9 Gbits/sec    0   1.66 MBytes
[ 13]   8.00-9.00   sec  5.21 GBytes  44.8 Gbits/sec    0   1.83 MBytes
[ 13]   9.00-10.00  sec  5.26 GBytes  45.2 Gbits/sec    0   1.92 MBytes
- - - - - - - - - - - - - - - - - - - - - - - - - - - - - - - -
[ ID] Interval           Transfer     Bitrate         Retr
[ 13]   0.00-10.00  sec  52.1 GBytes  44.7 Gbits/sec    0              sender
[ 13]   0.00-10.04  sec  52.1 GBytes  44.5 Gbits/sec                   receiver

iperf Done.
root@admin-pc:~# ▋
```

Fig. 51 Host h2 running on port 3250

```
X                                   "Host: h2"                        - ⤢ x
root@admin-pc:~# iperf3 -s
- - - - - - - - - - - - - - - - - - - - - - - - - - - - - - - - - - - - -
Server listening on 5201
- - - - - - - - - - - - - - - - - - - - - - - - - - - - - - - - - - - - -
▋
```

Fig. 52 Host h2 running iPerf3 as server

9.7   Export Results to JSON File

JSON (JavaScript Object Notation) is a lightweight data-interchange format. iPerf3 allows exporting the test results to a JSON file, which makes it easy for other applications to parse the file and interpret the results (e.g., plot the results).

Step 1 Start the iPerf3 server on host h2 (Fig. 52).

```
iperf3 -s
```

Step 2 Start the iPerf3 client on host h1. Specify the -J option to display the output in JSON format (Fig. 53).

```
iperf3 -c 10.0.0.2 -J
```

The -J option outputs JSON text to the screen through standard output (*stdout*) after the test is done (10 seconds by default). It is often useful to export the output

"Host: h1" — × ×
```
root@admin-pc:~# iperf3 -c 10.0.0.2 -J
```

Fig. 53 Host h1 using -J to output JSON to standard output (*stdout*)

"Host: h1" — × ×
```
root@admin-pc:~# iperf3 -c 10.0.0.2 -J > test_results.json
root@admin-pc:~#
```

Fig. 54 Host h1 using -J to output JSON and redirecting *stdout* to file

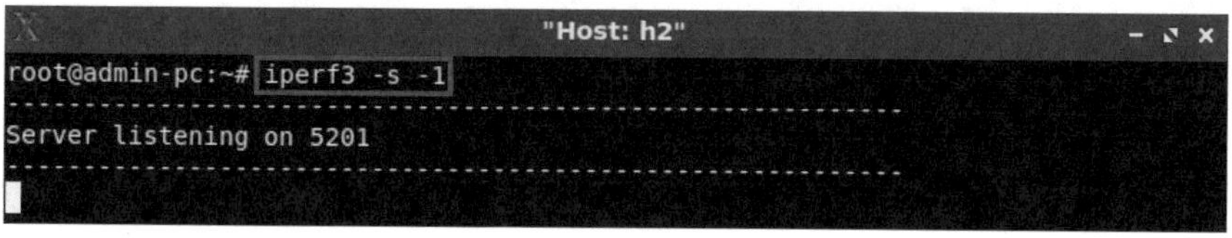

"Host: h2" — × ×
```
root@admin-pc:~# iperf3 -s -1
-------------------------------------------------
Server listening on 5201
-------------------------------------------------
```

Fig. 55 Host h2 running a server with one connection only

to a file that can be parsed later by other programs. This can be done by redirecting the standard output to a file using the redirection operator in Linux > (Fig. 54).

```
iperf3 -c 10.0.0.2 -J >test_results.json
```

After creating the JSON file, the ls command is used to verify that the file is created. The cat command can be used to display the file's contents.

Step 3 In order to stop the server, press Ctrl+c in host h2's terminal. The user can see the throughput results in the server side too.

9.8 Handle One Client

By default, an iPerf3 server keeps listening to incoming connections. To allow the server to handle one client and then stop, the -1 option is added to the server.

Step 1 Start the iPerf3 server on host h2. Use the -1 option to accept only one client (Fig. 55).

```
iperf3 -s -1
```

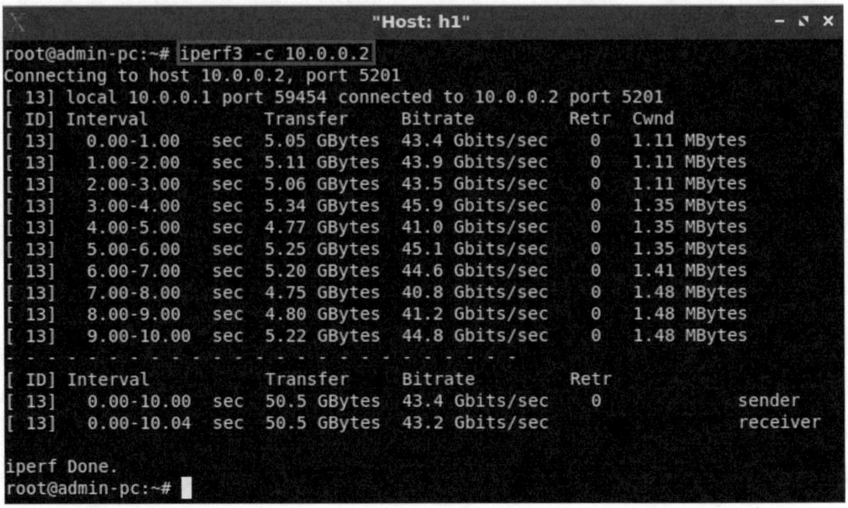

Fig. 56 Host h1 running an iPerf3 client

Fig. 57 Host h2 running iPerf3 as server

Step 2 Start the iPerf3 client on host h1 (Fig. 56).

```
iperf3 -c 10.0.0.2
```

After this test is finished, the server stops immediately.

10   Plotting iPerf3 Results

In Sect. 9.7, iPerf3's result was exported to a JSON file to be processed by other applications. A script called plot_iperf.sh is installed and configured on the Client's machine. It accepts a JSON file as input and generates PDF files plotting several variables produced by iPerf3.

Step 1 Start the iPerf3 server on host h2 (Fig. 57).

```
iperf3 -s
```

Fig. 58 Host h1 using `-J` to output JSON and redirecting *stdout* to file

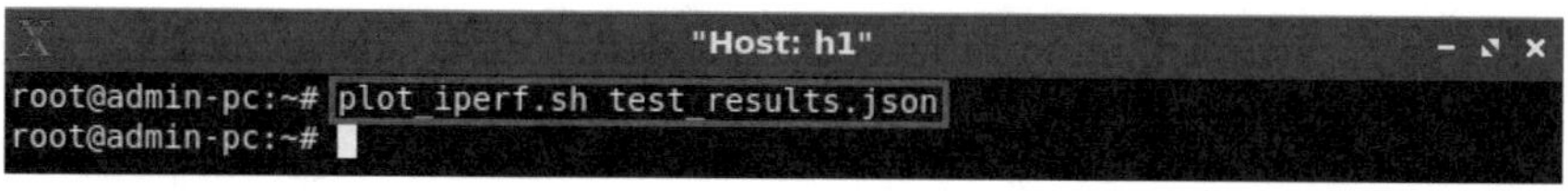

Fig. 59 `plot_iperf.sh` script generating output results

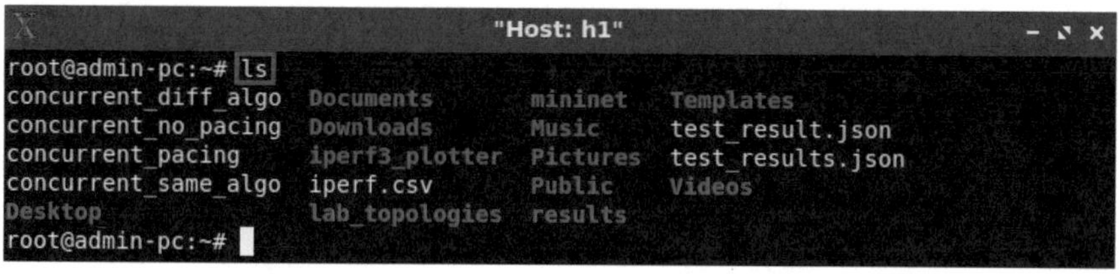

Fig. 60 Listing the current directory's contents using the `ls` command

Step 2 Start the iPerf3 client on host h1. Specify the `-J` option to produce the output in JSON format and redirect the output to the file *test_results.json*. Any data previously stored in this file will be replaced with current output as the `>` operator is being used here (Fig. 58).

```
iperf3 -c 10.0.0.2 -J >test_results.json
```

Step 3 To generate the output for iPerf3's JSON file run the following command (Fig. 59):

```
plot_iperf.sh test_results.json
```

This plotting script generates PDF files for the following fields: congestion window (*cwnd.pdf*), retransmits (*retransmits.pdf*), Round-Trip Time (*RTT.pdf*), Round-Trip Time variance (*RTT_Var.pdf*), throughput (*throughput.pdf*), maximum transmission unit (*MTU.pdf*), bytes transferred (*bytes.pdf*). The plotting script also generates a CSV file (1.dat), which can be used by other applications. These files are stored in a directory *results* created in the same directory where the script was executed as shown in the figure below (Fig. 60).

Step 4 Navigate to the results folder using the `cd` command (Fig. 61).

```
cd results/
```

Fig. 61 Entering the results directory using the cd command

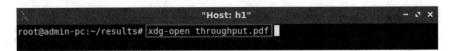

Fig. 62 Opening the *throughput.pdf* file using xdg-open

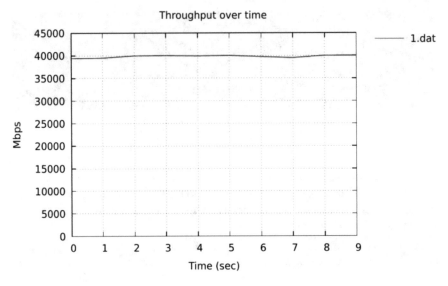

Fig. 63 *throughput.pdf* output

Step 5 To open any of the generated files, use the xdg-open command followed by the file name. For example, to open the throughput.pdf file, use the following command (Figs. 62 and 63):

```
xdg-open throughput.pdf
```

Step 6 In order to stop the server, press Ctrl+c in host h2's terminal. The user can see the throughput results in the server side too.

References

1. D. McNichol, *The roads that built America: the incredible story of the U.S. interstate system* (Sterling, 2003)
2. L. Farrell, Science DMZ: The fast path for science data. Sci Node J, May 2016. [Online]. Available: https://sciencenode.org/feature/science-dmz-a-data-highway-system.php
3. The energy science network. [Online]. Available: https://www.es.net
4. E. Dart, L. Rotman, B. Tierney, M. Hester, J. Zurawski, The science DMZ: a network design pattern for data-intensive science, in *Proceedings of the International Conference on High Performance Computing, Networking, Storage and Analysis* (2013, November)
5. A brief history of NSF and internet, Cyberinfrastructure Special Report, National Science Foundation. [Online]. Available https://www.nsf.gov/news/special_reports/cyber/internet.jsp
6. European organization for nuclear research. [Online]. Available: https://home.cern/about/computing
7. National radio astronomy observatory. [Online]. Available online: http://www.vla.nrao.edu/
8. J. Bashor, General atomics remote controls fusion experiments, bridges collaborators using ESnet-championed technology. ESnet News, Sep. 2015. [Online]. Available: https://es.net/news-and-publications/esnet-news/2015/science-dmz-fuels-fusion-research/
9. J. Van Horn, A. Toga, Human neuroimaging as a big data science. J. Brain Imaging Behav. **8**(2), 323–331 (2014)
10. S. Peisert, E. Dart, W. Barnett, J. Cuff, R. Grossman, E. Balas, A. Berman, A. Shankar, B. Tierney, The medical science DMZ: a network design pattern for data-intensive medical science. J. Am. Med. Inform. Assoc. (JAMIA) (2017). [Online]. Available: https://academic.oup.com/jamia/article/doi/10.1093/jamia/ocx104/4367749/The-medical-science-DMZ-a-network-design-pattern
11. S. Peisert, W. Barnett, E. Dart, J. Cuff, R. Grossman, E. Balas, A. Berman, A. Shankar, B. Tierney, The medical science DMZ. J. Am. Med. Inform. Assoc. **23**(6), 1199–1201 (2016)
12. General electric health care. [Online]. Available: http://www3.gehealthcare.com/en/global_gateway
13. G. Roberts, Big data and the X-ray laser, *Symmetry Magazine*, June 2013. [Online]. Available: https://www.symmetrymagazine.org/article/june-2013/big-data-and-the-x-ray-laser
14. SLAC national accelerator laboratory. [Online]. Available: https://www6.slac.stanford.edu/
15. E. Waltz, Portable DNA sequencer minion helps build the internet of living things. IEEE Spectr. Mag., Mar. 2016. [Online]. Available: https://spectrum.ieee.org/the-human-os/biomedical/devices/portable-dna-sequencer-minion-help-build-the-internet-of-living-things
16. Nanopore technologies. [Online]. Available: https://nanoporetech.com/

Network Cyberinfrastructure Aspects for Big Data Transfers

This chapter describes the elements of the cyberinfrastructure supporting Science DMZs and high-speed networks for large data transfers. They include friction-free network paths; dedicated, high-performance end devices, referred to as Data Transfer Nodes (DTNs); end-to-end performance measurement monitoring points; and security mechanisms suitable for high speed.

1 Limitations of Enterprise Networks and Motivation for Science DMZs

An enterprise network is composed of one or more interconnected local area networks (LANs). Common design goals are:

- To serve a large number of users and platforms: desktops, laptops, mobile devices, supercomputers, tablets, etc.
- To support a variety of applications: email, browsing, voice, video, procurement systems, and others.
- To provide security against the multiple threats that result from the large number of applications and platforms.
- To provide a level of Quality of Service (QoS) that satisfies user expectations.

To serve multiple applications and platforms, the network is designed for general purposes. To provide an adequate security level, the network may use multiple CPU-intensive appliances. Besides a centrally located firewall, internal firewalls are often used to add stringent filtering capability to sensitive subnetworks. The network may only provide a minimum level of QoS, which is often sufficient. The level of QoS does not need to be strict, as applications can improve on the service provided by the network. Moderate bandwidth, latency, and loss rates are most of the time acceptable, as flows have a small size (from few KBs to MBs) and a short

© The Author(s), under exclusive license to Springer Nature Switzerland AG 2022
J. Crichigno et al., *High-Speed Networks*, Practical Networking,
https://doi.org/10.1007/978-3-030-88841-1_2

duration. Rates of few Kbps to tens of Mbps can satisfy bandwidth requirements. Furthermore, most applications are elastic and can adapt to the bandwidth provided by the network. Similarly, packet losses can be repaired with retransmissions and jitter can be smoothed by buffering packets at the receiver.

Figure 1 shows a typical campus enterprise network. Packets coming from the WAN are inspected by multiple inline security appliances, including a firewall and an intrusion prevention system (IPS). Further processing is performed by a network address translator (NAT). Packets traverse through the network, from core-layer routers to access-layer switches. Important components of routers and switches, such as switching fabric, forwarding mechanism, size of memory buffers, etc., are adequate for small flows only. The devices also use processing techniques that yield poor performance when processing large flows, such as cut-through forwarding [1]. Additional security inspection by internal firewalls and distribution- and access-layer switches is common. These switches segregate LANs into virtual LANs (VLANs), requiring further frame processing and inter-VLAN routing. Further, end devices do not have the hardware nor software capabilities to send and receive data at high speed. The bandwidth of the network interface card (NIC) and the input/output and storage systems are often below 10 Gbps. Similarly, software applications perform poorly on WAN data transfers because of limitations such as small buffer size, excessive processing overhead, and inadequate flow and congestion control algorithms.

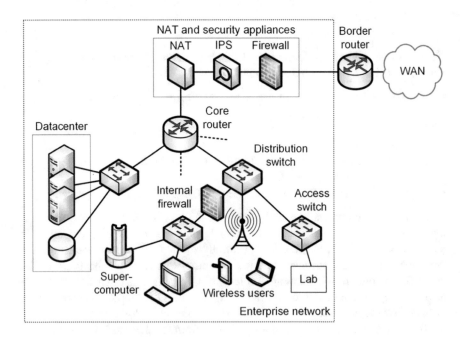

Fig. 1 A campus enterprise network

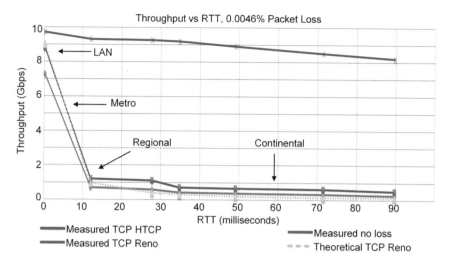

Fig. 2 Throughput vs. Round-Trip Time (RTT), for two devices connected via a 10 Gbps path. The performance of two TCP implementations are provided: Reno [3] (blue) and Hamilton TCP [4] (HTCP) (red). The theoretical performance with packet losses (green) and the measured throughput without packet losses (purple) are also shown [1]

Packet losses may occur at different locations in the enterprise network, including routers, switches, firewalls, IPS, etc. As a result of a packet loss, TCP reacts by drastically decreasing the rate at which packets are sent. The following example [1] illustrates the impact of a small packet loss rate. Figure 2 shows the TCP throughput of a data transfer across a 10 Gbps path. The packet loss rate is 1/22,000, or 0.0046%. The purple curve is the throughput in a loss-free environment; the green curve is the theoretical throughput computed according to the following equation [2]:

$$\text{throughput} = \frac{MSS}{RTT \cdot \sqrt{L}}. \tag{1}$$

Equation (1) indicates that the throughput of a TCP connection in steady state is directly proportional to the maximum segment size (MSS) and inversely proportional to the Round-Trip Time (RTT) and the square root of the packet loss rate (L). The red and blue curves are real measured throughput of two popular implementations of TCP: Reno [3] and Hamilton TCP (HTCP) [4]. Because TCP interprets losses as network congestion, it reacts by decreasing the rate at which packets are sent. This problem is exacerbated as the latency increases between the communicating hosts. Beyond LAN transfers, the throughput decreases rapidly to less than 1 Gbps. This is often the case when research collaborators sharing data are geographically distributed.

2 Science DMZ Architecture

The Science DMZ is designed to address the limitations of enterprise networks and is typically deployed near the main enterprise network. It is important to highlight, however, that the two networks, the Science DMZ and the enterprise network, are separated either physically or logically. There are important reasons for this choice. First, the path from the Science DMZ to the WAN must involve as few network devices as possible, to minimize the possibility of packet losses at intermediate devices. Second, the Science DMZ can also be considered as a security architecture, because it limits the application types and corresponding flows supported by end devices. While flows in enterprise networks are numerous and diverse, those in Science DMZs are usually well-identified, enabling security policies to be tied to those flows.

A Science DMZ example is illustrated in Fig. 3. The main characteristics of a Science DMZ are the deployment of a friction-free path between end devices across the WAN, the use of DTNs, the active performance measurement and monitoring of the paths between the Science DMZ and the collaborator networks, and the use of access-control lists (ACLs) and offline security appliances. Specifically:

- Friction-free network path: DTNs are connected to remote systems, such as collaborators' networks, via the WAN. The high-latency path is composed of

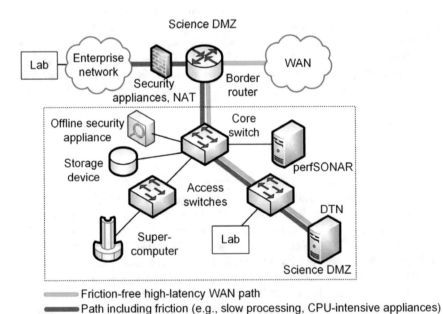

Fig. 3 A Science DMZ co-located with an enterprise network. Notice the absence of firewall or any stateful inline security appliance in the friction-free path

routers and switches, which have large buffer sizes to absorb transitory packet bursts and prevent losses. The path has no devices that may add excessive delays or cause the packet to be delivered out of order, e.g., firewall, IPS, NAT. The rationale for this design choice is to prevent any packet loss or retransmission, which can trigger a decrease in TCP throughput.

- Dedicated, high-performance DTNs: These devices are typically Linux devices built and configured for receiving WAN transfers at high speed. They use optimized data transfer tools such as Globus' GridFTP [5–7]. General-purpose applications (e.g., email clients, document editors, media players) are not installed. Having a narrow and specific set of applications simplifies the design and enforcement of security policies.

- Performance measurement and monitoring point: Typically, there is a primary high-capacity path connecting the Science DMZ with the WAN. An essential aspect is to maintain a healthy path. In particular, identifying and eliminating soft failures in the network is critical for large data transfers [1]. When soft failures occur, basic connectivity continues to exist but high throughput can no longer be achieved. Examples of soft failures include failing components and routers forwarding packets using the main CPU rather than the forwarding plane. Additionally, TCP was intentionally designed to hide transmission errors that may be caused by soft failures. As stated in RFC 793 [8], *As long as the TCPs continue to function properly and the internet system does not become completely partitioned, no transmission errors will affect the users.* The performance measurement and monitoring point provides an automated mechanism to actively measure end-to-end metrics such as throughput, latency, and packet loss. The most used tool is perfSONAR [9, 10].

- ACLs and offline security appliances: The primary method to protect a Science DMZ is via router's ACLs. Since ACLs are implemented in the forwarding plane of a router, they do not compromise the end-to-end throughput. Additional offline appliances include payload-based and flow-based intrusion detection systems (IDSs).

In Fig. 3, when data sets are transferred to a DTN from the WAN, they may be stored locally at the DTN or written into a storage device. DTNs can be dual-homed, with a second interface connected to the storage device. This approach allows the DTN to simultaneously receive data from the WAN and transfer the data to the storage device, avoiding double-copying it. Users located in a laboratory inside the Science DMZ have friction-free access to the data in the storage device. On the other hand, users from a laboratory located in the enterprise network are behind the security appliances protecting that network. These users may achieve reasonable performance accessing the stored data/Science DMZ. The reason here is that, because of the very low latency between the Science DMZ and enterprise users, the retransmissions caused by the security appliances have much less performance impact. TCP recovers from packet losses quickly at low latencies (discussed in Section IV), contrasting with the slow recovery observed when packet losses are experienced in high-latency WANs. The key is to provide the long-distance TCP connections with a friction-free service.

2.1 Addressing the Enterprise Network Limitations

The The Science DMZ addresses the limitations encountered in enterprise networks by using the coordinated set of resources shown in Fig. 4. At the physical layer/cyberinfrastructure, the WAN must be capable of handling large traffic volumes, with a predictable performance. Bit-error rates should be very low and congestion should not occur. The WAN path between end devices should include as few devices as possible. These requirements contrast with typical services delivered by commercial Internet Service Providers (ISPs), used in enterprise networks. ISPs often minimize operating costs at the expense of performance. For large data transfers and research purposes, many institutions are connected to regional or national backbones dedicated to supporting research and education, such as Internet2 [11].

At the data-link and network layers, the switches and routers must have a suitable architecture to forward frames/packets at a high speed (10 Gbps and above). Important attributes are the fabric, queueing, and forwarding techniques. These devices must also have large buffer sizes to absorb transient packet bursts generated by large flows. These requirements are opposite to those implemented by devices used in enterprise networks, which are driven by datacenter needs. The paths interconnecting devices inside a datacenter are characterized by a low latency. On the other hand, the paths interconnecting DTNs to remote networks are characterized by a high latency.

At the transport layer, the protocol must transfer a large amount of data between end devices without errors. TCP is the protocol used by most application-layer tools.

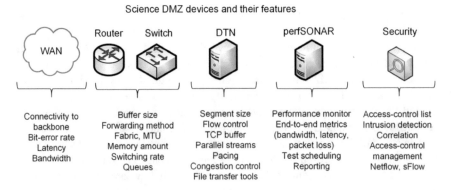

Fig. 4 Features of Science DMZ's devices

A large amount of memory must be allocated to the TCP buffer, which permits the sender to continuously send segments to fill up the WAN capacity. Otherwise, the TCP flow control mechanism leads to a stop-and-wait behavior. The transport layer should also permit the enabling or disabling of TCP extensions, the use of large segment sizes, and the selection of the congestion control algorithm. The segment size depends on the maximum transmission unit (MTU), which is defined by the layer-2 protocol. The congestion control algorithm must be suitable for high-throughput high-latency networks, as data transfers are often conducted over WANs.

At the application layer, applications are limited to data transfer tools at the DTN and perfSONAR at the measurement and monitoring point. The prevalent data transfer tool is Globus' GridFTP [5–7]. Globus implements features such as parallel streams and re-startable data transfer. perfSONAR [9, 10] provides an automated mechanism to actively measure and report end-to-end performance metrics.

With respect to security, by avoiding general-purpose applications and by separating the Science DMZ from the enterprise network, specific policies can be applied to the science traffic. Also, data transfer tools are relatively simple to monitor and to secure. Security policies are implemented with ACLs and offline appliances, such as IDSs. Routers and switches also provide functionality for collecting flow information, such as Netflow [12] and sFlow [13]. Netflow is a protocol used for collecting and exporting flow information that is increasingly used for monitoring big data transfers [14]. Similarly, sFlow uses sampling to decrease the amount of collected information. At high rates, inline security appliances such as firewalls and IPSs lead to packet losses and thus are not used in Science DMZs.

3 WAN Cyberinfrastructure

The Science DMZ can be treated as the portion of the cyberinfrastructure where the end devices are located. The second piece of the cyberinfrastructure is the WAN. In the U.S., there are multiple backbones and regional networks connecting institutions and corresponding Science DMZs. The primary backbone for science and engineering is Internet2 [11]. While most of this section focuses on the cyber-infrastructure needs for large flows using Internet2 as an example, the discussion is still applicable to other Research and Education Networks (RENs). A REN is a service provider network dedicated to supporting the needs of the research and education communities within a region. A particular REN that is deployed by a country is referred to as a National Research and Education Network (NREN). Examples of RENs include Internet2 in North America, GEANT [15] in Europe, UbuntuNet [16] in East and Southern Africa, APAN [17] in the Asia-Pacific region, and RedCLARA [18] in Latin America. Internet2 and RENs may contrast with commercial ISPs and Internet in several aspects, as summarized in Table 1.

Internet2 has multiple point of presences (POPs) distributed across the U.S., where institutions can connect to the network. While institutions located in the proximity of a POP can readily access a REN, others remotely located may only

Table 1 Differences between Internet and Internet2/REN

Feature	Internet	Internet2/REN
Traffic flows	Commercial flows: millions of small flows.	Research flows: smaller number of large flows.
Bandwidth	Limited, subject to ISP's policies/throttling.	Paths of up to 100 Gbps.
Network devices	Heterogeneous environment, routers and switches are not optimized for large flows.	Routers and switches with large buffer sizes suitable for accommodating large data transfers.
Bottlenecks	Congestion and outages are common.	Clear expectations, predictable WAN performance in terms of bandwidth, latency, and packet loss.
End-to-end path monitoring	Difficult to detect and solve soft failure problems. ISPs do not typically collaborate in keeping the internetwork healthy.	Easier to detect and solve soft failure problems. Active tools, such as perfSONAR, are used in Internet2 and partner networks.
Routing	Routing is achieved independently by each ISP. Routing decisions are based on policies that minimize operating costs at the expense of performance.	Routing is optimized for performance, leading to high-throughput, shorter paths.
Frame size	The maximum frame size in routers located in an ISP is typically 1500 bytes.	Routers within the Internet2 backbone support 9000-byte frames. Large frame sizes increase the throughput and the recovery speed from losses.
IPv6	Support for IPv6 is not ubiquitous.	Full IPv6 support.

connect to a REN indirectly. The connection of a Science DMZ to a REN can be accomplished in different ways, including a direct connection to the REN's POP, via a regional network, or via a commercial ISP.

3.1 Connecting a Science DMZ via an Internet2 POP

Many research institutions and universities connect directly to Internet2 via a direct link between the Science DMZ and an Internet2 POP. This connection type minimizes the number of devices or hops between the DTN and the WAN. Additionally, Internet2 is also optimized for throughput by avoiding the use of appliances that may reduce performance. Sometimes the POP is located in the institution campus, co-located with the border router. Alternatively, the institution campus may be located a few miles/kilometers away from the POP.

3.2 Connecting a Science DMZ via a Regional REN

A second option to access a major backbone/Internet2 is via a regional research network, which in turn is connected to Internet2. A representative example is the Western Regional Network (WRN) [19]. The WRN is a regional 100 Gbps REN in the western part of the U.S., as shown in Fig. 5. The interconnection with Internet2 is shown in blue. Connections to the Internet are achieved by peering with a tier-1 ISP, Level 3. The WRN is also connected to other research networks such as the Corporation for Education Network Initiatives in California (CENIC) network [20] and ESnet [21].

Figure 5 highlights the case of the University of Hawaii (UH), which has a link to the WRN. The WRN has access to Internet2 at several POPs. Although this alternative requires that flows traverse across two hierarchical levels (i.e., the WRN and Internet2), these research networks are typically optimized for performance.

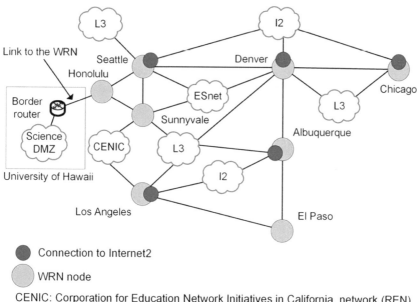

Connection to Internet2

WRN node

CENIC: Corporation for Education Network Initiatives in California network (REN)
ESnet: Energy Science network
I2: Internet2 network
L3: Level 3 network (tier-1 ISP)

Fig. 5 A Science DMZ connected to a REN, the Western Regional Network (WRN) [19]

3.3   Connecting a Science DMZ via a Commercial ISP

Most ISPs may have policies/throttling mechanisms that do not favor performance. Bottlenecks and congestion are common and clear performance expectations that cannot be established, because of the lack of collaborative monitoring between ISPs. Furthermore, policy criteria tend to dominate routing decisions rather than optimization criteria.

Figure 6a shows a use case of a campus enterprise network connected to the WAN via an ISP service. The lower level of the Internet hierarchy is the access ISP, whereas a second level provides connectivity to access ISPs, namely the regional ISP. Sometimes, the regional ISP can also provide connectivity to the end customer, i.e., the campus network. Each regional ISP then connects to a tier-1 ISP. Sometimes, the regional ISP can also provide connectivity to the end customer, i.e., the campus network. Each regional ISP then connects to a tier-1 ISP.

Figure 6b illustrates the communication between two Science DMZs in the state of New Mexico, U.S., located at Northern New Mexico College (NNMC) and at the University of New Mexico (UNM). The geographic distance between the two institutions is 90 miles (145 kilometers). NNMC is located in Espanola, where connectivity is provided by a commercial ISP. On the other hand, UNM is located in

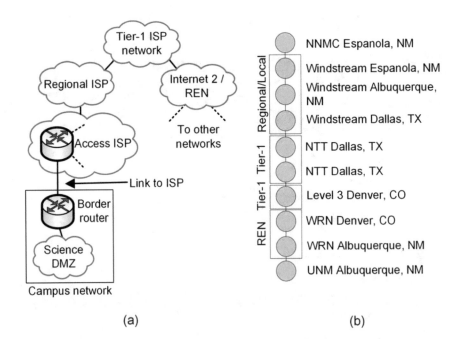

(a)   (b)

Fig. 6 Connecting a Science DMZ via an ISP. (**a**) A viewpoint of the connection in the Internet hierarchy. (**b**) The path between two Science DMZs, one attached to an ISP (NNMC) and another attached to a REN (UNM). NM, TX, and CO stand for New Mexico, Texas, and Colorado

Albuquerque and has a direct connection to a REN, namely the WRN. Note the long path between the two locations, which crosses a local/regional ISP (Windstreams), two tier-1 ISPs (NTT and Level 3), and a REN (WRN). The resulting RTT is approximately 60 milliseconds.

The above example illustrates that existing routing policies at ISPs can cause excessive delays. If instead NNMC was directly connected to a REN or Internet2, or the traffic was routed more efficiently when it entered Albuquerque, the delay would only be a few milliseconds.

3.4 Connecting a Science DMZ via a Commercial ISP Circuit

Science DMZs can be connected to Internet2 or a REN via layer-1 or layer-2 services provided by an ISP. A layer-1 service provides a dedicated wavelength on a fiber channel from the campus location to a POP of Internet2 or regional REN. A layer-2 service includes pseudowire emulation, virtual private LAN service (VPLS), and others. The advantage of this approach is that the terms of the service can be negotiated between the ISP and the institution, including a deterministic path to be followed by packets from the border router to the POP. Table 2 summarizes the four alternatives discussed in this section to connect Science DMZs to Internet2.

4 Current State: Science DMZ Deployment in the U.S.

The NSF recognizes the Science DMZ model as a proven operational best practice for university campuses supporting data-intensive science. This model has also been identified as eligible for funding through the NSF Campus Cyberinfrastructure program (CC*) [23]. Established in 2012, this program has funded more than 200 projects for network infrastructure deployment/Science DMZs. The locations of these institutions are shown in Fig. 7. Since a design goal of the Science DMZ is the establishment of a high-speed path across a WAN, the impact on improving the exchange of large data sets is significant. In essence, because of the data-sharing architecture of the Science DMZ, institutions implementing it have fast access to virtual co-location of large data that could reside anywhere in the world.

Academic Cloud and Virtual Laboratories
The book is accompanied by hands-on virtual laboratory experiments conducted in a cloud system, referred to as the Academic Cloud. Access to the Academic Cloud is available for a fee (6-month access) and includes all material needed to conduct the experiments. The URL is:

http://highspeednetworks.net/

Table 2 Alternative approaches to connect a Science DMZ to Internet2

Connection	Advantages	Disadvantages
SDMZ to Internet2 via a direct POP link	▷ Optimal technical approach; no additional hops from the Science DMZ to Internet2. ▷ Routing is optimized for performance. ▷ Internet2 has active performance monitoring. Thus, it is easier to detect soft failures.	▷ Based on location and providers, service may be more expensive than that of commercial service providers. ▷ Location; POPs to Internet2 may not be accessible to the client institution.
SDMZ to Internet2 via a regional research network	▷ If the regional research network is optimized for performance, there is a minimal performance degradation with respect to a direct link connection to Internet2 POP. ▷ Costs may be lower than that of establishing a direct link to an Internet2 POP.	▷ Additional hops are added to reach Internet2 backbone; packets must traverse at least two levels in the network hierarchy: to the research network and to Internet2.
SDMZ to Internet2 via commercial ISP circuit	▷ Resources are reserved in advance (bandwidth), and a more predictable quality of service is guaranteed (compared to regular commercial service).	▷ Additional hops and latency are added to reach Internet2 backbone; packets traverse at least two levels in the network. ▷ Soft failures may not be easy to detect if they occur within the network of the service provider.
SDMZ to Internet2 via a regular commercial ISP	▷ Costs are typically lower than that of connecting the Science DMZ to a research network or to an Internet2 POP.	▷ Performance is unpredictable due to congestion, latency, inadequate equipment for large flows, and bandwidth policies. ▷ End-to-end path monitoring and detection of soft failures are difficult.

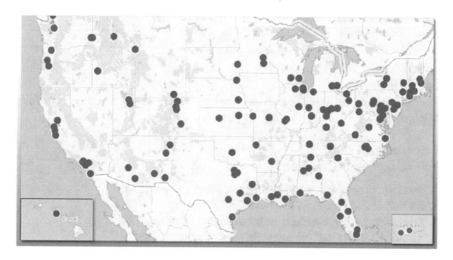

Fig. 7 Locations of institutions that have implemented cyberinfrastructure improvements and/or have deployed Science DMZs with the support of the NSF Campus Cyberinfrastructure program, as of 2016 [22]

Chapter 2—Lab 3: Emulating WAN with NETEM Part I—Latency and Jitter

Overview

To conduct the experiment described in this section, please login into the Academic Cloud at http://highspeednetworks.net/ and reserve a pod for Lab 3

This lab introduces NETEM and explains how it can be used to emulate real-world scenarios while having control on parameters that affect the performance of networks. Network parameters include latency, jitter, packet loss, reordering, and corruption. Correlation values between network parameters will also be set to provide a more realistic network environment.

Objectives

By the end of this lab, students should be able to:

1. Understand delay in networks and how to measure it.
2. Understand Linux queuing disciplines (qdisc) architecture.
3. Deploy emulated WANs characterized by large delays using NETEM and Mininet.
4. Perform measurements after introducing delays to an emulated WAN.
5. Deploy emulated WANs characterized by delays, jitters, and corresponding correlation values.
6. Modify the delay distribution of an emulated WAN.

Table 3 Credentials to
access Client1 machine

Device	Account	Password
Client1	admin	password

Lab Settings

The information in Table 3 provides the credentials of the machine containing Mininet.

Lab Roadmap

This lab is organized as follows:

1. Section 5: Introduction to network emulators and NETEM.
2. Section 6: Lab topology.
3. Section 7: Adding/changing delay to emulate a WAN.
4. Section 8: Restoring original values (deleting the rules).
5. Section 9: Adding jitter to emulated WAN.
6. Section 10: Adding correlation value for jitter and delay.
7. Section 11: Delay distribution.

5 Introduction to Network Emulators and NETEM

Network emulators play an important role for the research and development of network protocols and applications. Network emulators provide the ability to perform tests of realistic scenarios in a controlled manner, which is very difficult on production networks. This is particularly complex for researchers who develop and test tools for *Wide Area Networks (WANs)* and for multi-domain environments.

5.1 NETEM

One of the most popular network emulators is *NETEM*, a Linux network emulator for testing the performance of real applications over a virtual network. The virtual network may reproduce long-distance WANs in the lab environment. These scenarios facilitate the test and evaluation of protocols and devices from the application layer to the data-link layer under a variety of conditions. NETEM allows the user to modify parameters such as delay, jitter, packet loss, duplication, and re-ordering of packets.

NETEM is implemented in Linux and consists of two portions: a small kernel module for a queuing discipline and a command-line utility to configure it. Figure 8 shows the basic architecture of Linux queuing disciplines. The queuing disciplines exist between the IP protocol output and the network device. The default queuing discipline is a simple packet first-in first-out (FIFO) queue. A queuing discipline is a simple object with two interfaces. One interface queues packets to be sent and the

Fig. 8 Linux queueing
discipline

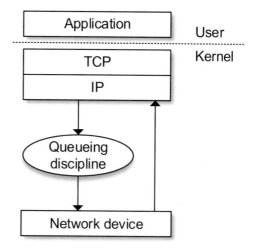

other interface releases packets to the network device. The queuing discipline makes
the policy decision of which packets to send, which packets to delay, and which
packets to drop. A classful queueing discipline, such as NETEM, has configurable
internal modules.

5.2 WANs and Delay

In networks, there are several processes and devices that contribute to the end-to-
end delay between a sender node and a destination node. Many times, the end-to-
end delay is dominated by the WAN's propagation delay. Consider two adjacent
switches A and B connected by a WAN. Once a bit is pushed onto the WAN by
switch A, it needs to propagate to switch B. The time required to propagate from
the beginning of the WAN to switch B is the propagation delay. The bit propagates
at the propagation speed of the WAN's link. The propagation speed depends on the
physical medium (that is, fiber optics, twisted-pair copper wire, etc.) and is in the
range of 2×10^8 m/s to 3×10^8 m/s, which is equal to, or a little less than, the speed
of light. The propagation delay is the distance between two switches divided by the
propagation speed. Once the last bit of the packet propagates to switch B, it and all
the preceding bits of the packet are stored in switch B.

Network tools usually estimate delay for troubleshooting and performance
measurements. For example, an estimate of end-to-end delay is the Round-Trip
Time (RTT), which is the time it takes for a small packet to travel from sender
to receiver and then back to the sender. The RTT includes packet-propagation
delays, packet-queuing delays in intermediate routers and switches, and packet-
processing delays. As mentioned above, if the propagation delay dominates other

delay components (as in the case of many WANs), then RTT is also an estimate of the propagation delay.

6   Lab Topology

Let's get started with creating a simple Mininet topology using MiniEdit. The topology uses 10.0.0.0/8, which is the default network assigned by Mininet (Fig. 9).

Step 1. A shortcut to MiniEdit is located on the machine's Desktop. Start MiniEdit by clicking on MiniEdit's shortcut (Fig. 10). When prompted for a password, type password .

Step 2. On MiniEdit's menu bar, click on *File* and then *Open* to load the lab's topology. Locate the *Lab 3.mn* topology file and click on *Open* (Fig. 11).

Step 3. Before starting the measurements between host h1 and host h2, the network must be started. Click on the *Run* button located at the bottom left of MiniEdit's window to start the emulation (Fig. 12).

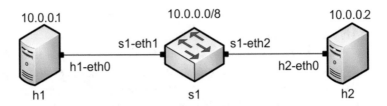

Fig. 9 Lab topology

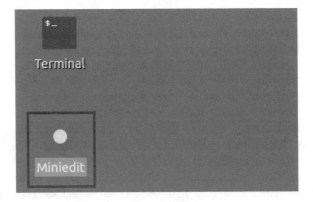

Fig. 10 MiniEdit shortcut

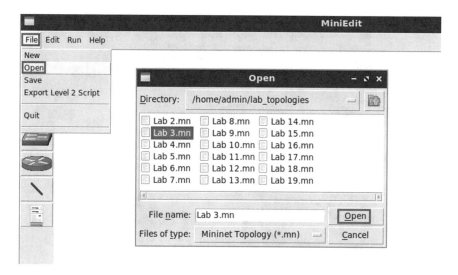

Fig. 11 MiniEdit's *Open* dialog

Fig. 12 Running the
emulation

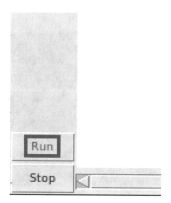

The above topology uses 10.0.0.0/8, which is the default network assigned by Mininet.

6.1 Starting Host h1 and Host h2

Step 1. Hold the right-click on host h1 and select *Terminal* (Fig. 13). This opens the terminal of host h1 and allows the execution of commands on host h1.

Step 2. Test connectivity between the end-hosts using the | ping | command. On host h1, type the command | ping 10.0.0.2 | (Fig. 14). This command tests the

File Edit Run Help
s1
h2
Host Options
Terminal

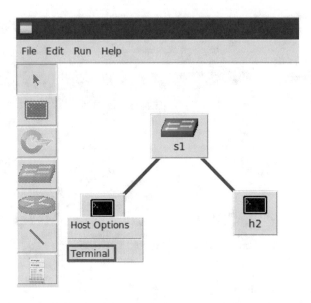

Fig. 13 Opening a terminal on host h1

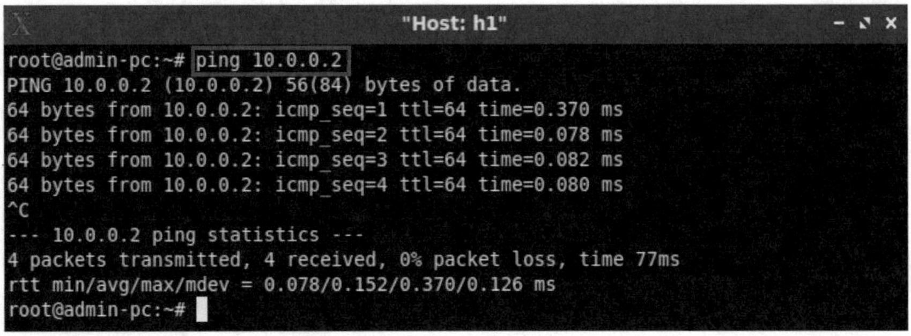

"Host: h1"
root@admin-pc:~# ping 10.0.0.2
PING 10.0.0.2 (10.0.0.2) 56(84) bytes of data.
64 bytes from 10.0.0.2: icmp_seq=1 ttl=64 time=0.370 ms
64 bytes from 10.0.0.2: icmp_seq=2 ttl=64 time=0.078 ms
64 bytes from 10.0.0.2: icmp_seq=3 ttl=64 time=0.082 ms
64 bytes from 10.0.0.2: icmp_seq=4 ttl=64 time=0.080 ms
^C
--- 10.0.0.2 ping statistics ---
4 packets transmitted, 4 received, 0% packet loss, time 77ms
rtt min/avg/max/mdev = 0.078/0.152/0.370/0.126 ms
root@admin-pc:~#

Fig. 14 Connectivity test using `ping` command

connectivity between host h1 and host h2. To stop the test, press `Ctrl+c`. The figure below shows a successful connectivity test.

The figure above indicates that there is connectivity between host h1 and host h2. Thus, we are ready to start the throughput measurement process.

7 Adding/Changing Delay to Emulate a WAN

The user invokes NETEM using the command-line utility called $\boxed{\text{tc}}$. With no additional parameters, NETEM behaves as a basic FIFO queue with no delay, loss, duplication, or reordering of packets. The basic $\boxed{\text{tc}}$ syntax used with NETEM is as follows:

```
sudo tc qdisc [add|del|replace|change|show] dev dev
id root netem opts
```

- $\boxed{\text{sudo}}$: enable the execution of the command with higher security privileges.
- $\boxed{\text{tc}}$: command used to interact with NETEM.
- $\boxed{\text{qdisc}}$: a queue discipline (qdisc) is a set of rules that determine the order in which packets arriving from the IP protocol output (see Fig. 8) are served. The queue discipline is applied to a packet queue to decide when to send each packet.
- $\boxed{\text{[add |del |replace |change |show]}}$: this is the operation on qdisc. For example, to add delay on a specific interface, the operation will be $\boxed{\text{add}}$. To change or remove delay on the specific interface, the operation will be $\boxed{\text{change}}$ or $\boxed{\text{del}}$
- $\boxed{\text{dev\_id}}$: this parameter indicates the interface to be subject to emulation.
- $\boxed{\text{opts}}$: this parameter indicates the amount of delay, packet loss, duplication, corruption, and others.

7.1 Identify Interface of Host h1 and Host h2

According to the previous section, we must identify the interfaces on the connected hosts.

Step 1. On host h1, type the command $\boxed{\text{ifconfig}}$ to display information related to its network interfaces and their assigned IP addresses (Fig. 15).

The output of the $\boxed{\text{ifconfig}}$ command indicates that host h1 has two interfaces: *h1-eth0* and *lo*. The interface *h1-eth0* at host h1 is configured with IP address 10.0.0.1 and subnet mask 255.0.0.0. This interface must be used in $\boxed{\text{tc}}$ when emulating the WAN.

Step 2. In host h2, type the command $\boxed{\text{ifconfig}}$ as well (Fig. 16).

The output of the $\boxed{\text{ifconfig}}$ command indicates that host h2 has two interfaces: *h2-eth0* and *lo*. The interface *h2-eth0* at host h1 is configured with IP address 10.0.0.2 and subnet mask 255.0.0.0. This interface must be used in $\boxed{\text{tc}}$ when emulating the WAN.

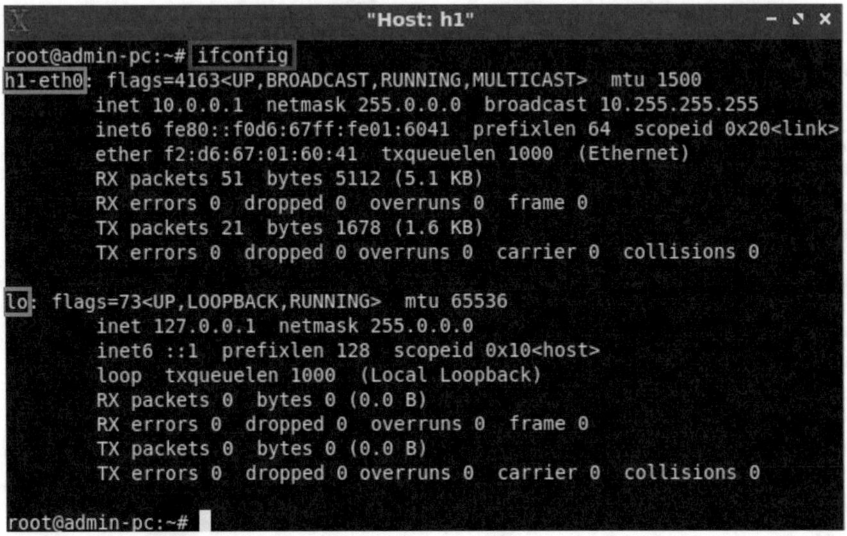

```
X                                   "Host: h1"                        -  ↙ ×
root@admin-pc:~# ifconfig
h1-eth0: flags=4163<UP,BROADCAST,RUNNING,MULTICAST>  mtu 1500
         inet 10.0.0.1  netmask 255.0.0.0  broadcast 10.255.255.255
         inet6 fe80::f0d6:67ff:fe01:6041  prefixlen 64  scopeid 0x20<link>
         ether f2:d6:67:01:60:41  txqueuelen 1000  (Ethernet)
         RX packets 51  bytes 5112 (5.1 KB)
         RX errors 0  dropped 0  overruns 0  frame 0
         TX packets 21  bytes 1678 (1.6 KB)
         TX errors 0  dropped 0 overruns 0  carrier 0  collisions 0

lo: flags=73<UP,LOOPBACK,RUNNING>  mtu 65536
         inet 127.0.0.1  netmask 255.0.0.0
         inet6 ::1  prefixlen 128  scopeid 0x10<host>
         loop  txqueuelen 1000  (Local Loopback)
         RX packets 0  bytes 0 (0.0 B)
         RX errors 0  dropped 0  overruns 0  frame 0
         TX packets 0  bytes 0 (0.0 B)
         TX errors 0  dropped 0 overruns 0  carrier 0  collisions 0

root@admin-pc:~#
```

Fig. 15 Output of ifconfig command on host h1

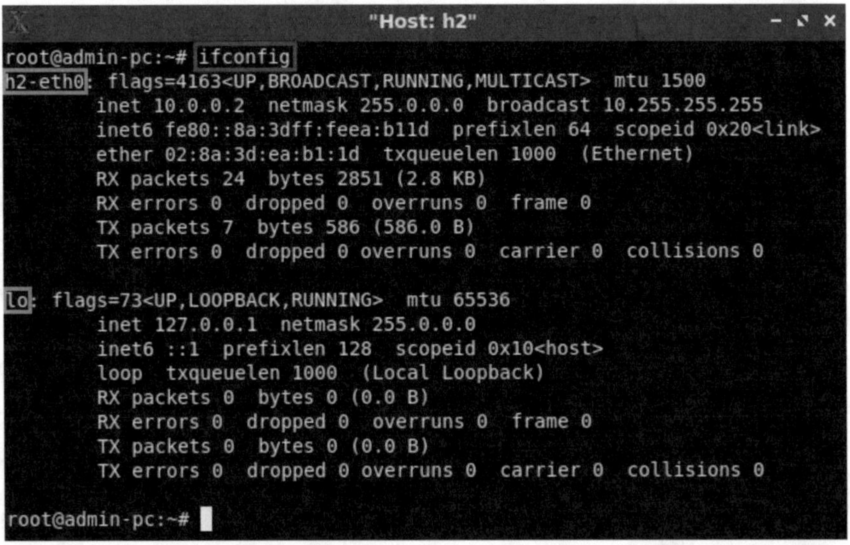

```
X                                   "Host: h2"                        -  ↙ ×
root@admin-pc:~# ifconfig
h2-eth0: flags=4163<UP,BROADCAST,RUNNING,MULTICAST>  mtu 1500
         inet 10.0.0.2  netmask 255.0.0.0  broadcast 10.255.255.255
         inet6 fe80::8a:3dff:feea:b11d  prefixlen 64  scopeid 0x20<link>
         ether 02:8a:3d:ea:b1:1d  txqueuelen 1000  (Ethernet)
         RX packets 24  bytes 2851 (2.8 KB)
         RX errors 0  dropped 0  overruns 0  frame 0
         TX packets 7  bytes 586 (586.0 B)
         TX errors 0  dropped 0 overruns 0  carrier 0  collisions 0

lo: flags=73<UP,LOOPBACK,RUNNING>  mtu 65536
         inet 127.0.0.1  netmask 255.0.0.0
         inet6 ::1  prefixlen 128  scopeid 0x10<host>
         loop  txqueuelen 1000  (Local Loopback)
         RX packets 0  bytes 0 (0.0 B)
         RX errors 0  dropped 0  overruns 0  frame 0
         TX packets 0  bytes 0 (0.0 B)
         TX errors 0  dropped 0 overruns 0  carrier 0  collisions 0

root@admin-pc:~#
```

Fig. 16 Output of ifconfig command on host h2

7.2   Add Delay to Interface Connecting to WAN

Network emulators emulate delays by introducing them to an interface. For example, the delay introduced to a switch A's interface that is connected to a switch B's

```
"Host: h1"                                             - ᴎ x
root@admin-pc:~# sudo tc qdisc add dev h1-eth0 root netem delay 100ms
root@admin-pc:~# █
```

Fig. 17 Adding 100 ms delay to the interface *h1-eth0*

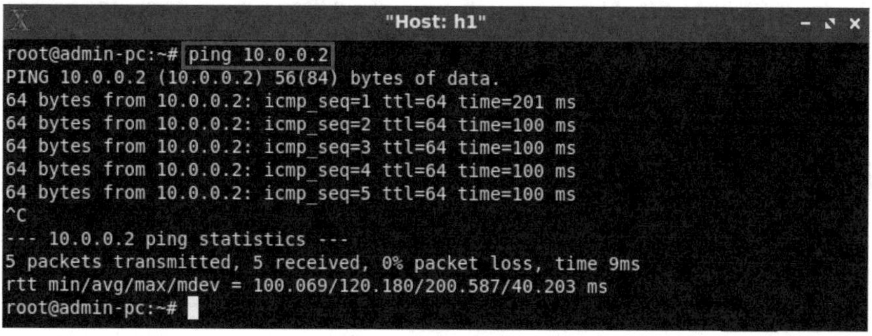

```
"Host: h1"                                             - ᴎ x
root@admin-pc:~# ping 10.0.0.2
PING 10.0.0.2 (10.0.0.2) 56(84) bytes of data.
64 bytes from 10.0.0.2: icmp_seq=1 ttl=64 time=201 ms
64 bytes from 10.0.0.2: icmp_seq=2 ttl=64 time=100 ms
64 bytes from 10.0.0.2: icmp_seq=3 ttl=64 time=100 ms
64 bytes from 10.0.0.2: icmp_seq=4 ttl=64 time=100 ms
64 bytes from 10.0.0.2: icmp_seq=5 ttl=64 time=100 ms
^C
--- 10.0.0.2 ping statistics ---
5 packets transmitted, 5 received, 0% packet loss, time 9ms
rtt min/avg/max/mdev = 100.069/120.180/200.587/40.203 ms
root@admin-pc:~# █
```

Fig. 18 Verifying latency after emulating delay using ping

interface may represent the propagation delay of a WAN connecting both switches. In this section, you will use netem command to insert delay to a network interface.

Step 1. In host h1, type the following command (Fig. 17):

```
sudo tc qdisc add dev h1-eth0 root netem delay 100 ms
```

This command can be summarized as follows:

- sudo : enable the execution of the command with higher security privileges.
- tc : invoke Linux's traffic control.
- qdisc : modify the queuing discipline of the network scheduler.
- add : create a new rule.
- dev h1-eth0 : specify the interface on which the rule will be applied.
- netem : use the network emulator.
- delay 100 ms : inject delay of 100 ms.

The above command adds a delay of 100 milliseconds (ms) to the output interface, exclusively.

Step 2. The user can verify now that the connection from host h1 to host h2 has a delay of 100 milliseconds by using the ping command from host h1 (Fig. 18).

```
ping 10.0.0.2
```

The result above indicates that all five packets were received successfully (0% packet loss) and that the minimum, average, maximum, and standard deviation of the Round-Trip Time (RTT) were 100.069, 120.180, 200.587, and 40.203 milliseconds, respectively.

Note that the above scenario emulates 100 milliseconds latency on the interface of host h1 connecting to the switch. In order to emulate a WAN where the delay is bidirectional, a delay of 100 milliseconds must also be added to the corresponding interface on host h2.

Step 3. In host h2's terminal, type the following command (Fig. 19):

```
sudo tc qdisc add dev h2-eth0 root netem delay 100 ms
```

Step 4. The user can verify now that the connection between host h1 and host h2 has an RTT of 200 milliseconds (100 ms from host h1 to host h2 plus 100 ms from host h2 to host h1) by retyping the ping command on host h's terminal (Fig. 20).

```
ping 10.0.0.2
```

The result above indicates that all five packets were received successfully (0% packet loss) and that the minimum, average, maximum, and standard deviation of the Round-Trip Time (RTT) were 200.078, 200.154, 204.447, and 0.511 milliseconds, respectively.

Fig. 19 Adding 100 ms delay to the interface *h2-eth0*

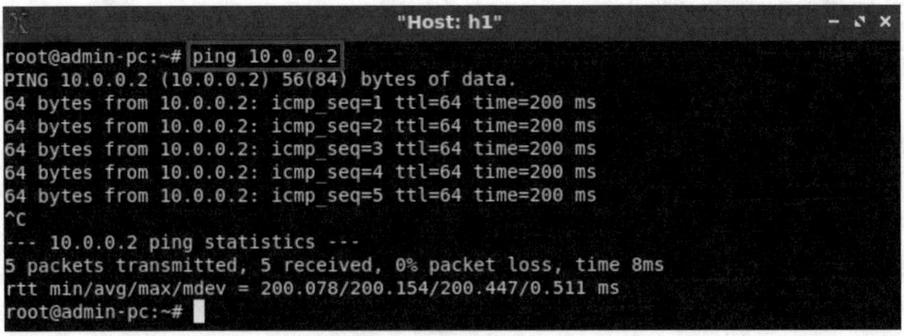

Fig. 20 Verifying latency after emulating delay on both host h1 and host h2 using ping

7.3   Changing the Delay in Emulated WAN

In this section, the user will change the delay from 100 milliseconds to 50 milliseconds in both sender and receiver. The RTT will be 100 milliseconds now.

Step 1. In host h1's terminal, type the following command (Fig. 21):

```
sudo tc qdisc change dev h1-eth0 root netem delay 50 ms
```

The new option added here is change , which changes the previously set delay to 50 milliseconds.

Step 2. Apply also the above step on host h2's terminal to change the delay to 50 ms (Fig. 22).

```
sudo tc qdisc change dev h2-eth0 root netem delay 50 ms
```

Step 3. The user can verify now that the connection from host h1 to host h2 has a delay of 100 milliseconds by using the ping command from host h's terminal (Fig. 23).

```
ping 10.0.0.2
```

The result above indicates that all five packets were received successfully (0% packet loss) and that the minimum, average, maximum, and standard deviation of the Round-Trip Time (RTT) were 100.079, 100.149, 100.411, and 0.131 milliseconds, respectively.

Fig. 21 Changing delay on the interface *h1-eth0*

Fig. 22 Changing delay to the interface *h2-eth0*

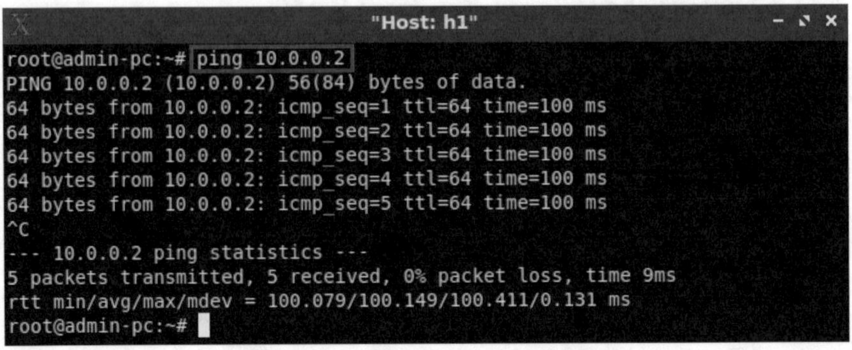

```
                              "Host: h1"                      — ↗ ✕
root@admin-pc:~# ping 10.0.0.2
PING 10.0.0.2 (10.0.0.2) 56(84) bytes of data.
64 bytes from 10.0.0.2: icmp_seq=1 ttl=64 time=100 ms
64 bytes from 10.0.0.2: icmp_seq=2 ttl=64 time=100 ms
64 bytes from 10.0.0.2: icmp_seq=3 ttl=64 time=100 ms
64 bytes from 10.0.0.2: icmp_seq=4 ttl=64 time=100 ms
64 bytes from 10.0.0.2: icmp_seq=5 ttl=64 time=100 ms
^C
--- 10.0.0.2 ping statistics ---
5 packets transmitted, 5 received, 0% packet loss, time 9ms
rtt min/avg/max/mdev = 100.079/100.149/100.411/0.131 ms
root@admin-pc:~#
```

Fig. 23 Verifying latency after emulating 100 ms delay using ping

```
                              "Host: h1"                      — ↗ ✕
root@admin-pc:~# sudo tc qdisc del dev h1-eth0 root netem
root@admin-pc:~#
```

Fig. 24 Deleting all rules on interface *h1-eth0*

```
                              "Host: h2"                      — ↗ ✕
root@admin-pc:~# sudo tc qdisc del dev h2-eth0 root netem
root@admin-pc:~#
```

Fig. 25 Deleting all rules on interface *h2-eth0*

8   Restoring Original Values (Deleting the Rules)

In this section, the user will restore the default configuration in both sender and receiver by deleting all the rules applied to the network scheduler of an interface.

Step 1. In host h1's terminal, type the following command (Fig. 24):

```
sudo tc qdisc del dev h1-eth0 root netem
```

The new option added here is del , which deletes the previously set rules on a given interface. As a result, the tc qdisc will restore its default values of the device *h1-eth0*.

Step 2. Apply the same steps to remove rules on host h2. In host h2's terminal, type the following command (Fig. 25):

```
sudo tc qdisc del dev h2-eth0 root netem
```

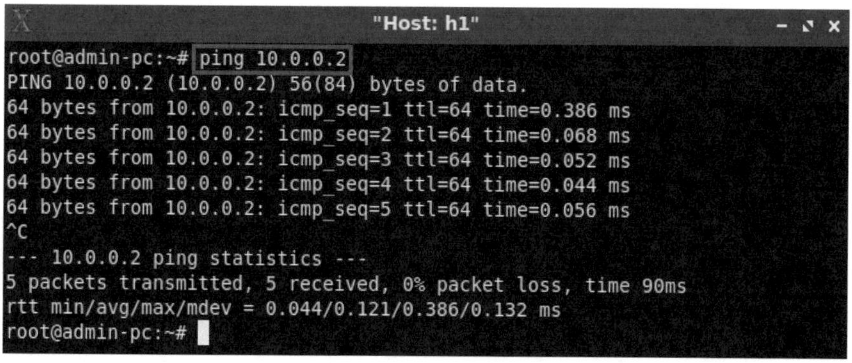

```
                              "Host: h1"                      —  ⬛ ✕
root@admin-pc:~# ping 10.0.0.2
PING 10.0.0.2 (10.0.0.2) 56(84) bytes of data.
64 bytes from 10.0.0.2: icmp_seq=1 ttl=64 time=0.386 ms
64 bytes from 10.0.0.2: icmp_seq=2 ttl=64 time=0.068 ms
64 bytes from 10.0.0.2: icmp_seq=3 ttl=64 time=0.052 ms
64 bytes from 10.0.0.2: icmp_seq=4 ttl=64 time=0.044 ms
64 bytes from 10.0.0.2: icmp_seq=5 ttl=64 time=0.056 ms
^C
--- 10.0.0.2 ping statistics ---
5 packets transmitted, 5 received, 0% packet loss, time 90ms
rtt min/avg/max/mdev = 0.044/0.121/0.386/0.132 ms
root@admin-pc:~# ▊
```

Fig. 26 Verifying latency after deleting all rules on both devices

As a result, the $\boxed{\text{tc}}$ queueing discipline will restore its default values of the device *h2-eth0*.

Step 3. The user can now verify that the connection from host h1 to host h2 has no explicit delay set by using the $\boxed{\text{ping}}$ command from host h's terminal (Fig. 26).

```
ping 10.0.0.2
```

The result above indicates that all five packets were received successfully (0% packet loss) and that the minimum, average, maximum, and standard deviation of the Round-Trip Time (RTT) were 0.044, 0.121, 0.386, and 0.132 milliseconds, respectively.

9   Adding Jitter to Emulated WAN

Networks do not exhibit constant delay; the delay may vary based on other traffic flows contending for the same path. Jitter is the variation of delay time. The delay parameters are described by the average value (μ), standard deviation (σ), and correlation. By default, NETEM uses a uniform distribution, so that the delay is within $\mu \pm \sigma$.

9.1   Add Jitter to Interface Connecting to WAN

In this section, the user will add delay of 100 milliseconds with a random variation of ± 10 milliseconds. Before doing so, make sure to restore the default configuration of the interfaces on host h1 and host h2 by applying the commands of Sect. 8. Then, apply the commands below:

Step 1. In host h1's terminal, type the following command (Fig. 27):

"Host: h1"
root@admin-pc:~# sudo tc qdisc add dev h1-eth0 root netem delay 100ms 10ms
root@admin-pc:~#

Fig. 27 Add 100 ms delay with ±10 millisecond

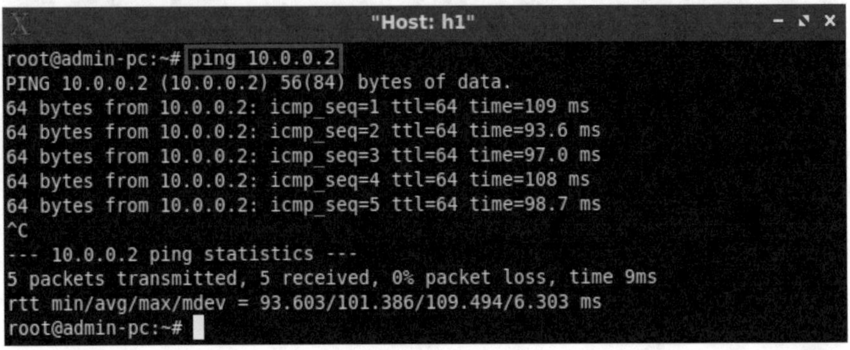
"Host: h1"
root@admin-pc:~# ping 10.0.0.2
PING 10.0.0.2 (10.0.0.2) 56(84) bytes of data.
64 bytes from 10.0.0.2: icmp_seq=1 ttl=64 time=109 ms
64 bytes from 10.0.0.2: icmp_seq=2 ttl=64 time=93.6 ms
64 bytes from 10.0.0.2: icmp_seq=3 ttl=64 time=97.0 ms
64 bytes from 10.0.0.2: icmp_seq=4 ttl=64 time=108 ms
64 bytes from 10.0.0.2: icmp_seq=5 ttl=64 time=98.7 ms
^C
--- 10.0.0.2 ping statistics ---
5 packets transmitted, 5 received, 0% packet loss, time 9ms
rtt min/avg/max/mdev = 93.603/101.386/109.494/6.303 ms
root@admin-pc:~#

Fig. 28 Verifying RTT after adding 100 millisecond delay and 10 millisecond jitter on interface *h1-eth0*

```
sudo tc qdisc add dev h1-eth0 root netem delay 100ms 10ms
```

The new value added here represents jitter, which defines the delay variation. Therefore, all packets leaving host h1 via interface *h1-eth0* will experience a delay of 100 ms, with a random variation of ±10 ms.

Step 2. The user can now verify that the connection from host h1 to host h2 has 100 ms delay with±10 millisecond random variation by using the `ping` command on host h's terminal (Fig. 28).

```
ping 10.0.0.2
```

The result above indicates that all five packets were received successfully (0% packet loss) and that the minimum, average, maximum, and standard deviation of the Round-Trip Time (RTT) were 93.603, 101.386, 109.494, and 6.303 milliseconds, respectively. Note that we are only adding jitter to the interface of host h1 at this point.

Step 3. In host h1's terminal, type the following command to delete previous configurations (Fig. 29):

```
sudo tc qdisc del dev h1-eth0 root netem
```

Fig. 29 Deleting all rules on interface *h1-eth0*

```
                              "Host: h1"                          − ↗ ×
root@admin-pc:~# sudo tc qdisc add dev h1-eth0 root netem delay 100ms 10ms 25%
root@admin-pc:~# █
```

Fig. 30 Adding a correlation value of 25%

10 Adding Correlation Value for Jitter and Delay

The correlation parameter controls the relationship between successive pseudo-random values. In this section, the user will add a delay of 100 milliseconds with a variation of ±10 milliseconds while adding a correlation value. Before doing so, make sure to restore the default configuration of the interfaces on host h1 and host h2 by applying the commands of Sect. 8. Then, apply the commands below:

Step 1. In host h1 terminal, type the following command (Fig. 30):

```
sudo tc qdisc add dev h1-eth0 root netem delay 100ms 10ms
25%
```

The new value added here represents the correlation value for jitter and delay. Therefore, all packets leaving the device host h1 on the interface *h1-eth0* will experience a 100 ms delay time, with a random variation of ±10 millisecond with the next random packet depending 25% on the previous one.

Step 2. Now, the user can test the connection from host h1 to host h2 by using the ping command on host h's terminal (Fig. 31).

```
ping 10.0.0.2
```

The result above indicates that all five packets were received successfully (0% packet loss) and that the minimum, average, maximum, and standard deviation of the Round-Trip Time (RTT) were 90.891, 101.007, 109.215, and 6.328 milliseconds, respectively.

Step 3. In host h1's terminal, type the following command to delete previous configurations (Fig. 32):

```
sudo tc qdisc del dev h1-eth0 root netem
```

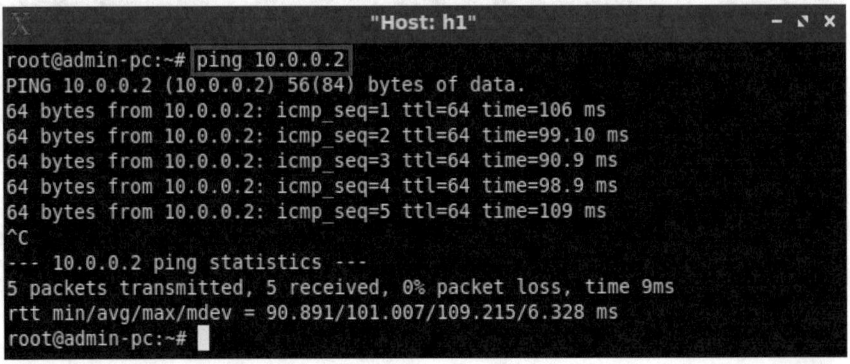

```
X                                    "Host: h1"                           -  ✕  ✕
root@admin-pc:~# ping 10.0.0.2
PING 10.0.0.2 (10.0.0.2) 56(84) bytes of data.
64 bytes from 10.0.0.2: icmp_seq=1 ttl=64 time=106 ms
64 bytes from 10.0.0.2: icmp_seq=2 ttl=64 time=99.10 ms
64 bytes from 10.0.0.2: icmp_seq=3 ttl=64 time=90.9 ms
64 bytes from 10.0.0.2: icmp_seq=4 ttl=64 time=98.9 ms
64 bytes from 10.0.0.2: icmp_seq=5 ttl=64 time=109 ms
^C
--- 10.0.0.2 ping statistics ---
5 packets transmitted, 5 received, 0% packet loss, time 9ms
rtt min/avg/max/mdev = 90.891/101.007/109.215/6.328 ms
root@admin-pc:~# █
```

Fig. 31 Verifying latency after setting the correlation value

```
X                                    "Host: h1"                           -  ✕  ✕
root@admin-pc:~# sudo tc qdisc del dev h1-eth0 root netem
root@admin-pc:~# █
```

Fig. 32 Deleting all rules on interface *h1-eth0*

11 Delay Distribution

NETEM permits user to specify a distribution that describes how delays vary in the network. Usually delays are not uniform, so it may be convenient to use a non-uniform distribution such as normal, pareto, or pareto-normal. For this test, the user will specify a normal distribution for the delay in the emulated network. Before doing so, make sure to restore the default configuration of the interfaces on host h1 and host h2 by applying the commands of Sect. 8. Then, apply the commands below:

Step 1. In host h1's terminal, type the following command (Fig. 33):

```
sudo tc qdisc add dev h1-eth0 root netem delay 100ms
20ms distribution normal
```

The new option added here (distribution) represents the delay distribution type. We define the delay to have a normal distribution, which provides a more realistic emulation of WAN networks. As a result, all packets leaving the host h1 on the interface *h1-eth0* will experience delay time that is normally distributed between the range of 100ms ±20ms.

Step 2. The user can now verify if the configuration was successfully done in the previous step (Step 1) by using the ping command on host h's terminal (Fig. 34)

```
"Host: h1"                                                        - ₰ ×
root@admin-pc:~# sudo tc qdisc add dev h1-eth0 root netem delay 100ms 20ms distribution normal
root@admin-pc:~#
```

Fig. 33 Adding normal distribution of delay

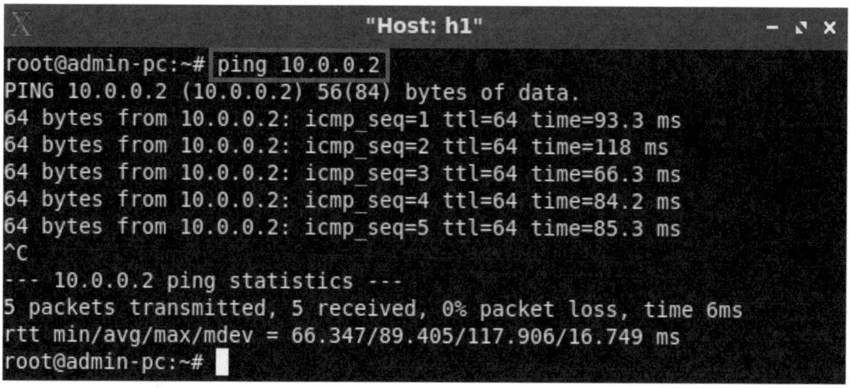

```
"Host: h1"                                              - ₰ ×
root@admin-pc:~# ping 10.0.0.2
PING 10.0.0.2 (10.0.0.2) 56(84) bytes of data.
64 bytes from 10.0.0.2: icmp_seq=1 ttl=64 time=93.3 ms
64 bytes from 10.0.0.2: icmp_seq=2 ttl=64 time=118 ms
64 bytes from 10.0.0.2: icmp_seq=3 ttl=64 time=66.3 ms
64 bytes from 10.0.0.2: icmp_seq=4 ttl=64 time=84.2 ms
64 bytes from 10.0.0.2: icmp_seq=5 ttl=64 time=85.3 ms
^C
--- 10.0.0.2 ping statistics ---
5 packets transmitted, 5 received, 0% packet loss, time 6ms
rtt min/avg/max/mdev = 66.347/89.405/117.906/16.749 ms
root@admin-pc:~#
```

Fig. 34 Verifying latency after using normal distribution

```
ping 10.0.0.2
```

The result above indicates that all five packets were received successfully (0% packet loss) and that the minimum, average, maximum, and standard deviation of the Round-Trip Time (RTT) were 66.347, 89.405, 117.906, and 16.749 milliseconds, respectively.

Chapter 2—Lab 4: Emulating WAN with NETEM II: Packet Loss, Duplication, Reordering, and Corruption

Overview
To conduct the experiment described in this section, please login into the Academic Cloud at http://highspeednetworks.net/ and reserve a pod for Lab 4.

This lab continues the description of NETEM and how to use it to emulate Wide Area Networks (WANs). Besides delay, this lab focuses on other parameters such as packet loss, packet duplication, reordering, and packet corruption. These parameters affect the performance of protocols and networks.

Objectives
By the end of this lab, students should be able to:

1. Deploy emulated WANs characterized by parameters such as delay, packet loss, packet corruption, packet reordering, and packet duplication.

Table 4 Credentials to
access Client1 machine

Device	Account	Password
Client1	admin	password

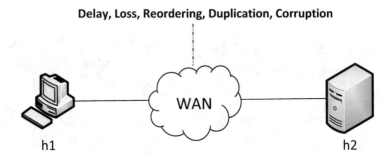

Delay, Loss, Reordering, Duplication, Corruption

Fig. 35 Parameters affecting throughput in a WAN

2. Measure the performance of WANs characterized by different parameter values.
3. Visualize WAN performance measures.

Lab Settings
The information in Table 4 provides the credentials of the machine containing Mininet.

Lab Roadmap
This lab is organized as follows:

1. Section 12: Introduction to network emulators and NETEM.
2. Section 13: Lab topology.
3. Section 14: Adding/changing packet loss.
4. Section 15: Adding packet corruption.
5. Section 16: Adding packet reordering.
6. Section 17: Adding packet duplication.

12   Introduction to Network Emulators and NETEM

Part I of Emulating WAN with NETEM described how to use NETEM to emulate WANs characterized by long delays. Part I also explained how the end-to-end delay can be dominated by the WAN's propagation delay and how the Round-Trip Time (RTT) estimates this delay.

In addition to delay, many WANs and LANs are subject to packet loss, reordering, corruption, and duplication (Fig. 35).

The above situations are described follows:

1. Packet loss: a condition that occurs when a packet travelling across a network fails to reach its destination. Packet loss may have a large impact on high-

throughput high-latency networks. A common cause of packet loss is the inability of routers to hold packets arriving at a rate higher than the departure rate. Even in cases where the high packet arrival rate is only temporary (e.g., short-term traffic bursts), the router is limited by the amount of buffer memory used to momentarily store packets. When packet loss occurs, TCP reduces the congestion window and consequently the throughput by half. Packet loss must be mitigated by using best-practice network designs, such as Science DMZ.

2. Packet reordering: a condition that occurs when packets are received in a different order from which they were sent. Packet reordering, also known as out-of-order packet delivery, is typically the result of packets following different routes to reach their destination. Packet reordering may deteriorate the throughput of TCP connections in high-throughput high-latency networks. For each segment received out of order, a TCP receiver sends an acknowledgement (ACK) for the last correctly received segment. Once the TCP sender receives three acknowledgements for the same segment (triple duplicate ACK), the sender considers that the receiver did not correctly receive the packet following the packet that is being acknowledged three times. It then proceeds to reduce the congestion window and throughput by half.

3. Packet corruption: corruption of bits comprising a packet may (mostly) occur at the physical layer. Two adjacent devices are connected by a physical channel (e.g., fiber, twisted-pair copper wire, etc.). The physical layer accepts a raw bit stream and delivers it to the data-link layer. If corruption occurs, some bits may have different values than those originally sent by the sender node. The receiver node then simply discards the packet. As a result, the TCP sender process will not receive an acknowledgement for the corresponding segment and will consider it as a lost segment. The TCP sender process will subsequently decrease the congestion window and throughput by half.

4. Packet duplication: a condition where multiple copies of a packet are present in the network and received by the destination. Packet duplication is the result of retransmissions, where a sender node retransmits unacknowledged (NACK) packets.

Packet loss, reordering, and corruption (the last two are interpreted as packet loss also by the TCP sender) lead to a drastic reduction of throughput. In this lab, we will use the NETEM tool to emulate these situations affecting end-to-end performance.

13 Lab Topology

Let's get started with creating a simple Mininet topology using MiniEdit. The topology uses 10.0.0.0/8, which is the default network assigned by Mininet (Fig. 36).

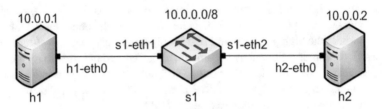

Fig. 36 Lab topology

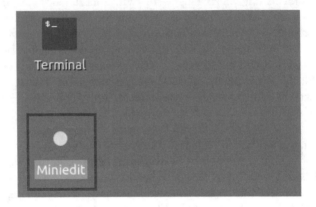

Fig. 37 MiniEdit shortcut

Step 1. A shortcut to MiniEdit is located on the machine's Desktop. Start MiniEdit by clicking on MiniEdit's shortcut (Fig. 37). When prompted for a password, type `password`.

Step 2. On MiniEdit's menu bar, click on *File* and then *Open* to load the lab's topology. Locate the *Lab 4.mn* topology file and click on *Open* (Fig. 38).

Step 3. Before starting the measurements between host h1 and host h2, the network must be started. Click on the *Run* button located at the bottom left of MiniEdit's window to start the emulation (Fig. 39).

The above topology uses 10.0.0.0/8, which is the default network assigned by Mininet.

13.1 Testing Connectivity Between Two Hosts

Step 1. Hold the right-click on host h1 and select *Terminal*. This opens the terminal of host h1 and allows the execution of commands on host h1 (Fig. 40).

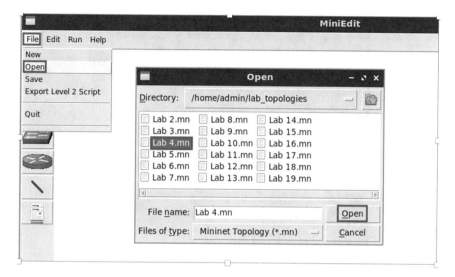

Fig. 38 MiniEdit's *Open* dialog

Fig. 39 Running the
emulation

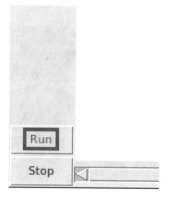

Step 2. Test connectivity between the end-hosts using the `ping` command. On host h1, type the command `ping 10.0.0.2`. This command tests the connectivity between host h1 and host h2. To stop the test, press `Ctrl+c`. The figure below shows a successful connectivity test (Fig. 41).

The figure above indicates that there is connectivity between host h1 and host h2. Thus, we are ready to start the throughput measurement process.

File Edit Run Help
s1
Host Options
Terminal
h2

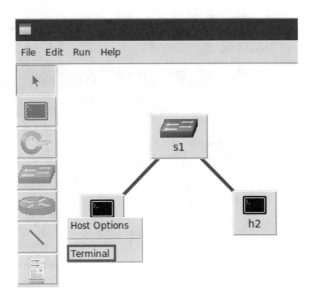

Fig. 40 Opening a terminal on host h1

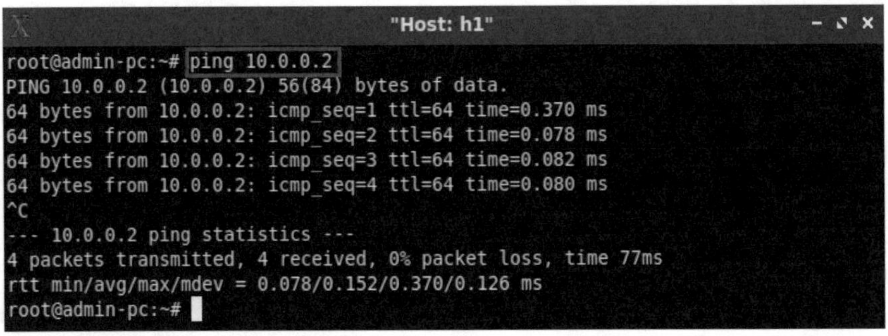

"Host: h1"
root@admin-pc:~# ping 10.0.0.2
PING 10.0.0.2 (10.0.0.2) 56(84) bytes of data.
64 bytes from 10.0.0.2: icmp_seq=1 ttl=64 time=0.370 ms
64 bytes from 10.0.0.2: icmp_seq=2 ttl=64 time=0.078 ms
64 bytes from 10.0.0.2: icmp_seq=3 ttl=64 time=0.082 ms
64 bytes from 10.0.0.2: icmp_seq=4 ttl=64 time=0.080 ms
^C
--- 10.0.0.2 ping statistics ---
4 packets transmitted, 4 received, 0% packet loss, time 77ms
rtt min/avg/max/mdev = 0.078/0.152/0.370/0.126 ms
root@admin-pc:~#

Fig. 41 Connectivity test using ping command

14 Adding/Changing Packet Loss

The user invokes NETEM using the command-line utility called tc. With no additional parameters, NETEM behaves as a basic FIFO queue with no delay, loss, duplication, or reordering of packets. The basic tc syntax used with NETEM is as follows:

```
sudo tc qdisc [add|del|replace|change|show] dev dev id root
netem opts
```

- sudo : enable the execution of the command with higher security privileges.
- tc : command used to interact with NETEM.
- qdisc : a queue discipline (qdisc) is a set of rules that determine the order in which packets arriving from the IP protocol output are served. The queue discipline is applied to a packet queue to decide when to send each packet.
- black lightgray!30[add |del |replace |change |show]: this is the operation on qdisc. For example, to add delay on a specific interface, the operation will be black lightgray!30add. To change or remove delay on the specific interface, the operation will be change or del .
- dev_id : this parameter indicates the interface to be subject to emulation.
- opts : this parameter indicates the amount of delay, packet loss, duplication, corruption, and others.

14.1   Identify Interface of Host h1 and Host h2

In this section, we must identify the interfaces on the connected hosts.

Step 1. On host h1, type the command ifconfig to display information related to its network interfaces and their assigned IP addresses (Fig. 42).

```
                              "Host: h1"                        -  ⌄ ✕
root@admin-pc:~# ifconfig
h1-eth0: flags=4163<UP,BROADCAST,RUNNING,MULTICAST>  mtu 1500
        inet 10.0.0.1  netmask 255.0.0.0  broadcast 10.255.255.255
        inet6 fe80::f0d6:67ff:fe01:6041  prefixlen 64  scopeid 0x20<link>
        ether f2:d6:67:01:60:41  txqueuelen 1000  (Ethernet)
        RX packets 51  bytes 5112 (5.1 KB)
        RX errors 0  dropped 0  overruns 0  frame 0
        TX packets 21  bytes 1678 (1.6 KB)
        TX errors 0  dropped 0 overruns 0  carrier 0  collisions 0

lo: flags=73<UP,LOOPBACK,RUNNING>  mtu 65536
        inet 127.0.0.1  netmask 255.0.0.0
        inet6 ::1  prefixlen 128  scopeid 0x10<host>
        loop  txqueuelen 1000  (Local Loopback)
        RX packets 0  bytes 0 (0.0 B)
        RX errors 0  dropped 0  overruns 0  frame 0
        TX packets 0  bytes 0 (0.0 B)
        TX errors 0  dropped 0 overruns 0  carrier 0  collisions 0

root@admin-pc:~#
```

Fig. 42 Output of ifconfig command on host h1

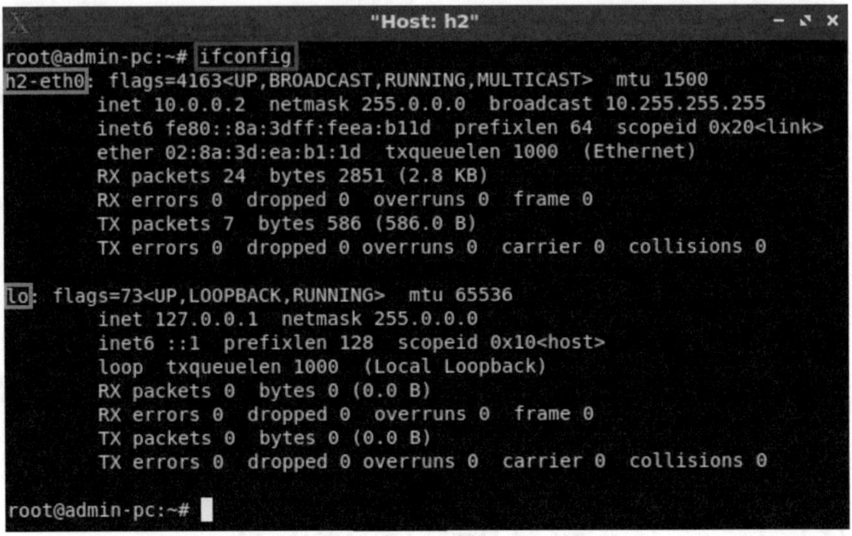

```
"Host: h2"                                    — ↙ ✕
root@admin-pc:~# ifconfig
h2-eth0: flags=4163<UP,BROADCAST,RUNNING,MULTICAST>  mtu 1500
        inet 10.0.0.2  netmask 255.0.0.0  broadcast 10.255.255.255
        inet6 fe80::8a:3dff:feea:b11d  prefixlen 64  scopeid 0x20<link>
        ether 02:8a:3d:ea:b1:1d  txqueuelen 1000  (Ethernet)
        RX packets 24  bytes 2851 (2.8 KB)
        RX errors 0  dropped 0  overruns 0  frame 0
        TX packets 7  bytes 586 (586.0 B)
        TX errors 0  dropped 0 overruns 0  carrier 0  collisions 0

lo: flags=73<UP,LOOPBACK,RUNNING>  mtu 65536
        inet 127.0.0.1  netmask 255.0.0.0
        inet6 ::1  prefixlen 128  scopeid 0x10<host>
        loop  txqueuelen 1000  (Local Loopback)
        RX packets 0  bytes 0 (0.0 B)
        RX errors 0  dropped 0  overruns 0  frame 0
        TX packets 0  bytes 0 (0.0 B)
        TX errors 0  dropped 0 overruns 0  carrier 0  collisions 0

root@admin-pc:~# ▌
```

Fig. 43 Output of ifconfig command on host h2

The output of the ifconfig command indicates that host h1 has two interfaces: *h1-eth0* and *lo*. The interface *h1-eth0* at host h2 is configured with IP address 10.0.0.1 and subnet mask 255.0.0.0. This interface must be used in tc when emulating the WAN.

Step 2. In host h2, type the command ifconfig as well (Fig. 43).

The output of the ifconfig command indicates that host h2 has two interfaces: *h2-eth0* and *lo*. The interface *h2-eth0* at host h1 is configured with IP address 10.0.0.2 and subnet mask 255.0.0.0. This interface must be used in tc when emulating the WAN.

14.2   Add Packet Loss to the Interface Connecting to the WAN

In a network, packets may be lost during transmission due to factors such as bit errors and network congestion. The rate of packets that are lost is often measured as a percentage of lost packets with respect to the number of sent packets. In this section, you will use netem command to insert packet loss on a network interface.

Step 1. In host h1's terminal, type the following command (Fig. 44):

```
sudo tc qdisc add dev h1-eth0 root netem loss 10%
```

```
"Host: h1"                                      - ⬞ ✕
root@admin-pc:~# sudo tc qdisc add dev h1-eth0 root netem loss 10%
root@admin-pc:~# █
```

Fig. 44 Adding 10% packet loss to host h1's interface *h1-eth0*

```
"Host: h1"                                      - ⬞ ✕
root@admin-pc:~# ping 10.0.0.2 -c 200
PING 10.0.0.2 (10.0.0.2) 56(84) bytes of data.
64 bytes from 10.0.0.2: icmp_seq=1 ttl=64 time=0.408 ms
64 bytes from 10.0.0.2: icmp_seq=3 ttl=64 time=0.060 ms
64 bytes from 10.0.0.2: icmp_seq=4 ttl=64 time=0.048 ms
64 bytes from 10.0.0.2: icmp_seq=5 ttl=64 time=0.044 ms
64 bytes from 10.0.0.2: icmp_seq=7 ttl=64 time=0.043 ms
64 bytes from 10.0.0.2: icmp_seq=8 ttl=64 time=0.069 ms
64 bytes from 10.0.0.2: icmp_seq=9 ttl=64 time=0.039 ms
64 bytes from 10.0.0.2: icmp_seq=11 ttl=64 time=0.048 ms
64 bytes from 10.0.0.2: icmp_seq=12 ttl=64 time=0.044 ms
64 bytes from 10.0.0.2: icmp_seq=13 ttl=64 time=0.039 ms
64 bytes from 10.0.0.2: icmp_seq=14 ttl=64 time=0.040 ms
64 bytes from 10.0.0.2: icmp_seq=15 ttl=64 time=0.040 ms
64 bytes from 10.0.0.2: icmp_seq=16 ttl=64 time=0.040 ms
64 bytes from 10.0.0.2: icmp_seq=18 ttl=64 time=0.044 ms
64 bytes from 10.0.0.2: icmp_seq=19 ttl=64 time=0.051 ms
64 bytes from 10.0.0.2: icmp_seq=20 ttl=64 time=0.047 ms
64 bytes from 10.0.0.2: icmp_seq=21 ttl=64 time=0.055 ms
64 bytes from 10.0.0.2: icmp_seq=22 ttl=64 time=0.053 ms
64 bytes from 10.0.0.2: icmp_seq=23 ttl=64 time=0.054 ms
64 bytes from 10.0.0.2: icmp_seq=24 ttl=64 time=0.051 ms
64 bytes from 10.0.0.2: icmp_seq=25 ttl=64 time=0.044 ms
64 bytes from 10.0.0.2: icmp_seq=26 ttl=64 time=0.041 ms
```

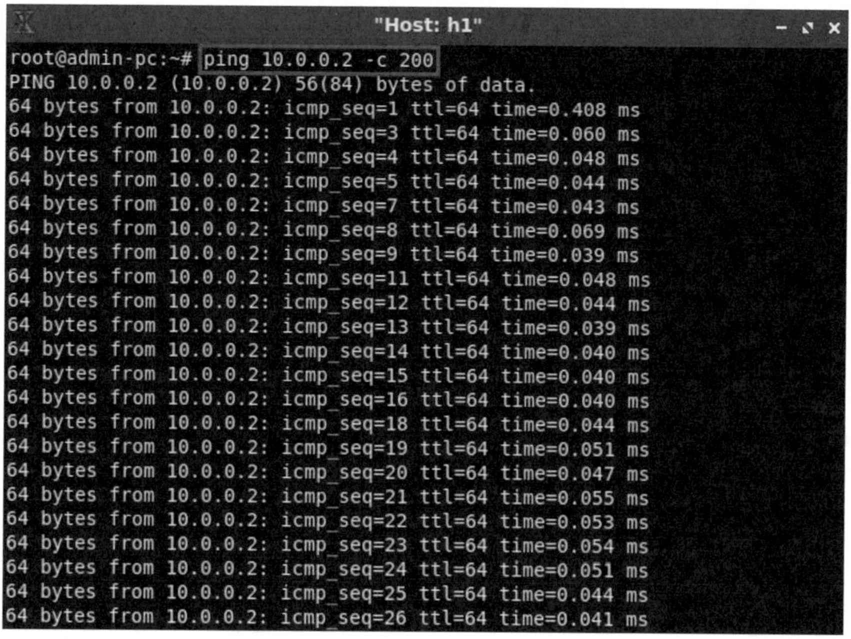

Fig. 45 `ping` command after introducing packet loss

The above command adds a 10% packet loss to host h1's interface *h1-eth0*.

Step 2. The user can verify now that the connection from host h1 to host h2 has packet losses by using the `ping` command from host h1's terminal (Fig. 45). The `-c` option specifies the total number of packets to send.

```
ping 10.0.0.2 -c 200
```

In Fig. 45, host h1 sends 200 ping packets to host h2. Note the *icmp_seq* values demonstrated in the figure below.

You can see that *icmp_seq*=2, 6, 10, and 17 are missing due to packet losses. Resulting packet loss will likely vary in each emulation.

```
--- 10.0.0.2 ping statistics ---
200 packets transmitted, 180 received, 10% packet loss, time 1154ms
rtt min/avg/max/mdev = 0.031/0.055/0.336/0.024 ms
root@admin-pc:~#
```

Fig. 46 ping summary report showing 10% packet loss

```
                                "Host: h2"                          –  ↖ ✕
root@admin-pc:~# sudo tc qdisc add dev h2-eth0 root netem loss 10%
root@admin-pc:~#
```

Fig. 47 Adding 10% packet loss to host h2's interface *h2-eth0*

Figure 46 shows the summary report of the previous command. By default, ping reports the percentage of packet loss after finishing the transmission. In our test, ping reported a packet loss rate of 10%. The measured packet loss rate will tend to become closer to the configured loss rate as more trials are performed.

Note that the above scenario emulates 10% packet loss on the unidirectional link from host h1 to host h2. If we want to emulate packet loss on both directions, a packet loss of 10% must also be added to host h2.

Step 3. In host h2's terminal, type the following command (Fig. 47):

```
sudo tc qdisc add dev h2-eth0 root netem loss 10%
```

Step 4. The user can verify now that the connection between host h1 and host h2 has more packets losses (10% from host h1 + 10% from host h2) by retyping the ping command on host h1's terminal (Fig. 48):

```
ping 10.0.0.2 -c 200
```

In Fig. 48, host h1 sends 200 ping packets to host h2. Note the *icmp_seq* values demonstrated in the figure below.

You can see that *icmp_seq*=3, 6, 10, 14, 23, and 27 are missing due to packet losses. Resulting packet loss will likely vary in each emulation.

Figure 49 shows the summary report of the previous command. By default, ping reports the percentage of packet loss after finishing the transmission. In our test, ping reported a packet loss rate of 10%. The measured packet loss rate will tend to become closer to the configured loss rate as more trials are performed.

The result above indicates that 159 out of 200 packets were received successfully (20.5% packet loss).

```
                                "Host: h1"
root@admin-pc:~# ping 10.0.0.2 -c 200
PING 10.0.0.2 (10.0.0.2) 56(84) bytes of data.
64 bytes from 10.0.0.2: icmp_seq=1 ttl=64 time=0.582 ms
64 bytes from 10.0.0.2: icmp_seq=2 ttl=64 time=0.074 ms
64 bytes from 10.0.0.2: icmp_seq=4 ttl=64 time=0.040 ms
64 bytes from 10.0.0.2: icmp_seq=5 ttl=64 time=0.069 ms
64 bytes from 10.0.0.2: icmp_seq=7 ttl=64 time=0.077 ms
64 bytes from 10.0.0.2: icmp_seq=8 ttl=64 time=0.074 ms
64 bytes from 10.0.0.2: icmp_seq=9 ttl=64 time=0.072 ms
64 bytes from 10.0.0.2: icmp_seq=10 ttl=64 time=0.066 ms
64 bytes from 10.0.0.2: icmp_seq=11 ttl=64 time=0.070 ms
64 bytes from 10.0.0.2: icmp_seq=12 ttl=64 time=0.075 ms
64 bytes from 10.0.0.2: icmp_seq=13 ttl=64 time=0.077 ms
64 bytes from 10.0.0.2: icmp_seq=16 ttl=64 time=0.082 ms
64 bytes from 10.0.0.2: icmp_seq=17 ttl=64 time=0.078 ms
64 bytes from 10.0.0.2: icmp_seq=18 ttl=64 time=0.086 ms
64 bytes from 10.0.0.2: icmp_seq=19 ttl=64 time=0.073 ms
64 bytes from 10.0.0.2: icmp_seq=20 ttl=64 time=0.076 ms
64 bytes from 10.0.0.2: icmp_seq=21 ttl=64 time=0.075 ms
64 bytes from 10.0.0.2: icmp_seq=22 ttl=64 time=0.083 ms
64 bytes from 10.0.0.2: icmp_seq=24 ttl=64 time=0.067 ms
64 bytes from 10.0.0.2: icmp_seq=25 ttl=64 time=0.076 ms
64 bytes from 10.0.0.2: icmp_seq=26 ttl=64 time=0.063 ms
64 bytes from 10.0.0.2: icmp_seq=28 ttl=64 time=0.072 ms
```

Fig. 48 ping command after introducing packet loss

```
--- 10.0.0.2 ping statistics ---
200 packets transmitted, 159 received, 20.5% packet loss, time 966ms
rtt min/avg/max/mdev = 0.028/0.054/0.345/0.026 ms
root@admin-pc:~#
```

Fig. 49 ping summary report showing 20.5% packet loss

14.3   Restore Default Values

To remove the packet loss added in Sect. 7.2 and restore the default configuration, you must delete the rules of the interfaces on host h1 and host h2 (Fig. 50).

Step 1. In host h1's terminal, type the following command (Fig. 23):

```
sudo tc qdisc del dev h1-eth0 root netem
```

Step 2. Apply the same steps to remove rules on host h2. In host h2's terminal, type the following command (Fig. 51):

```
                              "Host: h1"                    - ⟋ ✕
root@admin-pc:~# sudo tc qdisc del dev h1-eth0 root netem
root@admin-pc:~# ▮
```

Fig. 50 Deleting all rules on interface *h1-eth0*

```
                              "Host: h2"                    - ⟋ ✕
root@admin-pc:~# sudo tc qdisc del dev h2-eth0 root netem
root@admin-pc:~# ▮
```

Fig. 51 Deleting all rules on interface *h2-eth0*

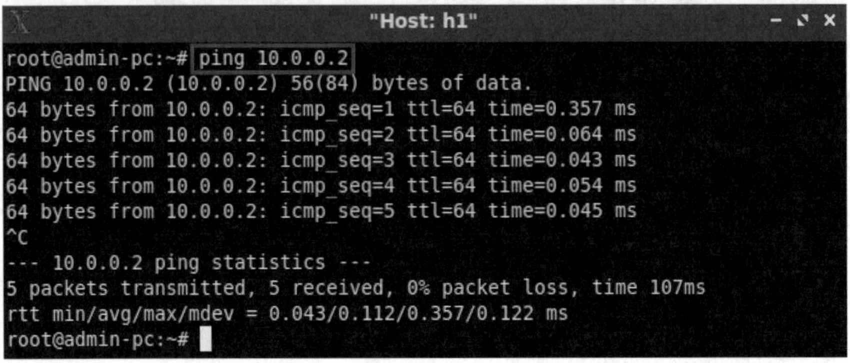

```
                              "Host: h1"                    - ⟋ ✕
root@admin-pc:~# ping 10.0.0.2
PING 10.0.0.2 (10.0.0.2) 56(84) bytes of data.
64 bytes from 10.0.0.2: icmp_seq=1 ttl=64 time=0.357 ms
64 bytes from 10.0.0.2: icmp_seq=2 ttl=64 time=0.064 ms
64 bytes from 10.0.0.2: icmp_seq=3 ttl=64 time=0.043 ms
64 bytes from 10.0.0.2: icmp_seq=4 ttl=64 time=0.054 ms
64 bytes from 10.0.0.2: icmp_seq=5 ttl=64 time=0.045 ms
^C
--- 10.0.0.2 ping statistics ---
5 packets transmitted, 5 received, 0% packet loss, time 107ms
rtt min/avg/max/mdev = 0.043/0.112/0.357/0.122 ms
root@admin-pc:~# ▮
```

Fig. 52 Verifying latency after deleting all rules on both devices

```
sudo tc qdisc del dev h2-eth0 root netem
```

As a result, the ⟨tc⟩ queueing discipline will restore its default values of the device *h2-eth0*.

Step 3. Now, the user can verify that the connection from host h1 to host h2 has no explicit packet loss configured by using the ⟨ping⟩ command from host h1's terminal; press ⟨Ctrl+c⟩ to stop the test (Fig. 52):

```
ping 10.0.0.2
```

The result above indicates that all five packets were received successfully (0% packet loss) and that the minimum, average, maximum, and standard deviation of the Round-Trip Time (RTT) were 0.043, 0.112, 0.357, and 0.122 milliseconds, respectively.

Fig. 53 Emulating packet losses with a correlation value

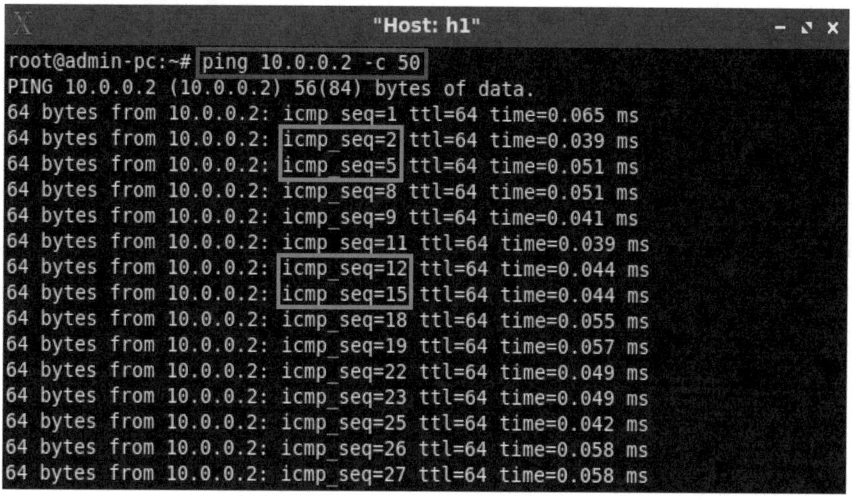

Fig. 54 ⎢ping⎢ in progress showing successive packet loss

14.4 Add Correlation Value for Packet Loss to Interface Connecting to WAN

An optional correlation may be added. Adding correlation causes the random number generator to be less random and can be used to emulate packet burst losses.

Step 1. In host h1's terminal, type the following command (Fig. 53):

```
sudo tc qdisc add dev h1-eth0 root netem loss 50% 50%
```

The above command introduces a packet loss rate of 50%, and each successive probability depends 50% on the last one. Note that a packet loss rate this high is unlikely.

Step 2. The user can verify now that the connection from host h1 to host h2 has packet losses by using the ⎢ping⎢ command from host h1's terminal (Fig. 54).

Fig. 55 Deleting all rules on interface *h1-eth0*

```
ping 10.0.0.2 -c 50
```

The result above shows an example where successive packets were dropped: [3, 4, 6, 10,], [13, 14, 16, 17, 20, 21], etc.

Step 3. In host h1's terminal, type the following command to delete previous configurations (Fig. 55):

```
sudo tc qdisc del dev h1-eth0 root netem
```

15   Adding Packet Corruption

Besides packet loss, packet corruption can be introduced with NETEM.

15.1   Add Packet Corruption to an Interface Connected to the WAN

Step 1. In host h1's terminal, type the following command (Fig. 56):

```
sudo tc qdisc add dev h1-eth0 root netem corrupt 0.01%
```

The new value added here represents packet corruption percentage (0.01%).

Step 2. The user can now verify the previous configuration by using the iperf3 tool to check the retransmissions. To launch iPerf3 in server mode, run the command iperf3 -s in host h2's terminal (Fig. 57).

```
iperf3 -s
```

Step 3. To launch iPerf3 in client mode, run the command iperf3 -c 10.0.0.2 in host h1's terminal (Fig. 58).

Fig. 56 Adding packets corruption (0.01%) to interface *h1-eth0*

Fig. 57 Host h2 running iPerf3 as server

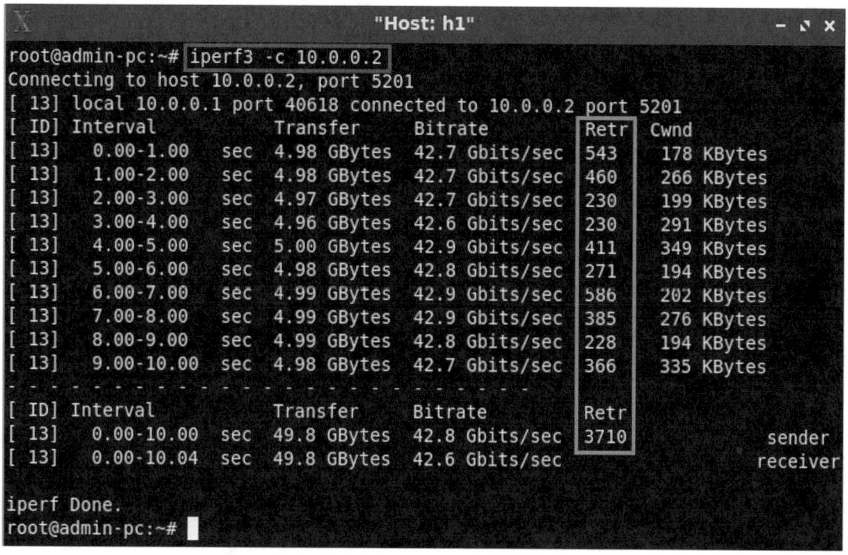

Fig. 58 Retransmissions after packets corruption

```
iperf3 -c 10.0.0.2
```

The figure above shows the retransmission values on each time interval (1 s). The total number of retransmitted packets, due to packet corruption, is 3710. This verifies that packet corruption was indeed applied to the interface on host h1.

Step 4. In host h1's terminal, type the following command to delete previous configurations (Fig. 59):

```
"Host: h1"                                          – ⟋ ×
root@admin-pc:~# sudo tc qdisc del dev h1-eth0 root netem
root@admin-pc:~# ▮
```

Fig. 59 Deleting all rules on interface *h1-eth0*

```
"Host: h1"                                          – ⟋
root@admin-pc:~# sudo tc qdisc add dev h1-eth0 root netem delay 10ms reorder 25% 50%
root@admin-pc:~# ▮
```

Fig. 60 Adding packet reordering

```
sudo tc qdisc del dev h1-eth0 root netem
```

Step 5. In order to stop the server, press Ctrl+c in host h2's terminal. The user can see the throughput results in the server side too. The summarized data on the server is similar to that of the client side's and must be interpreted in the same way.

16   Add Packet Reordering

Packets are sometimes not delivered in the same order they were sent. In order to emulate reordering in NETEM, the reorder option is used. Proceed with the steps below:

Step 1. In host h1's terminal, type the following command (Fig. 60):

```
sudo tc qdisc add dev h1-eth0 root netem delay 10ms
reorder 25% 50%
```

In this command, 25% of the packets (with a correlation value of 50%) will be sent immediately, while the remainder 75% will be delayed by 10 ms.

Step 2. The user can verify the effect of packet reorder by using the ping command on host h1's terminal; press Ctrl+c to stop the test (Fig. 61):

```
ping 10.0.0.2
```

Consider the first four packets of the figure above. The first and second packets did not experience delay (one out of four, or 25%), while the next three packets experienced a delay of ~10 milliseconds (three out of four, or 75%). The measured reordering rate will tend to become closer to the configured reordering rate as more trials are performed.

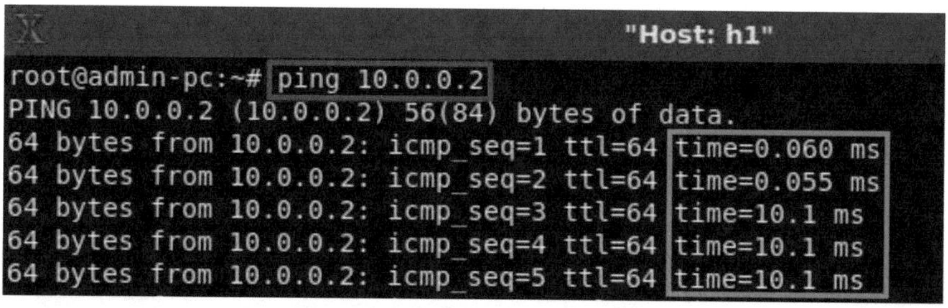

```
                                                    "Host: h1"
root@admin-pc:~# ping 10.0.0.2
PING 10.0.0.2 (10.0.0.2) 56(84) bytes of data.
64 bytes from 10.0.0.2: icmp_seq=1 ttl=64 time=0.060 ms
64 bytes from 10.0.0.2: icmp_seq=2 ttl=64 time=0.055 ms
64 bytes from 10.0.0.2: icmp_seq=3 ttl=64 time=10.1 ms
64 bytes from 10.0.0.2: icmp_seq=4 ttl=64 time=10.1 ms
64 bytes from 10.0.0.2: icmp_seq=5 ttl=64 time=10.1 ms
```

Fig. 61 | ping | test illustrating the effect of packet reordering

```
                               "Host: h1"                            – ⤢ ✕
root@admin-pc:~# sudo tc qdisc del dev h1-eth0 root netem
root@admin-pc:~# ▮
```

Fig. 62 Deleting all rules on interface *h1-eth0*

It is possible that your first packet will experience delay, but this effect will eventually occur in future tests.

Step 3. In host h1's terminal, type the following command to delete previous configurations (Fig. 62):

```
sudo tc qdisc del dev h1-eth0 root netem
```

17   Add Packet Duplication

Duplicate packets may be present in networks as a result of retransmissions. NETEM provides the option duplicate to inject | duplicate | packets. Before introducing packet corruption, make sure to restore the default configuration of the interfaces on host h1 and host h2 by applying the commands of Sect. 14.3. Then, proceed with the following steps:

Step 1. In host h1's terminal, type the following command (Fig. 63):

```
sudo tc qdisc change dev h1-eth0 root netem duplicate 50%
```

The above command will produce a duplication of 50% (i.e., 50% of the packets will be received twice at the destination).

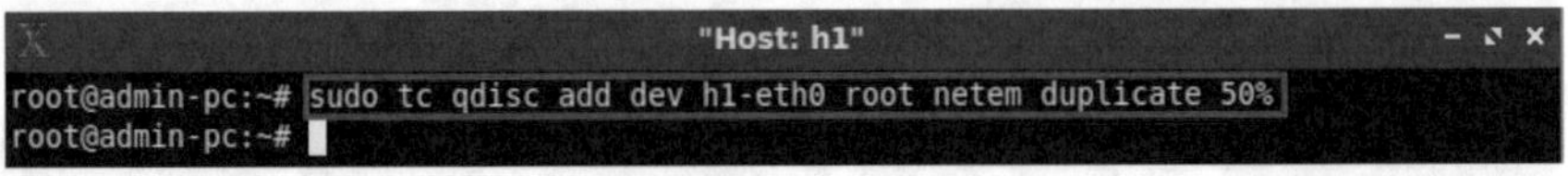

```
X                                    "Host: h1"                              - ⬞ x
root@admin-pc:~# sudo tc qdisc add dev h1-eth0 root netem duplicate 50%
root@admin-pc:~#
```

Fig. 63 Adding packet duplication

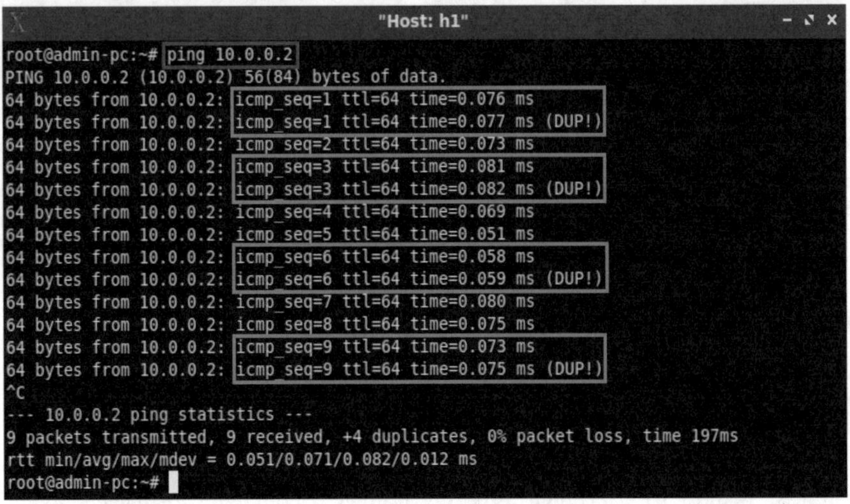

```
X                                    "Host: h1"                              - ⬞ x
root@admin-pc:~# ping 10.0.0.2
PING 10.0.0.2 (10.0.0.2) 56(84) bytes of data.
64 bytes from 10.0.0.2: icmp_seq=1 ttl=64 time=0.076 ms
64 bytes from 10.0.0.2: icmp_seq=1 ttl=64 time=0.077 ms (DUP!)
64 bytes from 10.0.0.2: icmp_seq=2 ttl=64 time=0.073 ms
64 bytes from 10.0.0.2: icmp_seq=3 ttl=64 time=0.081 ms
64 bytes from 10.0.0.2: icmp_seq=3 ttl=64 time=0.082 ms (DUP!)
64 bytes from 10.0.0.2: icmp_seq=4 ttl=64 time=0.069 ms
64 bytes from 10.0.0.2: icmp_seq=5 ttl=64 time=0.051 ms
64 bytes from 10.0.0.2: icmp_seq=6 ttl=64 time=0.058 ms
64 bytes from 10.0.0.2: icmp_seq=6 ttl=64 time=0.059 ms (DUP!)
64 bytes from 10.0.0.2: icmp_seq=7 ttl=64 time=0.080 ms
64 bytes from 10.0.0.2: icmp_seq=8 ttl=64 time=0.075 ms
64 bytes from 10.0.0.2: icmp_seq=9 ttl=64 time=0.073 ms
64 bytes from 10.0.0.2: icmp_seq=9 ttl=64 time=0.075 ms (DUP!)
^C
--- 10.0.0.2 ping statistics ---
9 packets transmitted, 9 received, +4 duplicates, 0% packet loss, time 197ms
rtt min/avg/max/mdev = 0.051/0.071/0.082/0.012 ms
root@admin-pc:~#
```

Fig. 64 ping test illustrating the effect of packet duplication

```
X                                    "Host: h1"                              - ⬞ x
root@admin-pc:~# sudo tc qdisc del dev h1-eth0 root netem
root@admin-pc:~#
```

Fig. 65 Deleting all rules on interface *h1-eth0*

Step 2. The user can verify the effect of packet duplication by using the ping command on host h1's terminal; press Ctrl+c to stop the test (Fig. 64):

```
ping 10.0.0.2
```

The result above indicates that five duplicate packets were received. Duplicate packets are also marked with (DUP!). The measured rate of duplicate packets will tend to become closer to the configured rate as more trials are performed.

Step 3. In host h1's terminal, type the following command to delete previous configurations (Fig. 65):

```
sudo tc qdisc del dev h1-eth0 root netem
```

Table 5 Credentials to
access Client1 machine

Device	Account	Password
Client1	admin	password

Chapter 2—Lab 5: Setting WAN Bandwidth with Token Bucket Filter (TBF)

Overview
To conduct the experiment described in this section, please login into the Academic Cloud at http://highspeednetworks.net/ and reserve a pod for Lab 5.

This lab explains the Token Bucket Filter (TBF) queuing discipline that shapes incoming/outgoing traffic to limit the bandwidth. Throughput measurements are also conducted in this lab to verify the bandwidth-limiting configuration with TBF.

Objectives
By the end of this lab, students should be able to:

1. Understand the token bucket algorithm.
2. Use Token Bucket Filter (*tbf*), which is a Linux implementation of the token bucket algorithm on network interfaces.
3. Understand how to combine queueing disciplines in Linux Traffic Control (*tc*).
4. Combine *tbf* and *NETEM*.
5. Emulate WAN properties in Mininet.
6. Visualize iPerf3's output after modifying the network's parameters.

Lab Settings
The information in Table 5 provides the credentials of the machine containing Mininet.

Lab Roadmap
This lab is organized as follows:

1. Section 18: Introduction to token bucket algorithm.
2. Section 19: Lab topology.
3. Section 20: Rate limiting on end-hosts.
4. Section 21: Rate limiting on switches.
5. Section 22: Combining NETEM and TBF.

18   Introduction to Token Bucket Algorithm

When simulating a Wide Area Network (WAN), it is sometimes necessary to limit the bandwidth of devices (end-hosts and networking devices) to observe the network's behavior in different conditions.

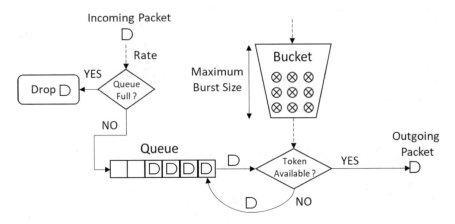

Fig. 66 Token bucket filter

The *Token Bucket* is an algorithm used in packet-switching networks to limit the bandwidth and the burstiness of the traffic. In summary, token bucket consists of adding tokens (represented as packets or packets' bytes) at a fixed rate to a fixed-capacity bucket. When a new packet arrives, the bucket is inspected to check the number of available tokens; if at least *n* tokens are available, *n* tokens are removed from the bucket, and the packet is sent to the network. Else, no tokens are removed, and the packet is considered *non-conformant*. In such case, the packet might be dropped, enqueued, or transmitted but marked as non-conformant. This algorithm is illustrated in Fig. 66.

The *rate,* which is the transmission speed, is determined by the frequency at which tokens are added to the bucket.

Another important property of the token bucket algorithm is *burstiness*; when the bucket becomes completely occupied (i.e., no packets are consuming tokens), new packets will consume tokens right away, without being limited. Burstiness is defined as the number of tokens that can fit in the bucket, or the bucket size.

To provide limits and control over the bursts, token bucket implementations often create another smaller bucket with a size equal to the *Maximum Transmission Unit (MTU)*, and a rate much faster than the original bucket (the *peak rate*). Its rate defines the maximum speed of bursts.

The token bucket algorithm implemented in Linux is the Token Bucket Filter (*tbf*), which is a queuing discipline used in conjunction with the Linux Traffic Control (*tc*) to shape traffic.

Figure 67 depicts the main parameters used by tbf.

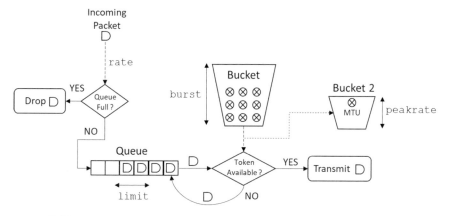

Fig. 67 | tbf | parameters and architecture

The basic | tbf | syntax used with | tc | is as follows:

```
tc qdisc [add |...] dev [dev id] root tbf limit [BYTES]
burst [BYTES] rate [BPS] [mtu BYTES] [ peakrate BPS ]
[ latency TIME ]
```

- | tc | : Linux traffic control tool.
- | qdisc | : a queue discipline (qdisc) is a set of rules that determine the order in which packets arriving from the IP protocol output are served. The queue discipline is applied to a packet queue to decide when to send each packet.
- | [add |del |replace |change |show] | : this is the operation on qdisc. For example, to add the token bucket algorithm on a specific interface, the operation will be | add |. To change or remove it, the operation will be | change | or | del |, respectively.
- | dev [dev_id] | : this parameter indicates the interface is to be subject to emulation.
- | tbf | : this parameter specifies the Token Bucket Filter algorithm.
- | limit [BYTES] | : size of the packet queue in bytes.
- | burst [BYTES] | : number of bytes that can fit in the bucket.
- | rate [BPS] | : transmission speed, determined by the frequency at which tokens are added to the bucket.
- | mtu [BYTES] | : maximum transmission unit in bytes.
- | peak rate [BPS] | : the maximum speed of a burst.
- | latency [TIME] | : the maximum time a packet can wait in the queue.

In this lab, we will use the $\boxed{\text{tbf}}$ queueing discipline to emulate the aforementioned parameters affecting the network behavior.

19 Lab Topology

Let's get started with creating a simple Mininet topology using MiniEdit. The topology uses 10.0.0.0/8, which is the default network assigned by Mininet (Fig. 68).

Step 1. A shortcut to MiniEdit is located on the machine's Desktop. Start MiniEdit by clicking on MiniEdit's shortcut (Fig. 69). When prompted for a password, type $\boxed{\text{password}}$.

Step 2. On MiniEdit's menu bar, click on *File* and then *Open* to load the lab's topology. Locate the *Lab 5.mn* topology file and click on *Open* (Fig. 70).

Step 3. Before starting the measurements between host h1 and host h2, the network must be started. Click on the *Run* button located at the bottom left of MiniEdit's window to start the emulation (Fig. 71).

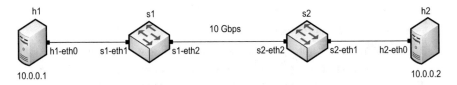

Fig. 68 Lab topology

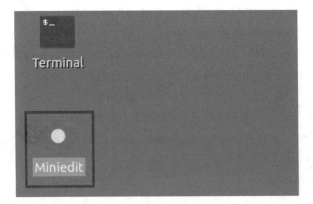

Fig. 69 MiniEdit shortcut

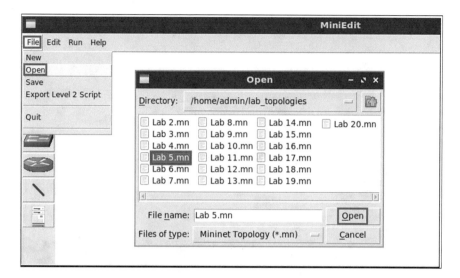

Fig. 70 MiniEdit's *Open* dialog

Fig. 71 Running the
emulation

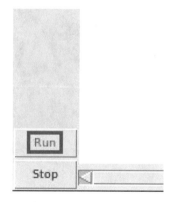

The above topology uses 10.0.0.0/8, which is the default network assigned by Mininet.

19.1   Starting Host h1 and Host h2

Step 1. Hold right-click on host h1 and select *Terminal*. This opens the terminal of host h1 and allows the execution of commands on that host (Fig. 72).

Step 2. Apply the same steps on host h2 and open its *Terminal*.

MiniEdit
File Edit Run Help
s1
s2
Host Options
Terminal
h2

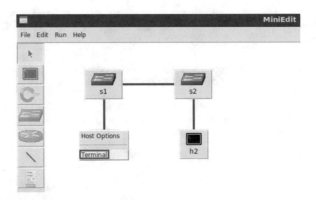

Fig. 72 Opening a terminal on host h1

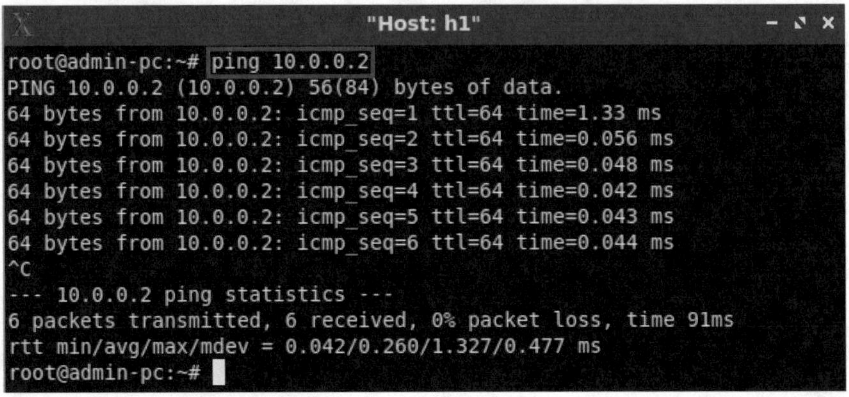

"Host: h1"
root@admin-pc:~# ping 10.0.0.2
PING 10.0.0.2 (10.0.0.2) 56(84) bytes of data.
64 bytes from 10.0.0.2: icmp_seq=1 ttl=64 time=1.33 ms
64 bytes from 10.0.0.2: icmp_seq=2 ttl=64 time=0.056 ms
64 bytes from 10.0.0.2: icmp_seq=3 ttl=64 time=0.048 ms
64 bytes from 10.0.0.2: icmp_seq=4 ttl=64 time=0.042 ms
64 bytes from 10.0.0.2: icmp_seq=5 ttl=64 time=0.043 ms
64 bytes from 10.0.0.2: icmp_seq=6 ttl=64 time=0.044 ms
^C
--- 10.0.0.2 ping statistics ---
6 packets transmitted, 6 received, 0% packet loss, time 91ms
rtt min/avg/max/mdev = 0.042/0.260/1.327/0.477 ms
root@admin-pc:~#

Fig. 73 Connectivity test using `ping` command

Step 3. Test connectivity between the end-hosts using the `ping` command. On host h1, type the command `ping 10.0.0.2`. This command tests the connectivity between host h1 and host h2. To stop the test, press `Ctrl+c`. The figure below shows a successful connectivity test (Fig. 73).

Figure 73 indicates that there is connectivity between host h1 and host h2.

20 Rate Limiting on End-Hosts

The `tc` command can be applied on the network interface of a device to shape egress traffic. In this section, the user will limit the sending rate of an end-host using the Token Bucket Filter (tbf), which is an implementation of the token bucket algorithm.

20.1   Identify Interface of Host h1 and Host h2

According to the previous section, we must identify the interfaces on the connected hosts.

Step 1. On host h1, type the command ifconfig to display information related to its network interfaces and their assigned IP addresses (Fig. 74).

The output of the ifconfig command indicates that host h1 has two interfaces: *h1-eth0* and *lo*. The interface *h1-eth0* at host h1 is configured with IP address 10.0.0.1 and subnet mask 255.0.0.0. This interface must be used in tc when emulating the network.

Step 2. In host h2's command line, type the command ifconfig as well (Fig. 75).

The output of the ifconfig command indicates that host h2 has two interfaces: *h2-eth0* and *lo*. The interface *h2-eth0* at host h1 is configured with IP address 10.0.0.2 and subnet mask 255.0.0.0. This interface must be used in tc when emulating the network.

```
                            "Host: h1"                    – ⤢ ✕
root@admin-pc:~# ifconfig
h1-eth0: flags=4163<UP,BROADCAST,RUNNING,MULTICAST>  mtu 1500
        inet 10.0.0.1  netmask 255.0.0.0  broadcast 10.255.255.255
        inet6 fe80::f0d6:67ff:fe01:6041  prefixlen 64  scopeid 0x20<link>
        ether f2:d6:67:01:60:41  txqueuelen 1000  (Ethernet)
        RX packets 51  bytes 5112 (5.1 KB)
        RX errors 0  dropped 0  overruns 0  frame 0
        TX packets 21  bytes 1678 (1.6 KB)
        TX errors 0  dropped 0 overruns 0  carrier 0  collisions 0

lo: flags=73<UP,LOOPBACK,RUNNING>  mtu 65536
        inet 127.0.0.1  netmask 255.0.0.0
        inet6 ::1  prefixlen 128  scopeid 0x10<host>
        loop  txqueuelen 1000  (Local Loopback)
        RX packets 0  bytes 0 (0.0 B)
        RX errors 0  dropped 0  overruns 0  frame 0
        TX packets 0  bytes 0 (0.0 B)
        TX errors 0  dropped 0 overruns 0  carrier 0  collisions 0
```

Fig. 74 Output of ifconfig command on host h1

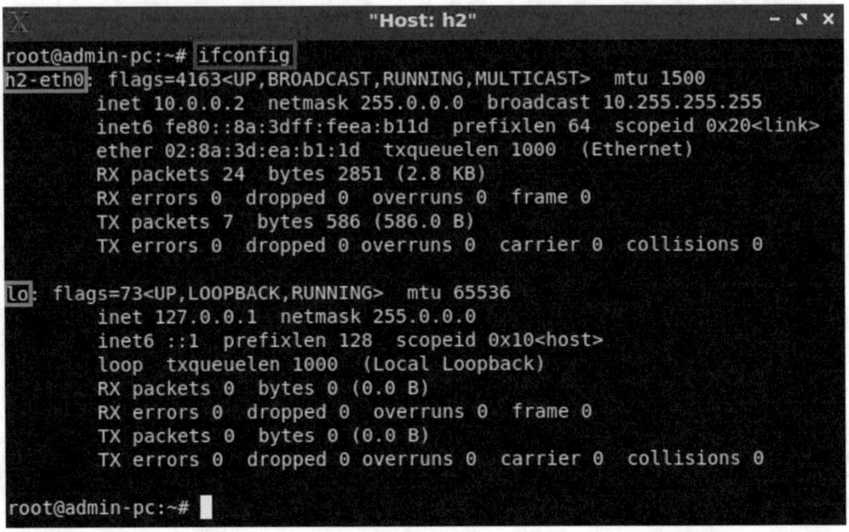

```
X                                    "Host: h2"                          -  ꞏ  X
root@admin-pc:~# ifconfig
h2-eth0: flags=4163<UP,BROADCAST,RUNNING,MULTICAST>  mtu 1500
        inet 10.0.0.2  netmask 255.0.0.0  broadcast 10.255.255.255
        inet6 fe80::8a:3dff:feea:b11d  prefixlen 64  scopeid 0x20<link>
        ether 02:8a:3d:ea:b1:1d  txqueuelen 1000  (Ethernet)
        RX packets 24  bytes 2851 (2.8 KB)
        RX errors 0  dropped 0  overruns 0  frame 0
        TX packets 7  bytes 586 (586.0 B)
        TX errors 0  dropped 0 overruns 0  carrier 0  collisions 0

lo: flags=73<UP,LOOPBACK,RUNNING>  mtu 65536
        inet 127.0.0.1  netmask 255.0.0.0
        inet6 ::1  prefixlen 128  scopeid 0x10<host>
        loop  txqueuelen 1000  (Local Loopback)
        RX packets 0  bytes 0 (0.0 B)
        RX errors 0  dropped 0  overruns 0  frame 0
        TX packets 0  bytes 0 (0.0 B)
        TX errors 0  dropped 0 overruns 0  carrier 0  collisions 0

root@admin-pc:~# █
```

Fig. 75 Output of ifconfig command on host h2

```
X                                    "Host: h1"                          -  ꞏ  X
root@admin-pc:~# sudo tc qdisc add dev h1-eth0 root tbf rate 10gbit burst 5000000 limit 15000000
root@admin-pc:~# █
```

Fig. 76 Limiting rate with TBF to 10 Gbps

20.2 Emulating 10 Gbps High-Latency WAN

In this section, you will use tbf command on a network interface to control the egress rate.

Step 1. Modify the bandwidth of host h1 by typing the command below. This command sets the bandwidth to 10 Gbps on host h1's *h1-eth0* interface (Fig. 76). The tbf parameters are the following:

- rate : 10gbit
- burst : 5,000,000
- limit : 15,000,000

```
sudo tc qdisc add dev h1-eth0 root tbf rate 10gbit burst
5000000 limit 15000000
```

This command can be summarized as follows:

- sudo : enable the execution of the command with higher security privileges.

- tc : invoke Linux's traffic control.
- qdisc : modify the queuing discipline of the network scheduler.
- add : create a new rule.
- dev h1-eth0 root : specify the interface on which the rule will be applied.
- tbf : use the token bucket filter algorithm.
- rate : specify the transmission rate (10 Gbps).
- burst : number of bytes that can fit in the bucket (*5,000,000*).
- limit : queue size in bytes (15,000,000).

Burst calculation: *tbf* requires setting a burst value when limiting the rate. This value must be high enough to allow your configured rate. Specifically, it must be at least the specified rate/HZ, where HZ is clock rate, configured as a kernel parameter, and can be extracted using the command shown below (Fig. 77):

```
egrep '^CONFIG_HZ_[0-9]+' /boot/config-$(uname -r)
```

The HZ on Client1 is 250. Thus, to calculate the burst, we divide 10 Gbps by 250:

10 Gbps = 10,000,000,000 bps

Burst = $\frac{10,000,000,000}{250}$ = 40, 000, 000 bits

Burst = 40,000,000 bits = 5,000,000 bytes

The resulting value is to be used in the command as the burst value.

Step 2. The user can now verify the previous configuration by using the iperf3 tool to measure throughput. To launch iPerf3 in server mode, run the command iperf3 -s in host h2's terminal as shown in the figure below (Fig. 78):

```
iperf3 -s
```

Step 3. Now to launch iPerf3 in client mode, run the command iperf3 -c 10.0.0.2 in host h1's terminal as shown below (Fig. 79):

```
iperf3 -c 10.0.0.2
```

```
X                                    "Host: h1"
root@admin-pc:~# egrep '^CONFIG_HZ_[0-9]+' /boot/config-$(uname -r)
CONFIG_HZ_250=y
root@admin-pc:~#
```

Fig. 77 Retrieving system's HZ

```
X                                    "Host: h2"                              -  ▵  ✕
root@admin-pc:~# iperf3 -s
--------------------------------------------------------------------------------
Server listening on 5201
--------------------------------------------------------------------------------
```

Fig. 78 Host h2 running iPerf3 as server

```
X                                    "Host: h1"                              -  ▵  ✕
root@admin-pc:~# iperf3 -c 10.0.0.2
Connecting to host 10.0.0.2, port 5201
[ 15] local 10.0.0.1 port 34924 connected to 10.0.0.2 port 5201
[ ID] Interval           Transfer     Bitrate         Retr  Cwnd
[ 15]   0.00-1.00   sec  1.12 GBytes  9.62 Gbits/sec    0    564 KBytes
[ 15]   1.00-2.00   sec  1.11 GBytes  9.57 Gbits/sec    0    701 KBytes
[ 15]   2.00-3.00   sec  1.11 GBytes  9.56 Gbits/sec    0    740 KBytes
[ 15]   3.00-4.00   sec  1.11 GBytes  9.56 Gbits/sec    0    775 KBytes
[ 15]   4.00-5.00   sec  1.11 GBytes  9.57 Gbits/sec    0    854 KBytes
[ 15]   5.00-6.00   sec  1.11 GBytes  9.56 Gbits/sec    0   1.01 MBytes
[ 15]   6.00-7.00   sec  1.11 GBytes  9.56 Gbits/sec    0   1.01 MBytes
[ 15]   7.00-8.00   sec  1.11 GBytes  9.56 Gbits/sec    0   1.01 MBytes
[ 15]   8.00-9.00   sec  1.11 GBytes  9.56 Gbits/sec    0   1.01 MBytes
[ 15]   9.00-10.00  sec  1.11 GBytes  9.57 Gbits/sec    0   1.01 MBytes
- - - - - - - - - - - - - - - - - - - - - - - - - - - - - -
[ ID] Interval           Transfer     Bitrate         Retr
[ 15]   0.00-10.00  sec  11.1 GBytes  9.57 Gbits/sec    0         sender
[ 15]   0.00-10.04  sec  11.1 GBytes  9.53 Gbits/sec              receiver

iperf Done.
root@admin-pc:~# █
```

Fig. 79 iPerf3's report after limiting the rate on host h1 to 10 Gbps

The figure above shows the iPerf3 report after limiting the rate on host h1 using
tbf . The average achieved throughputs are 9.57 Gbps (sender) and 9.53 Gbps
(receiver). Since we executed the command on host h1's terminal, the rule was
applied to host h1's network interface. However, it is also possible to limit the rate
on the switch interfaces as explained next.

Step 4. In order to stop the server, press Ctrl+c in host h2's terminal. The user
can see the throughput results in the server side too.

21 Rate Limiting on Switches

The previous section explained how to use the token bucket filter on end-hosts'
network interfaces. In this section, we will explain how to apply the filter on switch
interfaces. By limiting the rate on switch S1's *s1-eth2* interface, all communication
sessions between switch S1 and switch S2 will be filtered by the applied rule(s).

In previous tests, we applied the commands on host h1's terminal; switches,
however, do not have terminals where commands can be set and applied. Recall that

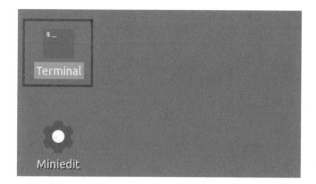

Fig. 80 Shortcut to open a Linux terminal

we are using Mininet for this emulation, which creates virtual interfaces emulating the switch functionality. Therefore, these virtual interfaces can be identified using the ⟨ifconfig⟩ command, but this time, it should be issued on the client's terminal (e.g., the terminal located on the Desktop) and not on end-hosts (host h1 or host h2).

Step 1. Launch a Linux terminal by holding the ⟨Ctrl+Alt+T⟩ keys or by clicking on the Linux terminal icon (Fig. 80).

The Linux terminal is a program that opens a window and permits you to interact with a command-line interface (CLI). A CLI is a program that takes commands from the keyboard and sends them to the operating system for execution.

Step 2. Type in the terminal the command ⟨ifconfig⟩ to display information related to its network interfaces (Fig. 81).

Figure 81 shows the network interfaces of the client:

- *s1-eth1* is the interface connecting switch S1 to host h1.
- *s1-eth2* is the interface connecting switch S1 to switch S2.
- *s2-eth1* is interface connecting switch S2 to host h2.
- *s2-eth2* is interface connecting switch S2 to switch S1.

Step 3. Remove the previous configuration on host h1. Write the following command on host h1's terminal (Fig. 82):

```
sudo tc qdisc del dev h1-eth0 root
```

Step 4. Apply ⟨tbf⟩ rate limiting rule on switch S1's interface, which connects it to switch S2 (*s1-eth2*). In the Client1's terminal, type the command below (Fig. 83). When prompted for a password, type ⟨password⟩ and hit enter. The ⟨tbf⟩ parameters are the following:

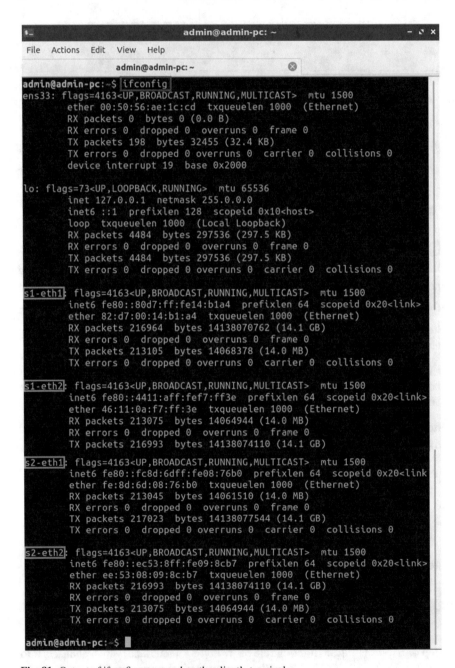

Fig. 81 Output of ifconfig command on the client's terminal

Fig. 82 Deleting all rules on host h1's network scheduler

Fig. 83 Limiting rate with TBF to 10 Gbps on switch S1's interface

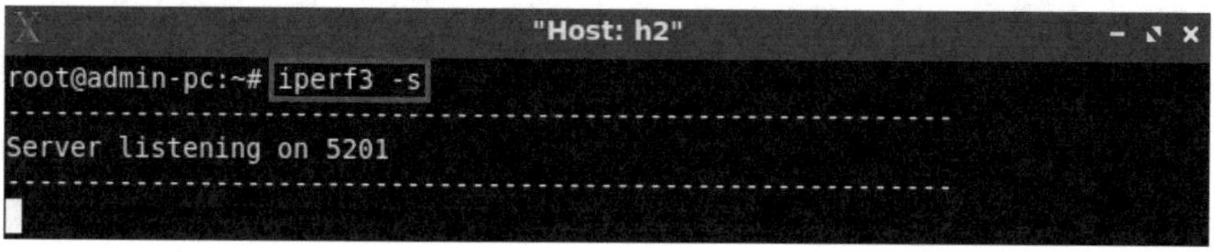

Fig. 84 Host h2 running iPerf3 as server

- rate : 10gbit
- burst : 5,000,000
- limit : 15,000,000

```
sudo tc qdisc add dev s1-eth2 root tbf rate 10gbit burst
5000000 limit 15000000
```

Step 5. The user can now verify the previous configuration by using the iperf3 tool to measure throughput. To launch iPerf3 in server mode, run the command iperf3 -s in host h2's terminal as shown in Fig. 84:

```
iperf3 -s
```

Step 6. Now to launch iPerf3 in client mode, run the command iperf3 -c 10.0.0.2 in host h1's terminal as shown in the figure below (Fig. 85):

```
iperf3 -c 10.0.0.2
```

Again, the reported values match the desired throughput (10 Gbps). In practice, the reported throughput will not achieve the target (10 Gbps) but will achieve a throughput slightly less than the target.

```
X                                  "Host: h1"                        -  ⬚  x
root@admin-pc:~# iperf3 -c 10.0.0.2
Connecting to host 10.0.0.2, port 5201
[ 15] local 10.0.0.1 port 34932 connected to 10.0.0.2 port 5201
[ ID] Interval           Transfer     Bitrate       Retr  Cwnd
[ 15]   0.00-1.00   sec  1.13 GBytes  9.69 Gbits/sec    0    8.27 MBytes
[ 15]   1.00-2.00   sec  1.11 GBytes  9.56 Gbits/sec    0    8.27 MBytes
[ 15]   2.00-3.00   sec  1.11 GBytes  9.56 Gbits/sec    0    8.27 MBytes
[ 15]   3.00-4.00   sec  1.11 GBytes  9.57 Gbits/sec    0    8.27 MBytes
[ 15]   4.00-5.00   sec  1.11 GBytes  9.56 Gbits/sec    0    8.27 MBytes
[ 15]   5.00-6.00   sec  1.11 GBytes  9.56 Gbits/sec    0    8.27 MBytes
[ 15]   6.00-7.00   sec  1.11 GBytes  9.57 Gbits/sec    0    8.27 MBytes
[ 15]   7.00-8.00   sec  1.11 GBytes  9.56 Gbits/sec    0    8.27 MBytes
[ 15]   8.00-9.00   sec  1.11 GBytes  9.55 Gbits/sec    0    8.27 MBytes
[ 15]   9.00-10.00  sec  1.11 GBytes  9.57 Gbits/sec    0    8.27 MBytes
- - - - - - - - - - - - - - - - - - - - - - - - - - - -
[ ID] Interval           Transfer     Bitrate       Retr
[ 15]   0.00-10.00  sec  11.1 GBytes  9.58 Gbits/sec    0           sender
[ 15]   0.00-10.05  sec  11.1 GBytes  9.53 Gbits/sec              receiver

iperf Done.
root@admin-pc:~# █
```

Fig. 85 iPerf3's report after limiting the rate on switch S1 to 10 Gbps

Step 7. In order to stop the server, press Ctrl+c in host h2's terminal. The user can see the throughput results in the server side too.

22 Combining NETEM and TBF

NETEM is used to introduce delay, jitter, packet corruption, etc. TBF on the other hand can be used to limit the rate. However, this is not enough for emulating real networks, particularly WANs. Therefore, it is also possible to combine multiple impairments and activate them at the same time.

As shown in Fig. 86, the first *qdisc* (*qdisc₁*) is attached to the *root* label. Then, subsequent *qdiscs* can be attached to their *parents* by specifying the correct label. In this section, we will look at how to combine NETEM and TBF in order to have more properties emulated in our network. Specifically, we will introduce delay, jitter, and packet corruption while specifying the rate on switch S1's interface.

Step 1. In the Client's terminal, type the following command to remove the previous configuration on switch S1 (Fig. 87):

```
sudo tc qdisc del dev s1-eth2 root
```

Fig. 86 Chaining *qdiscs* hierarchy

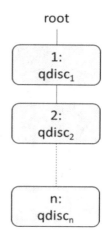

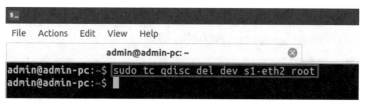

Fig. 87 Deleting all rules on switch S1's *s1-eth2*

Fig. 88 Adding delay of 10 ms to switch S1's *s1-eth2* interface

Step 2. In the client's terminal, type the command below. When prompted for a password, type password and hit *Enter* (Fig. 88):

```
sudo tc qdisc add dev s1-eth2 root handle 1: netem delay
10 ms
```

The new keyword in this command is *handle* and its value reflects the number shown in Fig. 86 each *qdisc*. This means that our NETEM *qdisc* is attached to the root with the handle 1:

Step 3. The user can now verify the previous configuration by using the ping tool to measure the Round-Trip Time (RTT). On the terminal of host h1, type ping 10.0.0.2. To stop the test, press Ctrl+c. The figure below shows a successful

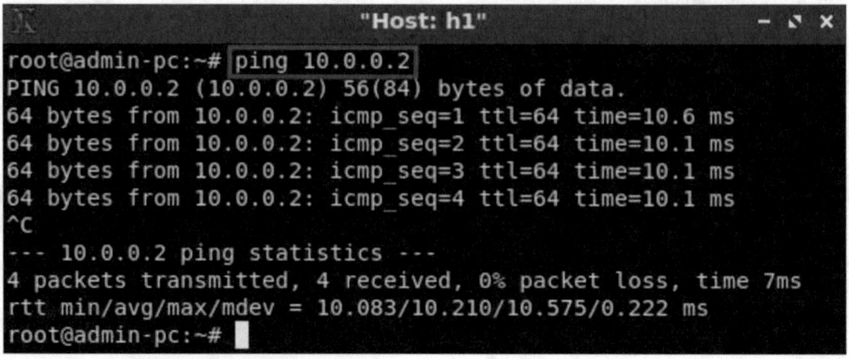

```
                              "Host: h1"                     -  ⩘ ✗
root@admin-pc:~# ping 10.0.0.2
PING 10.0.0.2 (10.0.0.2) 56(84) bytes of data.
64 bytes from 10.0.0.2: icmp_seq=1 ttl=64 time=10.6 ms
64 bytes from 10.0.0.2: icmp_seq=2 ttl=64 time=10.1 ms
64 bytes from 10.0.0.2: icmp_seq=3 ttl=64 time=10.1 ms
64 bytes from 10.0.0.2: icmp_seq=4 ttl=64 time=10.1 ms
^C
--- 10.0.0.2 ping statistics ---
4 packets transmitted, 4 received, 0% packet loss, time 7ms
rtt min/avg/max/mdev = 10.083/10.210/10.575/0.222 ms
root@admin-pc:~# █
```

Fig. 89 Output of ping 10.0.0.2 command

connectivity test. Host h1 (10.0.0.1) sent four packets to host h2 (10.0.0.2), successfully receiving responses back (Fig. 89).

```
ping 10.0.0.2
```

The result above indicates that all four packets were received successfully (0% packet loss) and that the minimum, average, maximum, and standard deviation of the Round-Trip Time (RTT) were 10.083, 10.210, 10.575, and 0.222 milliseconds, respectively. Essentially, the standard deviation is an average of how far each ping RTT is from the average RTT. The higher the standard deviation, the more variable the RTT is.

Step 4. Now to add the second rule that applies rate limiting using tbf, issue the command shown below on the client's terminal (Fig. 90). The tbf parameters are the following:

- rate : 2gbit
- burst : 1,000,000
- limit : 2,500,000

```
sudo tc qdisc add dev s1-eth2 parent 1: handle 2: tbf rate
2gbit burst 1000000 limit 2500000
```

Step 5. The user can now verify the previous configuration by using the iperf3 tool to measure throughput. To launch iPerf3 in server mode, run the command iperf3 -s in host h2's terminal as shown in Fig. 91:

```
iperf3 -s
```

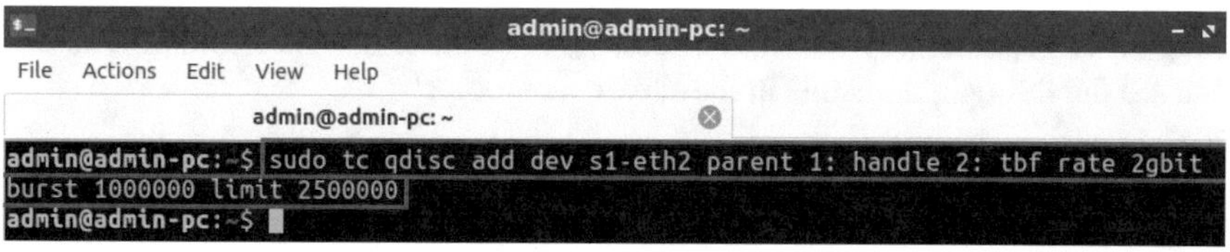

Fig. 90 Adding a new rule while combining it with the previous

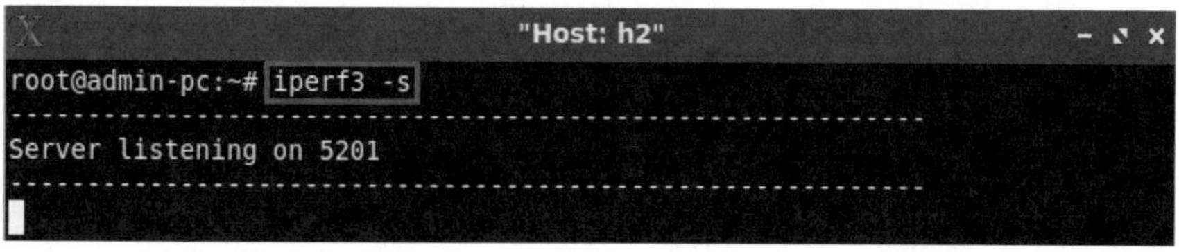

Fig. 91 Host h2 running iPerf3 as server

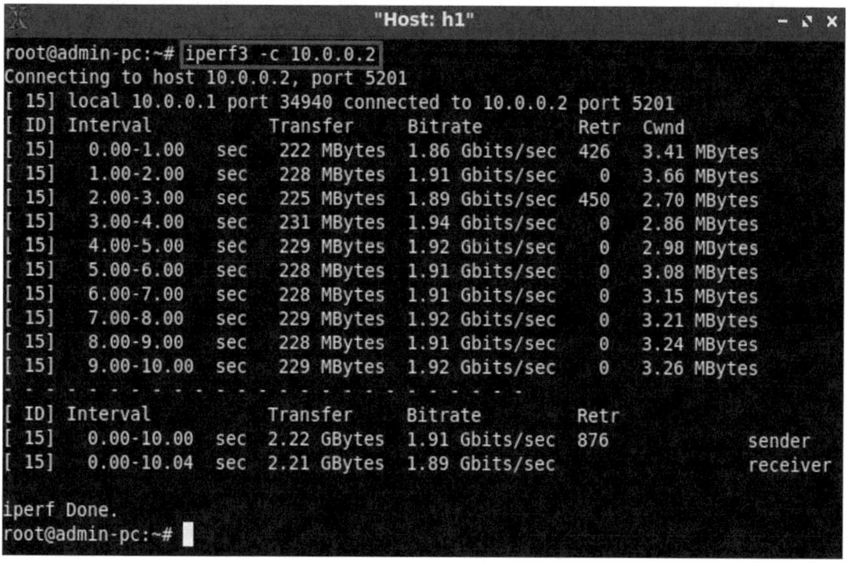

Fig. 92 iPerf3 throughput test after combining *qdiscs*

Step 6. Now to launch iPerf3 in client mode again by running the command iperf3 -c 10.0.0.2 in host h1's terminal as shown in Fig. 92:

```
iperf3 -c 10.0.0.2
```

The figure above shows the iPerf3 test output report. The average achieved throughputs are 1.86 Gbps (sender) and 1.84 Gbps (receiver).

Step 7. In order to stop the server, press $\boxed{\text{Ctrl+c}}$ in host h2's terminal. The user can see the throughput results in the server side too.

References

1. E. Dart, L. Rotman, B. Tierney, M. Hester, J. Zurawski, The science DMZ: a network design pattern for data-intensive science, in *Proceedings of the International Conference on High Performance Computing, Networking, Storage and Analysis* (2013)
2. M. Mathis, J. Semke, J. Mahdavi, T. Ott, The macroscopic behavior of the TCP congestion avoidance algorithm. ACM Comput. Commun. Rev. **27**(3), 67–82 (1997)
3. K. Fall, S. Floyd, Simulation-based comparisons of Tahoe, Reno, and SACK TCP. Comput. Commun. Rev. **26**(3), 5–21 (1996)
4. D. Leith, R. Shorten, Y. Lee, H-TCP: a framework for congestion control in high-speed and long-distance networks. Hamilton Institute Technical Report (2005). http://www.hamilton.ie/net/htcp2005.pdf
5. K. Chard, S. Tuecke, I. Foster, Globus: recent enhancements and future plans, in *Proceedings of the XSEDE16 Conference on Diversity, Big Data, and Science at Scale* (2016)
6. W. Allcock, J. Bresnahan, R. Kettimuthu, M. Link, The Globus striped GridFTP framework and server, in *Proceedings of the 2005 ACM/IEEE Conference on Supercomputing* (2005)
7. B. Radic, V. Kajic, E. Imamagic, Optimization of data transfer for grid using GridFTP, in *Proceedings of the International Conference on Information Technology Interfaces* (2008)
8. J. Postel, Transmission control protocol (TCP), in *Internet Request for Comments, RFC Editor, RFC 793* (1981). https://tools.ietf.org/html/rfc793
9. J. Zurawski, S. Balasubramanian, A. Brown, E. Kissel, A. Lake, M. Swany, B. Tierney, M. Zekauskas, perfSONAR: on-board diagnostics for big data, in *Workshop on Big Data and Science: Infrastructure and Services* (2013)
10. A. Hanemann, J. Boote, E. Boyd, J. Durand, L. Kudarimoti, R. Lapacz, D. Swany, J. Zurawski, S. Trocha, perfSONAR: a service oriented architecture for multi-domain network monitoring, in *Proceedings of the Third international conference on Service-Oriented Computing* (2005), pp. 241–254
11. Internet2. https://www.internet2.edu/
12. B. Claise, Cisco Systems NetFlow Services Export Version 9. Internet Request for Comments, RFC Editor, RFC 3954 (2004). https://www.ietf.org/rfc/rfc3954.txt
13. K. Miller, DDOS mitigation with sFlow. http://www.rn.psu.edu/2014/07/25/ddos-mitigation-with-sflow/
14. R. Hofstede, A. Pras, A. Sperotto, G. Rodosek, Flow-based compromise detection: lessons learned. IEEE Secur. Priv. **16**(1), 82–89 (2018)
15. F. Farina, P. Szegedi, J. Sobieski, GEANT world testbed facility: federated and distributed testbeds as a service facility of GEANT, in *International Tele-traffic Congress* (2014)
16. UbuntuNet. https://ubuntunet.net/
17. Asia Pacific Advanced Network. https://apan.net/
18. RedCLARA network. https://www.redclara.net/index.php/en/
19. The western regional network. http://nets.ucar.edu/nets/ongoing-activities/wrn/wrnroot/
20. The corporation for education network initiatives in California. http://cenic.org
21. The energy science network. https://www.es.net
22. K. Thompson, Campus cyberinfrastructure, in *Principal Investigators Workshop, NSF Campus Cyberinfrastructure Program* (2016). https://www.thequilt.net/wp-content/uploads/CC_PIMeeting2016_KLT.pdf
23. National science foundation (NSF) campus cyberinfrastructure program, in *National Science Foundation*. https://www.nsf.gov/funding/pgm_summ.jsp?pims_id=504748

Data-Link and Network Layer Considerations for Large Data Transfers

One of the main functions of routers and switches is forwarding. Forwarding refers to the switching of a packet from the input port to the appropriate output port. This chapter reviews the architecture and forwarding-related attributes of switches and routers. Attributes include forwarding rates, memory for buffering packets, forwarding methods such as store-and-forward and cut-through, queue management, and maximum transmission units.

1 Data-Link and Network-Layer Devices

Two essential functions performed by routers are routing and forwarding. Routing refers to the determination of the route taken by packets. Forwarding refers to the switching of a packet from the input port to the appropriate output port. The term switching is also used interchangeably with forwarding.

Traditional routing approaches such as static and dynamic routing (e.g., Open Shortest Path First (OSPF) [1], BGP [2]) are used in the implementation of Science DMZs. Routing events, such as routing table updates, occur at the millisecond, second, or minute timescale, and best practices used in regular enterprise networks are applicable to Science DMZs. On the other hand, with transmission rates of 10 Gbps and above, the forwarding operation occurs at the nanosecond timescale. Since forwarding functionality is common in both routers and switches, this section reviews the architecture and forwarding-related attributes of switches. These attributes are applicable to routers as well; thus, for the remainder of this section, the terms switch and router are used interchangeably. Switching attributes discussed in this section are illustrated in Fig. 1.

Fig. 1 Switching attributes requiring consideration in a Science DMZ

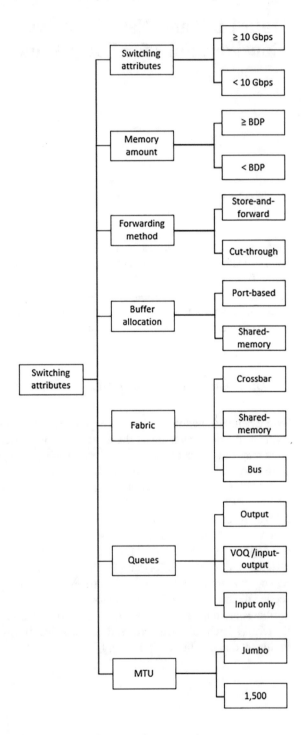

2   Switching Review

A generic router architecture is shown in Fig. 2. Modern routers may have a network processor (NP) and a table derived from the routing table in each port, which is referred to as the forwarding table (FT) or forwarding information base (FIB). The router in Fig. 2 has two input ports, iP1 and iP2, with their respective queues. iP1 has three packets in its queue, which will be forwarded to output ports oP1 (green packets) and oP2 (blue packet) by the fabric.

Router queues/buffers absorb traffic fluctuations. Even in the absence of congestion, fluctuations are present, resulting mostly from coincident traffic bursts [3].

Consider an input buffer implemented as a first-in first-out in the router of Fig. 2. As iP1 and iP2 both have one packet to be forwarded to oP1 at the front of the buffer, only one of them, say the packet at iP2, will be forwarded to oP1. The consequence of this is that not only the first packet at iP1 must wait, so too must the second packet that is queued at iP1 wait, even though there is no contention for oP2. This phenomenon is known as head-of-line (HOL) blocking [4]. To avoid HOL blocking, many switches use output buffering, a mixture of internal and output buffering, or techniques emulating output buffering such as Virtual Output Queueing (VOQ).

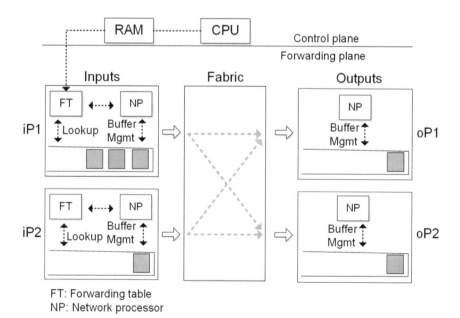

Fig. 2   A generic router architecture

3   Switching Considerations for Science DMZs

There are critical switching attributes that must be considered for a well-designed Science DMZ. These attributes are related to the characteristics of the science traffic and the role of switches in mitigating packet losses. Key considerations are now presented.

3.1   *Traffic Profile*

At a switch, buffer size, forwarding or switching rate, and queues should be selected based on the traffic profile to be supported by the network. Enterprise networks and Science DMZs are subject to different traffic profiles, as listed in Table 1. In a typical enterprise network, a very large number of flows consume a relatively small amount of bandwidth each. Figure 3 shows an example of a traffic profile at a small campus enterprise network serving approximately 1000 hosts. The number of flows observed in a week-long period is approximately 33 million, 81% TCP, 18% UDP, and 1% other protocols. According to the cumulative distribution function (CDF) of the flow duration, more than 90% of these flows have a duration of less than 200 s. Similarly, approximately 90% of the flows have a size of 10 KBs or less. This traffic profile is very different from that of a science flow, which may last several hours and consume the total available bandwidth. For example, transferring 100 TBs at 10 Gbps takes over 24 h. In this type of flow, bursts occur occasionally but are not the norm. When both small and large flows are transported across the same network, smaller flows do not saturate ports. However, when bursts associated with a science flow occur, then these events can cause the starvation of the small flows [5].

Enterprise flows are less sensitive than Science DMZ flows to packet loss and throughput requirements. Typically, the size of files in enterprise applications

Table 1 Comparison between enterprise network and Science DMZ flows

Feature	Enterprise network flow	Science DMZ flow
Duration	Short	Long
Data size	KBs, MBs	TBs, PBs
Nature of the data	Large variety: web, email, media content, database-related, mobile applications, streaming	Files
Bursty	Yes	No
Packet loss	Less sensitive	Very sensitive
Latency	Sensitive	Less sensitive
Throughput	Less sensitive	Very sensitive
Concurrent flows	Thousands of flows per second	Few flows per second

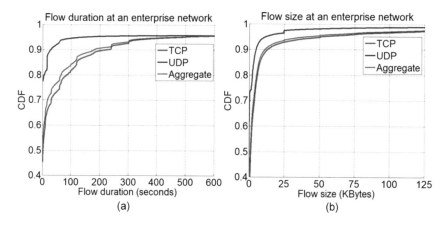

Fig. 3 A week-long (Apr. 16-22, 2018) measurement data for a small campus enterprise network. The total number of observed flows is approximately 33 million; 81% of flows are TCP, 18% UDP, and 1% other protocols. (**a**) Cumulative distribution function (CDF) of the flow duration and (**b**) the flow size. The flow duration is the time interval between the first and last packets of the flow observed in the network, whereas the flow size is the aggregate number of bytes contained in the packets of that flow

is small. Even though packet losses reduce the TCP throughput, from a user perspective this reduction results in a modest increase of the data transfer time. On the other hand, Science DMZ applications typically transfer terabyte-scale files. Hence, even a very small packet loss rate can cause the TCP throughput to collapse below 1 Gbps. As a result, a terabyte-scale data transfer requires many additional hours or days to complete.

A well-designed Science DMZ is minimally sensitive to latency. One of the goals of the Science DMZ is to prevent packet loss and thus to sustain high throughput over high-latency WANs. Hence, the Science DMZ uses dedicated DTNs and switches capable of absorbing transient bursts. It also avoids inline security appliances that may cause packets to be dropped or delivered out of order. By fulfilling these requirements, the achievable throughput can approach the full network capacity. For example, with no packet losses, the throughput is high. Note that the throughput is only slightly sensitive to latency.

3.2 Maximum Transmission Unit

The MTU has a prominent impact on TCP throughput. The throughput is directly proportional to the MSS. Congestion control algorithms perform multiple probes to see how much the network can handle. With high-speed networks, using half a dozen or so small probes to see how the network responds wastes a huge amount of bandwidth. Similarly, when a packet loss is detected, the rate is decreased by a factor of two. TCP can only recover slowly from this rate reduction. The speed at which the recovery occurs is proportional to the MTU. Thus, for Science DMZs, it is recommended to use large frames.

3.3   Buffer Size of Output or Transmission Ports

The buffer size of a router's output port must be large enough, since packets from coincident arrivals from different input ports may be forwarded to the same output port. Additionally, buffers prevent packet losses when traffic bursts occur. A key question is how large should buffers be to absorb the fluctuations generated by large flows. The rule of thumb has been that the amount of buffering (in bits) in a router's port should equal the RTT (in seconds) multiplied by the capacity C (in bits per second) of the port [6, 7]:

$$\text{buffer size} = C \cdot RTT. \tag{1}$$

The above quantity is also known as the bandwidth-delay product (BDP). The rationale behind this quantity is explained in Fig. 4 [8]. In a TCP connection, a sender can have at most W_{max} in-flight or outstanding bits (or the equivalent in segments), where W_{max} is the TCP buffer size dictated by the receiver. Assume that the output port of the router is the bottleneck link of the end-to-end connection. Due to the additive increase behavior of TCP, the sender will keep increasing the rate. The number of queued packets at the router will also increase, until it becomes full and a packet is dropped. At that point, TCP decreases the congestion window to $\frac{W_{max}}{2}$. In order to maximize the throughput of the connection, the bottleneck link should always be utilized. With sufficient buffering, the window size is always above the critical threshold $\frac{W_{max}}{2}$. Since the buffer size is equal to the height of the TCP sawtooth [9], then the size needs to be equal to BDP as well. Notice that the buffer absorbs the changes observed in the TCP window size.

Appenzeller et al. [8] demonstrated that when there is a large number of TCP flows passing through a link, say N, the amount of buffering can be reduced to:

$$\text{buffer size} \approx \frac{C \cdot RTT}{\sqrt{N}}. \tag{2}$$

This result is observed when there is no dominant flow and the router aggregates thousands of flows.

Empirical results [10, 11] suggest that the buffer size of a router in a Science DMZ should equal the bandwidth-delay product. However, a formal proof remains an open research problem. The main challenge in finding an analytical solution is the mathematical complexity of queueing systems with complex packet inter-arrival times. Specifically, the network traffic exhibits high levels of burstiness and self-similarity. A critical characteristic of self-similar traffic is that there is no natural

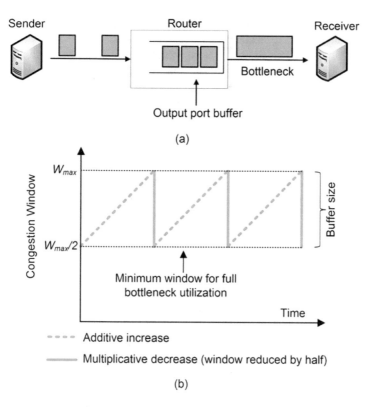

Sender   Router   Receiver

Bottleneck

Output port buffer

(a)

(b)

Fig. 4 TCP viewpoint of a connection and its behavior. (**a**) A simplified TCP interpretation of the connection. (**b**) The congestion control behavior characterized by the additive increase and multiplicative decrease

length of a burst; at every time scale ranging from a few milliseconds to minutes and hours, similar-looking traffic bursts are present. Thus, the results predicted by the $M/M/1$ model from queueing theory (which models packet arrivals as a Poisson process) deviate from the actual performance [12].

In this context, consider again Fig. 4a. Assume that the router behaves like an $M/M/1$ queue, and X is the number of packets in the system. The utilization factor is defined as:

$$\rho = \frac{\text{packet arrival rate at the input port/s}}{\text{packet departure rate at the output port}}. \tag{3}$$

Note that ρ can be interpreted as the utilization of the bottleneck link. According to the $M/M/1$ model, the expected number of packets in the system is $\mathbb{E}(X) = \frac{\rho}{1-\rho}$, and the probability that at least B packets are in the system is given by ρ^B. For a link utilization of $\rho = 0.8$, the expected number of packets in the system is small, namely $\mathbb{E}(X) = 4$. Thus, with a modest buffer size, say 60 packets, the packet drop rate would be less than $\rho^{60} = 0.0000015$. By contrast, the buffer size of a modern 10 Gbps router interface can be over 1,000,000 packets.

Modeling packet arrivals as a Poisson process severely underestimates the traffic burstiness. Traditional TCP congestion control algorithms typically send as many packets as possible at once. Hence, a potential approach to reduce the traffic burstiness (which would permit to reduce the buffer size of a router as predicted by the $M/M/1$ model) is to space out or *pace* packets at sender nodes. The pacing technique can be accomplished by requiring sender nodes to send packets at a fixed rate, so that they are spread over an RTT interval. Results by Beheshti et al. [13] indicate that high throughput can be achieved with small buffer sizes, provided that short-term bursts are minimized. Notably, the first TCP congestion control algorithm based upon pacing has been recently proposed, namely the Bottleneck Bandwidth and Round-Trip Time (BBR) algorithm [3]. Thus, studying the impact of BBR on routers' buffer size is a promising open research direction.

3.4 Bufferbloat

While allocating sufficient memory for buffering is desirable, it is also important to note that the term RTT in Eq. (1) depends upon the use case at hand. Hence, allocating additional unneeded buffer space may result in more latency. This undesirable latency phenomenon is known as bufferbloat [14, 15] and can be mitigated by avoiding the over-allocation of buffers. Controlling excess delay is an active research area. For example, new active-queue management techniques based on control theory have been recently proposed in [16].

3.5 Routers and Switches in a Hierarchical Network

Figure 5 illustrates a typical hierarchical network. The access layer represents the network edge, where traffic enters or exits the network. In Science DMZs, usually DTNs, supercomputer, and research labs have access to the network through access-layer switches. The distribution layer interfaces between the access layer and the core layer, aggregating traffic from the access layer. The core layer is the network backbone. Core routers forward traffic at very high speeds. In this simplified topology, the core is also the border router, connecting the network to the WAN.

Access-layer switches must support a range of traffic capacity needs, sometimes starting as low as 10 Mbps and reaching to as much as 100 Gbps. This wide mix can strain the choice of buffers required, particularly on output switch ports connecting to the distribution layer [17]. Specifically, buffer sizes must be large enough to absorb bursts from the end devices (DTNs, supercomputer, lab devices).

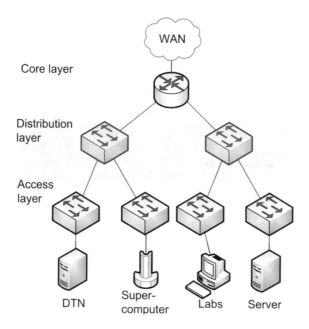

Fig. 5 Hierarchical network

Distribution- and core-layer switches must have as much buffer space as possible to handle the bursts coming from the access-layer switches and from the WAN. Hence, attention must be paid to bandwidth capacity changes (e.g., aggregation of multiple smaller input ports into a larger output port).

Switches manufactured for datacenters may not be a good choice for Science DMZs. They often use fabrics based upon shared memory designs. In these designs, the size of the output buffers may not be tunable, which may become a key performance limitation during the transfer of large flows.

4   Switches in Enterprise Networks and Science DMZs

Table 2 compares switches for enterprise networks and Science DMZs. In general, the crossbar switch fabric is suggested for Science DMZs, because of its high bandwidth. A crossbar switch is also non-blocking; a packet being forwarded to an output port will not be blocked from reaching the output port as long as no other packet is currently being forwarded to that output port. The shared memory technology usually does not allow the allocation of per-port memory for buffering. In Science DMZs, ideally output ports will be statically allocated enough memory for buffering, as suggested by Eq. (1). Although the bus technology still provides sufficient bandwidth for enterprise networks, e.g., Cisco Catalysts 6500 switches

Table 2 Comparison between enterprise network and Science DMZ switches

Feature		Enterprise network switch	Science DMZ switch
Fabric	Crossbar	Recommended.	Recommended.
	Shared memory[a]	Suitable for low-latency datacenters.	Not recommended; buffers usually cannot be allocated on a per-port basis.
	Bus	Suitable for small enterprise networks.	Not recommended; low switching capacity.
Queues	Input queue only	Not recommended; it suffers HOL blocking.	Not recommended; it suffers HOL blocking.
	Input and output queues	Adequate performance.	Adequate performance.
	VOQ	Adequate, attainable throughput approximates 100% of total capacity.	Adequate, attainable throughput approximates 100% of total capacity.
Forwarding	Cut-through	Preferred for low-latency enterprise networks.	Not recommended.
	Store-and-forward	Adequate performance.	Recommended.
Output buffer size	$\frac{RTT \cdot C}{\sqrt{N}}$	Adequate for enterprise flows.	Not sufficient to accommodate large flows.
	$RTT \cdot C$	Not needed.	Recommended; adequate to absorb large flows' bursts and changes in TCP window size.
Buffer allocation	Port-based	Adequate performance.	Recommended.
	Dynamic shared memory[a]	Adequate performance.	Not recommended.
Jumbo frame		Minimum impact for small, short duration flows.	Recommended.

[a] A shared memory fabric often implies dynamic shared memory allocation for ports.

[18], its underlying time-sharing operation is not appropriate for Science DMZs. Consider now buffering; HOL blocking limits the throughput of an input-buffered switch to 59% of the theoretical maximum (which is the sum of the link bandwidths for the switch) [19]. While this technology may be acceptable for small enterprise networks, it should not be used in high-throughput high-latency environments. Science DMZs should use switches that implement output buffering, a mixture of input and output buffering, or techniques emulating output buffering such as VOQ [19].

Forwarding techniques include cut-through and store-and-forward. Cut-through switches start forwarding a packet before the entire packet has been received, normally, as soon as the destination address is processed. They are designed to avoid buffering packets and to minimize latency. Store-and-forward switches buffer the entire packet before it is forwarded to the output port.

Store-and-forward switches provide flexibility to support any mix of speeds. Consider an incoming packet traveling at 10 Gbps that must be forwarded to a 100 Gbps output port. The bit time at the input port is 10 times longer than that at the output port. In a cut-through switch, as incoming bits are processed, they are transmitted to the output port. As soon as a bit is sent out, the 100 Gbps output port is idle waiting for the next bit, which is still being received by the 10 Gbps input port. Hence, much of the 100 Gbps bandwidth would be wasted. Thus, in order to optimize the use of the available bandwidth, the cut-through switch would have to change its operation mode to store-and-forward. However, a significant throughput degradation has been observed when a cut-through switch operates as a store-and-forward switch [20]. This degradation is partially attributed to the small buffer size of a typical cut-through switch. On the other hand, a store-and-forward switch provides automatic buffering of all incoming packets. The forwarding process from a slower interface to a faster interface is made easier, as the reception process at the input port and transmission process at the output port are decoupled.

For Science DMZs, port-based buffer allocation is highly recommended. To absorb transient bursts formed by large flows, or when traffic streams are merged and multiplexed to the same output port, the amount of memory allocated to that port is recommended to be equal to the bandwidth-delay product. Many enterprise networks use switches based on dynamic shared memory. These switches deposit packets into a common memory that is shared by all ports. With dynamic shared memory, there is no guarantee that a port will be allocated an appropriate amount of memory, as this is dynamically allocated.

Academic Cloud and Virtual Laboratories

The book is accompanied by hands-on virtual laboratory experiments conducted in a cloud system, referred to as the Academic Cloud. Access to the Academic Cloud is available for a fee (6-month access) and includes all material needed to conduct the experiments. The URL is

http://highspeednetworks.net/

Chapter 3—Lab 6: Router's Buffer Size

Overview

To conduct the experiment described in this section, please login into the Academic Cloud at http://highspeednetworks.net/ and reserve a pod for Lab 6.

This lab reviews the internal architecture of routers and switches. These devices are essential in high-speed networks, as they must be capable of absorbing transient packet bursts generated by large flows and thus avoid packet loss. The lab describes

Table 3 Credentials to access Client1 machine

Device	Account	Password
Client1	admin	password

the buffer requirements to absorb such traffic fluctuations, which are then validated by experimental results.

Objectives

By the end of this lab, students should be able to:

1. Describe the internal architecture of routers and switches.
2. Understand the importance of buffers of routers and switches to prevent packet loss.
3. Conduct experiments with routers and switches of variable buffer sizes.
4. Calculate the buffer size required by routers and switches to absorb transient bursts.
5. Use experimental results to draw conclusions and make appropriate decision related to routers' and switches' buffers.

Lab Settings

The information in Table 3 provides the credentials of the machine containing Mininet.

Lab Roadmap

This lab is organized as follows:

1. Section 5: Introduction.
2. Section 6: Lab topology.
3. Section 7: Testing throughput with 100*MTU switch's buffer size.
4. Section 8: Testing throughput with one BDP switch's buffer size.
5. Section 9: Emulating high-latency WAN with packet loss.

5   Introduction

5.1   Introduction to Switching

Two essential functions performed by routers are routing and forwarding. Routing refers to the determination of the route taken by packets. Forwarding refers to the switching of a packet from the input port to the appropriate output port. The term switching is also used interchangeably with forwarding. Traditional routing approaches such as static and dynamic routing (e.g., Open Shortest Path First [OSPF], BGP) are used in the implementation of high-speed networks, e.g., Science DMZs. Routing events, such as routing table updates, occur at the millisecond, second, or minute timescale, and best practices used in regular enterprise networks

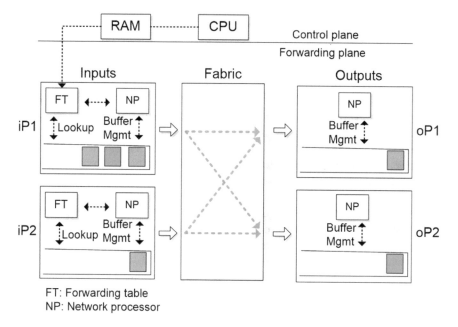

Fig. 6 A generic router architecture

are applicable to high-speed networks as well. These functions are sometimes collectively referred to as the control plane and are usually implemented in software and execute on the routing processor (typically a traditional CPU); see Fig. 6. On the other hand, with transmission rates of 10 Gbps and above, the forwarding operations related to moving packets from input to output interfaces at very high speed must occur at the nanosecond timescale. Thus, forwarding operations, collectively referred to as forwarding or data plane, are executed in specialized hardware and optimized for performance.

Since forwarding functionality is common in both routers and switches, this lab reviews the architecture and forwarding-related attributes of switches. These attributes are applicable to routers as well; thus, for this lab, the terms switch and router are used interchangeably.

5.2 Router Architecture

Consider the generic router architecture that is shown in Fig. 6. Modern routers may have a network processor (NP) and a table derived from the routing table in each port, which is referred to as the forwarding table (FT) or forwarding information base (FIB). The router in Fig. 6 has two input ports, iP1 and iP2, with their respective queues. iP1 has three packets in its queue, which will be forwarded to output ports oP1 (green packets) and oP2 (blue packet) by the fabric. A switch fabric moves packets from input to output ports. Switch fabric designs are shared memory, crossbar network, and bus. In shared memory switches, packets are written into a memory location by an input port and then read from that memory location by the output port. Crossbar switches implement a matrix of pathways that can be configured to connect any input port to any output port. Bus switches use a shared bus to move packets from the input ports to the output ports.

Router queues/buffers absorb traffic fluctuations. Even in the absence of congestion, fluctuations are present, resulting mostly from coincident traffic bursts. Consider an input buffer implemented as a first-in first-out in the router of Fig. 6. As iP1 and iP2 both have one packet to be forwarded to oP1 at the front of the buffer, only one of them, say the packet at iP2, will be forwarded to oP1. The consequence of this is that not only the first packet must wait at iP1. Also, the second packet that is queued at iP1 must wait, even though there is no contention for oP2. This phenomenon is known as Head-Of-Line (HOL) blocking. To avoid HOL blocking, many switches use output buffering, a mixture of internal and output buffering, or techniques emulating output buffering such as Virtual Output Queueing (VOQ).

5.3   Where does Packet Loss Occur?

Packet queues may form at both the input ports and the output ports. The location and extent of queueing (either at the input port queues or the output port queues) will depend on the traffic load, the relative speed of the switching fabric, and the line speed. However, in modern switches with large switching rate capability, queues are commonly formed at output or transmission ports. A main contributing factor is the coincident arrivals of traffic bursts from different input ports that must be forwarded to the same output port. If transmission rates of input and output ports are the same, then packets from coincident arrivals must be momentarily buffered.

Note, however, that buffers will only prevent packet losses in case of transient traffic bursts. If those were not transient but permanent, such as approximately constant bit rates from large file transfers, the aggregate rate of input ports will surpass the rate of the output port. Thus, the output buffer would be permanently full, and the router would drop packets.

Packet loss occurs when a router drops the packet. It is the queues within a router, where such packets are dropped and lost.

5.4   Buffer Size

From the above observation, a key question is how large should buffers be to absorb the fluctuations generated by TCP flows. The rule of thumb has been that the amount of buffering (in bits) in a router's port should equal the average Round-Trip Time (RTT) (in seconds) multiplied by the capacity C (in bits per second) of the port.

Router's buffer size $= C \cdot RTT$ [bits] (single/small number of flows)

Note that RTT is the average of individual RTTs. For example, if there are five TCP flows sharing a router's link (port), the RTT used in the equation above is the average value of the five flows, and the capacity C is the router's port capacity. For example, for 250 millisecond connections and a 10 Gbps port, the router's buffer size equals 2.5 Gbits. The above quantity is a conservative value that can be used in high-throughput high-latency networks.

In 2004, Appenzeller et al. presented a study that suggests that when there is a large number of TCP flows passing through a link, say N (e.g., hundreds, thousands, or more), the amount of buffering can be reduced to:

Router's buffer size $= \frac{C \cdot RTT}{\sqrt{N}}$ [bits] (large number of flows N)

This result is observed when there is no dominant flow and the router aggregates hundreds, thousands, or more flows. The observed effect is that the fluctuations of the sum of congestion windows are smoothed, and the buffer size at an output port can be reduced to the expression given above. Note that N can be very large for campus and backbone networks, and the reduction in needed buffer size can become considerable.

6 Lab Topology

Let's get started with creating a simple Mininet topology using MiniEdit. The topology uses 10.0.0.0/8, which is the default network assigned by Mininet (Fig. 7).

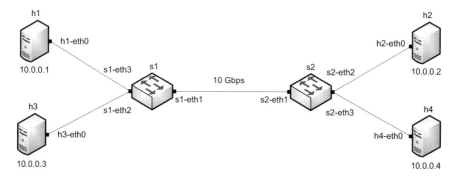

Fig. 7 Lab topology

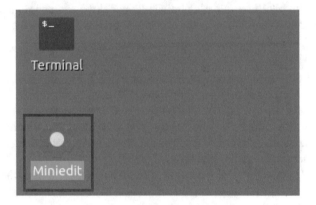

Fig. 8 MiniEdit shortcut

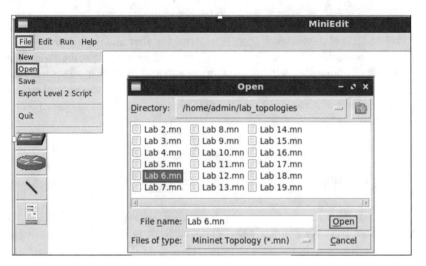

Fig. 9 MiniEdit's *Open* dialog

Step 1. A shortcut to MiniEdit is located on the machine's Desktop. Start MiniEdit by clicking on MiniEdit's shortcut (Fig. 8). When prompted for a password, type ⌷ password ⌷.

Step 2. On MiniEdit's menu bar, click on *File* and then *Open* to load the lab's topology. Locate the *Lab 11.mn* topology file and click on *Open* (Fig. 9).

Step 3. Before starting the measurements between host h1 and host h2, the network must be started. Click on the *Run* button located at the bottom left of MiniEdit's window to start the emulation (Fig. 10).

Fig. 10 Running the
emulation

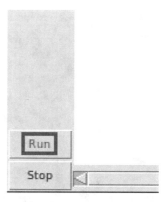

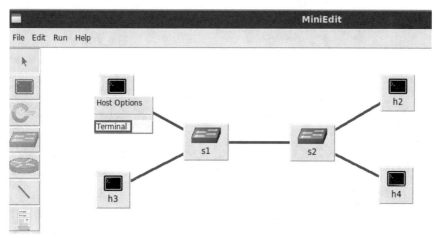

Fig. 11 Opening a terminal on host h1

The above topology uses 10.0.0.0/8, which is the default network assigned by Mininet.

6.1   *Starting Host h1, Host h2, Host h3, and Host h4*

Step 1.   Hold the right-click on host h1 and select *Terminal*. This opens the terminal of host h1 and allows the execution of commands on that host (Fig. 11).

Step 2.   Apply the same steps on host h2 and open its *Terminal*.

Step 3.   Test connectivity between the end-hosts using the ping command. On host h1, type the command ping 10.0.0.2 . This command tests the connectivity

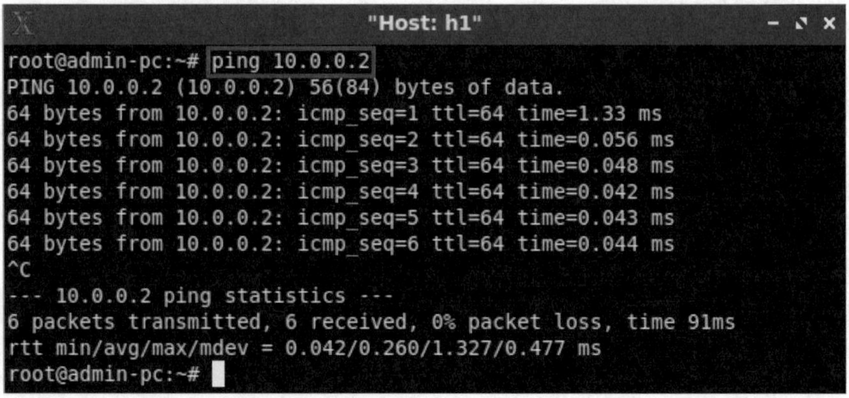

```
X                              "Host: h1"                          -  ⬦  x
root@admin-pc:~# ping 10.0.0.2
PING 10.0.0.2 (10.0.0.2) 56(84) bytes of data.
64 bytes from 10.0.0.2: icmp_seq=1 ttl=64 time=1.33 ms
64 bytes from 10.0.0.2: icmp_seq=2 ttl=64 time=0.056 ms
64 bytes from 10.0.0.2: icmp_seq=3 ttl=64 time=0.048 ms
64 bytes from 10.0.0.2: icmp_seq=4 ttl=64 time=0.042 ms
64 bytes from 10.0.0.2: icmp_seq=5 ttl=64 time=0.043 ms
64 bytes from 10.0.0.2: icmp_seq=6 ttl=64 time=0.044 ms
^C
--- 10.0.0.2 ping statistics ---
6 packets transmitted, 6 received, 0% packet loss, time 91ms
rtt min/avg/max/mdev = 0.042/0.260/1.327/0.477 ms
root@admin-pc:~#
```

Fig. 12 Connectivity test using | ping | command

```
X                              "Host: h3"                          -  ⬦  x
root@admin-pc:~# ping 10.0.0.4
PING 10.0.0.4 (10.0.0.4) 56(84) bytes of data.
64 bytes from 10.0.0.4: icmp_seq=1 ttl=64 time=0.075 ms
64 bytes from 10.0.0.4: icmp_seq=2 ttl=64 time=0.089 ms
64 bytes from 10.0.0.4: icmp_seq=3 ttl=64 time=0.071 ms
64 bytes from 10.0.0.4: icmp_seq=4 ttl=64 time=0.069 ms
64 bytes from 10.0.0.4: icmp_seq=5 ttl=64 time=0.064 ms
64 bytes from 10.0.0.4: icmp_seq=6 ttl=64 time=0.061 ms
^C
--- 10.0.0.4 ping statistics ---
6 packets transmitted, 6 received, 0% packet loss, time 110ms
rtt min/avg/max/mdev = 0.061/0.071/0.089/0.012 ms
root@admin-pc:~#
```

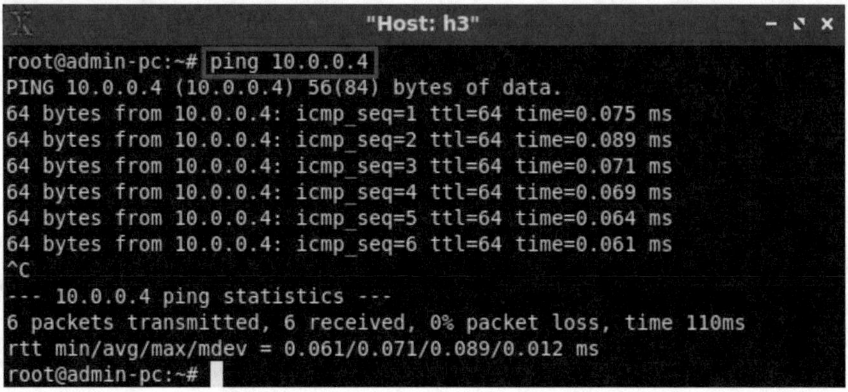

Fig. 13 Connectivity test using | ping | command

between host h1 and host h2 (Fig. 12). To stop the test, press | Ctrl+c |. The figure below shows a successful connectivity test.

Step 4. Test connectivity between the end-hosts using the | ping | command. On host h3, type the command | ping 10.0.0.4 |. This command tests the connectivity between host h3 and host h4 (Fig. 13). To stop the test, press | Ctrl+c |. The figure below shows a successful connectivity test.

```
                              "Host: h1"
root@admin-pc:~# sysctl -w net.ipv4.tcp_rmem='10240 87380 131072000'
net.ipv4.tcp_rmem = 10240 87380 131072000
root@admin-pc:~# █
```

Fig. 14 Receive window change in $\boxed{\text{sysctl}}$

6.2   Modifying Hosts' Buffer Size

In the following tests the bandwidth is limited to 10 Gbps, and the RTT (delay or latency) is 20 ms.

In order to have enough TCP buffer size, we will set the sending and receiving buffer to 5 · BDP in all hosts.

$$BW = 10,000,000,000 \, \text{bits/second}$$

$$RTT = 0.02 \, \text{seconds}$$

$$BDP = 10,000,000,000 \cdot 0.02 = 200,000,000 \, \text{bits}$$

$$= 25,000,000 \, \text{bytes} \approx 25 \, \text{Mbytes}$$

The send and receive buffer sizes should be set to 5 •BDP. We will use the 25 Mbytes value for the BDP instead of 25,000,000 bytes.

$$1 \, \text{Mbyte} = 1024^2 \, \text{bytes}$$

$$BDP = 25 \, \text{Mbytes} = 25 \cdot 1024^2 \, \text{bytes} = 26,214,400 \, \text{bytes}$$

$$5 \cdot BDP = 5 \cdot 26,214,400 \, \text{bytes} = 131,072,000 \, \text{bytes}$$

Step 1. Now, we have calculated the maximum value of the TCP sending and receiving buffer size. In order to change the receiving buffer size, on host h1's terminal type the command shown below. The values set are: 10,240 (minimum), 87,380 (default), and 131,072,000 (maximum) (Fig. 14).

```
sysctl -w net.ipv4.tcp_rmem='10240 87380 131072000'
```

The returned values are measured in bytes. 10,240 represents the minimum buffer size that is used by each TCP socket. 87,380 is the default buffer that is allocated

```
                                    "Host: h1"
root@admin-pc:~# sysctl -w net.ipv4.tcp_wmem='10240 87380 131072000'
net.ipv4.tcp_wmem = 10240 87380 131072000
root@admin-pc:~# █
```

Fig. 15 Send window change in sysctl

```
                                    "Host: h2"
root@admin-pc:~# sysctl -w net.ipv4.tcp_rmem='10240 87380 131072000'
net.ipv4.tcp_rmem = 10240 87380 131072000
root@admin-pc:~# █
```

Fig. 16 Receive window change in sysctl

```
                                    "Host: h2"
root@admin-pc:~# sysctl -w net.ipv4.tcp_wmem='10240 87380 131072000'
net.ipv4.tcp_wmem = 10240 87380 131072000
root@admin-pc:~# █
```

Fig. 17 Send window change in sysctl

when applications create a TCP socket. 131,072,000 is the maximum receive buffer that can be allocated for a TCP socket.

Step 2. To change the current send-window size value(s), use the following command on host h1's terminal. The values set are: 10,240 (minimum), 87,380 (default), and 131,072,000 (maximum) (Fig. 15).

```
sysctl -w net.ipv4.tcp_wmem='10240 87380 131072000'
```

Next, the same commands must be configured on host h2, host h3, and host h4.

Step 3. To change the current receiver-window size value(s), use the following command on host h2's terminal. The values set are: 10,240 (minimum), 87,380 (default), and 131,072,000 (maximum) (Fig. 16).

```
sysctl -w net.ipv4.tcp_rmem='10240 87380 131072000'
```

Step 4. To change the current send-window size value(s), use the following command on host h2's terminal. The values set are: 10,240 (minimum), 87,380 (default), and 131,072,000 (maximum) (Fig. 17).

```
sysctl -w net.ipv4.tcp_wmem='10240 87380 131072000'
```

Fig. 18 Receive window change in sysctl

Fig. 19 Send window change in sysctl

Fig. 20 Receive window change in sysctl

Step 5. To change the current receiver-window size value(s), use the following command on host h3's terminal. The values set are: 10,240 (minimum), 87,380 (default), and 131,072,000 (maximum) (Fig. 18).

```
sysctl -w net.ipv4.tcp_rmem='10240 87380 131072000'
```

Step 6. To change the current send-window size value(s), use the following command on host h3's terminal. The values set are: 10,240 (minimum), 87,380 (default), and 131,072,000 (maximum) (Fig. 19).

```
sysctl -w net.ipv4.tcp_wmem='10240 87380 131072000'
```

Step 7. To change the current receiver-window size value(s), use the following command on host h4's terminal. The values set are: 10,240 (minimum), 87,380 (default), and 131,072,000 (maximum) (Fig. 20).

```
sysctl -w net.ipv4.tcp_rmem='10240 87380 131072000'
```

Fig. 21 Send window change in `sysctl`

Fig. 22 Shortcut to open a Linux terminal

Step 8. To change the current send-window size value(s), use the following command on host h4's terminal. The values set are: 10,240 (minimum), 87,380 (default), and 131,072,000 (maximum) (Fig. 21).

```
sysctl -w net.ipv4.tcp_wmem='10240 87380 131072000'
```

6.3   Emulating High-Latency WAN

This section emulates a high-latency WAN. We will first emulate 20 ms delay between switches, setting 10 ms delay on switch S1 and 10 ms delay on switch S2, resulting in 20 ms of Round-Trip Time (*RTT*).

Step 1. Launch a Linux terminal by holding the Ctrl+Alt+T keys or by clicking on the Linux terminal icon (Fig. 22).

The Linux terminal is a program that opens a window and permits you to interact with a command-line interface (CLI). A CLI is a program that takes commands from the keyboard and sends them to the operating system to perform.

Step 2. In the terminal, type the command below. When prompted for a password, type password and hit *Enter*. This command introduces 10 ms delay to switch S1's *s1-eth1* interface (Fig. 23).

Fig. 23 Adding delay of 10 ms to switch S1's *s1-eth1* interface

Fig. 24 Adding delay of 10 ms to switch S2's *s2-eth1* interface

```
sudo tc qdisc add dev s1-eth1 root handle 1: netem delay
10 ms
```

Step 3. Similarly, repeat again the previous step to set a 10 ms delay to switch S2's interface. When prompted for a password, type | password | and hit *Enter*. This command introduces 10 ms delay on switch S2's *s2-eth1* interface (Fig. 24).

```
sudo tc qdisc add dev s2-eth1 root handle 1: netem delay
10 ms
```

6.4 Testing Connection

To test connectivity, you can use the command | ping |.

Step 1. On the terminal of host h1, type | ping 10.0.0.2 |. To stop the test, press | Ctrl+c |. The figure below shows a successful connectivity test. Host h1 (10.0.0.1) sent four packets to host h2 (10.0.0.2), successfully receiving responses back (Fig. 25).

The result above indicates that all four packets were received successfully (0deviation of the Round-Trip Time (RTT) were 20.096, 20.110, 20.135, and 0.101 milliseconds, respectively. The output above verifies that delay was injected successfully, as the RTT is approximately 20 ms.

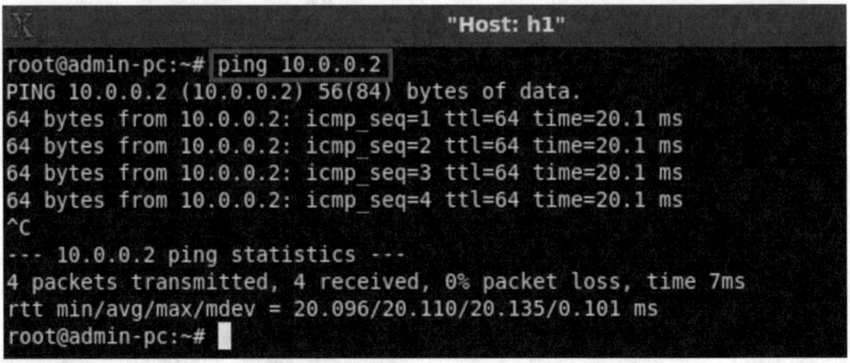

```
                              "Host: h1"
root@admin-pc:~# ping 10.0.0.2
PING 10.0.0.2 (10.0.0.2) 56(84) bytes of data.
64 bytes from 10.0.0.2: icmp_seq=1 ttl=64 time=20.1 ms
64 bytes from 10.0.0.2: icmp_seq=2 ttl=64 time=20.1 ms
64 bytes from 10.0.0.2: icmp_seq=3 ttl=64 time=20.1 ms
64 bytes from 10.0.0.2: icmp_seq=4 ttl=64 time=20.1 ms
^C
--- 10.0.0.2 ping statistics ---
4 packets transmitted, 4 received, 0% packet loss, time 7ms
rtt min/avg/max/mdev = 20.096/20.110/20.135/0.101 ms
root@admin-pc:~#
```

Fig. 25 Output of ping 10.0.0.2 command

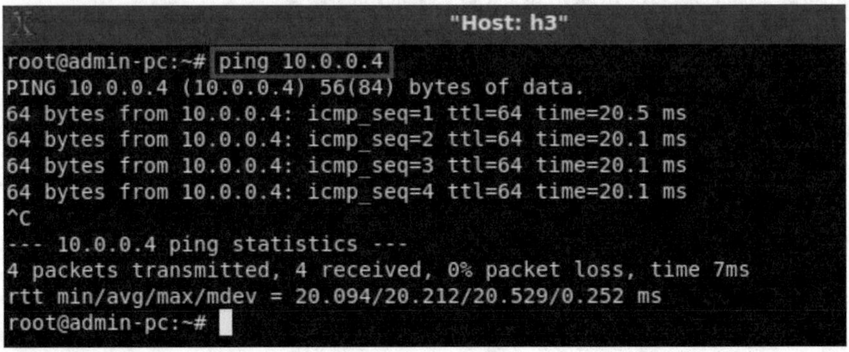

```
                              "Host: h3"
root@admin-pc:~# ping 10.0.0.4
PING 10.0.0.4 (10.0.0.4) 56(84) bytes of data.
64 bytes from 10.0.0.4: icmp_seq=1 ttl=64 time=20.5 ms
64 bytes from 10.0.0.4: icmp_seq=2 ttl=64 time=20.1 ms
64 bytes from 10.0.0.4: icmp_seq=3 ttl=64 time=20.1 ms
64 bytes from 10.0.0.4: icmp_seq=4 ttl=64 time=20.1 ms
^C
--- 10.0.0.4 ping statistics ---
4 packets transmitted, 4 received, 0% packet loss, time 7ms
rtt min/avg/max/mdev = 20.094/20.212/20.529/0.252 ms
root@admin-pc:~#
```

Fig. 26 Output of ping 10.0.0.4 command

Step 2. On the terminal of host h3, type ping 10.0.0.4 . The ping output in this test should be relatively similar to the results of the test initiated by host h1 in Step 1 (Fig. 26). To stop the test, press Ctrl+c .

The result above indicates that all four packets were received successfully (0deviation of the Round-Trip Time (RTT) were 20.094, 20.212, 20.529, and 0.252 milliseconds, respectively. The output above verifies that delay was injected successfully, as the RTT is approximately 20 ms.

7 Testing Throughput with 100 · MTU Switch's Buffer Size

In this section, you are going to change the switch S1's buffer size to 100·MTU and emulate a 10 Gbps Wide Area Network (*WAN*) using the Token Bucket Filter (tbf). Then, you will test the throughput between host h1 and host h2 while there is another

Fig. 27 Limiting rate to 10 Gbps and setting the buffer size to 100·MTU on switch S1's interface

TCP flow between host h3 and host h4. On each test, you will modify the congestion control algorithm in host h1, namely *cubic*, *reno*, and *bbr*. The congestion control algorithm will still be *cubic* in host h3 for all tests. In this section, the MTU is 1600 bytes; thus, the tbf limit value will be set to 100 · MTU = 160,000 bytes.

7.1 Setting Switch S1's Buffer Size to 100 · MTU

Step 1. Apply tbf rate limiting rule on switch S1's *s1-eth1* interface. In the client's terminal, type the command below (Fig. 27). When prompted for a password, type password and hit *Enter*.

- rate : 10gbit
- burst : 5,000,000
- limit : 160,000

```
sudo tc qdisc add dev s1-eth1 parent 1: handle 2: tbf rate
10gbit burst 5000000 limit 160000
```

7.2 TCP Cubic

The default congestion avoidance algorithm in the following test is *cubic*; thus, there is no need to specify it manually.

Step 1. Launch iPerf3 in server mode on host h2's terminal (Fig. 28).

```
iperf3 -s
```

Step 2. Launch iPerf3 in server mode on host h4's terminal (Fig. 29).

Fig. 28 Starting iPerf3 server on host h2

Fig. 29 Starting iPerf3 server on host h4

Fig. 30 Typing iPerf3 client command on host h1

```
iperf3 -s
```

The following two steps should be executed almost simultaneously; thus, you will type the commands displayed in Step 3 and Step 4, and then in Step 5 you will execute them.

Step 3. Type the following iPerf3 command in host h1's terminal without executing it (Fig. 30):

```
iperf3 -c 10.0.0.2 -t 90
```

Step 4. Type the following iPerf3 command in host h3's terminal without executing it (Fig. 31):

```
iperf3 -c 10.0.0.2 b-t 90
```

Step 5. Press *Enter* to execute the commands, first in host h1 terminal and then in host h3 terminal (Fig. 32).

Fig. 31 Typing iPerf3 client command on host h3

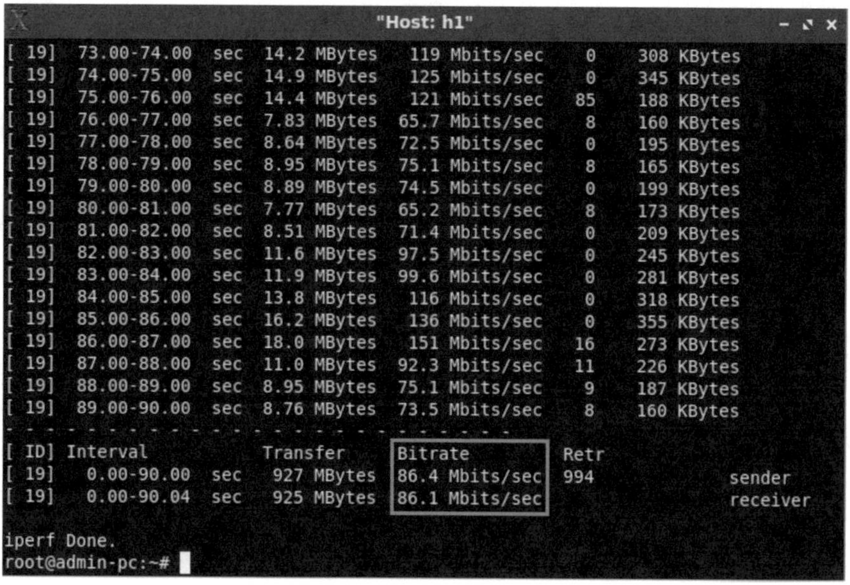

Fig. 32 Running iPerf3 client on host h1

The figure above shows the iPerf3 test output report by the last 20 s. The average achieved throughput is 86.4 Mbps (sender) and 86.1 Mbps (receiver), and the number of retransmissions is 994. Host h3's results are similar to the above; however, we are just focused on host h1's results.

Step 6. In order to stop the server, press ⎪Ctrl+c⎪ in host h2's and host h4's terminals. The user can see the throughput results in the server side too.

7.3 TCP Reno

Step 1. In host h1's terminal, change the TCP congestion control algorithm to Reno by typing the following command (Fig. 33):

```
sysctl -w net.ipv4.tcp_congestion_control=reno
```

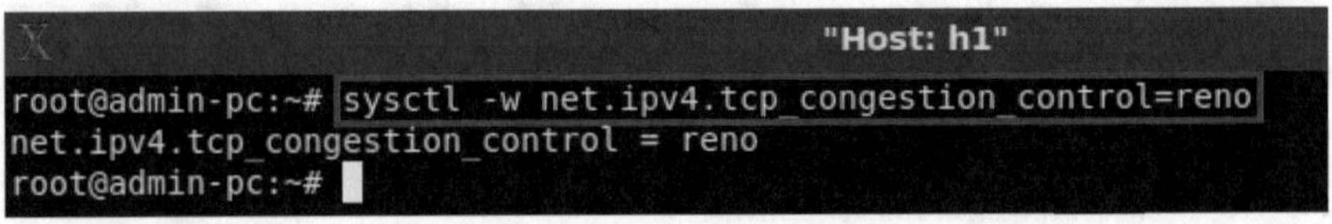

```
X                                                    "Host: h1"
root@admin-pc:~# sysctl -w net.ipv4.tcp_congestion_control=reno
net.ipv4.tcp_congestion_control = reno
root@admin-pc:~#
```

Fig. 33 Changing TCP congestion control algorithm to reno in host h1

```
X                                "Host: h2"                        -  ᵔ  ✕
root@admin-pc:~# iperf3 -s
-----------------------------------------------------------------------
Server listening on 5201
-----------------------------------------------------------------------
```

Fig. 34 Starting iPerf3 server on host h2

```
X                                "Host: h4"                        -  ᵔ  ✕
root@admin-pc:~# iperf3 -s
-----------------------------------------------------------------------
Server listening on 5201
-----------------------------------------------------------------------
```

Fig. 35 Starting iPerf3 server on host h4

Note that host h3's congestion control algorithm is cubic by default.

Step 2. Launch iPerf3 in server mode on host h2's terminal (Fig. 34).

```
iperf3 -s
```

Step 3. Launch iPerf3 in server mode on host h4's terminal (Fig. 35).

```
iperf3 -s
```

The following two steps should be executed almost simultaneously; thus, you will type the commands displayed in Step 3 and Step 4, and then in Step 5 you will execute them.

Step 4. Type the following iPerf3 command in host h1's terminal without executing it (Fig. 36):

```
iperf3 -c 10.0.0.2 -t 90
```

Fig. 36 Typing iPerf3 client command on host h1

Fig. 37 Typing iPerf3 client command on host h3

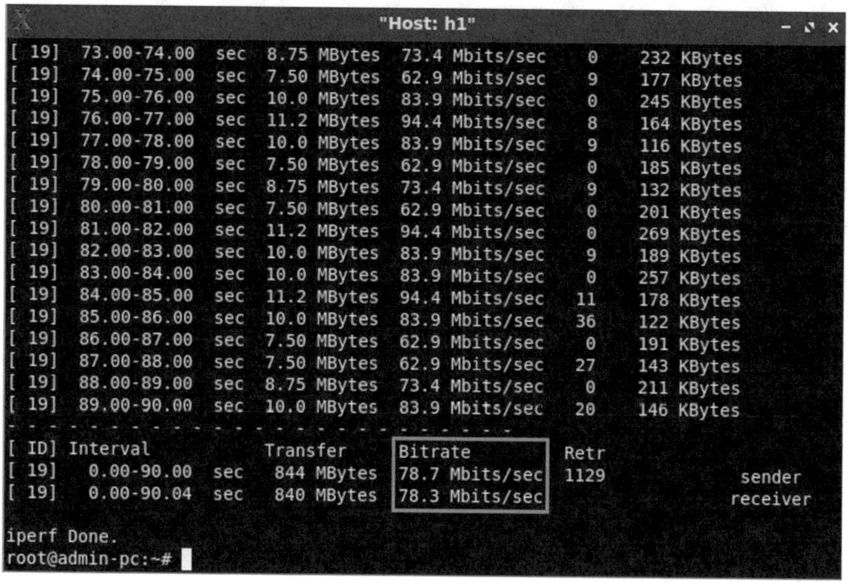

Fig. 38 Running iPerf3 client on host h1

Step 5. Type the following iPerf3 command in host h3's terminal without executing it (Fig. 37):

```
iperf3 -c 10.0.0.2 -t 90
```

Step 6. Press *Enter* to execute the commands, first in host h1 terminal and then in host h3 terminal (Fig. 38).

The figure above shows the iPerf3 test output report by the last 20 s. The average achieved throughput is 78.7 Mbps (sender) and 78.3 Mbps (receiver), and the number of retransmissions is 1129. Host h3's results are similar to the figure above; however, we are just focused on host h1's results.

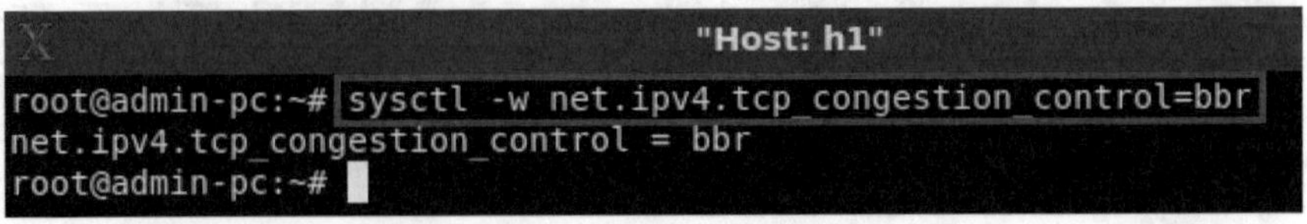

```
X                               "Host: h1"
root@admin-pc:~# sysctl -w net.ipv4.tcp_congestion_control=bbr
net.ipv4.tcp_congestion_control = bbr
root@admin-pc:~# ▌
```

Fig. 39 Changing TCP congestion control algorithm to bbr in host h1

```
X                               "Host: h2"                          —  ↙ ×
root@admin-pc:~# iperf3 -s
- - - - - - - - - - - - - - - - - - - - - - - - - - - - - - - - - - - - - - - -
Server listening on 5201
- - - - - - - - - - - - - - - - - - - - - - - - - - - - - - - - - - - - - - - -
▌
```

Fig. 40 Starting iPerf3 server on host h2

Step 7. In order to stop the server, press ⎡Ctrl+c⎤ in host h2's and host h4's terminals. The user can see the throughput results in the server side too.

7.4 TCP BBR

Step 1. In host h1's terminal, change the TCP congestion control algorithm to BBR by typing the following command (Fig. 39):

```
sysctl -w net.ipv4.tcp_congestion_control=bbr
```

Note that host h3's congestion control algorithm is cubic by default.

Step 2. Launch iPerf3 in server mode on host h2's terminal (Fig. 40).

```
iperf3 -s
```

Step 3. Launch iPerf3 in server mode on host h4's terminal (Fig. 41).

```
iperf3 -s
```

The following two steps should be executed almost simultaneously; thus, you will type the commands displayed in Step 3 and Step 4, and then in Step 5 you will execute them.

Step 4. Type the following iPerf3 command in host h1's terminal without executing it (Fig. 42):

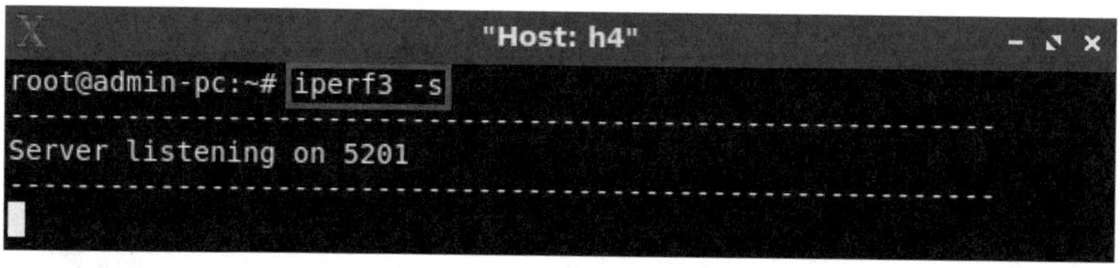

Fig. 41 Starting iPerf3 server on host h4

Fig. 42 Typing iPerf3 client command on host h1

Fig. 43 Typing iPerf3 client command on host h3

```
iperf3 -c 10.0.0.2 -t 90
```

Step 5. Type the following iPerf3 command in host h3's terminal without executing it (Fig. 43):

```
iperf3 -c 10.0.0.2 -t 90
```

Step 6. Press *Enter* to execute the commands, first in host h1 terminal and then in host h3 terminal (Fig. 44).

The figure above shows the iPerf3 test output report by the last 20 s. The average achieved throughput is 3.48 Gbps (sender) and 3.47 Gbps (receiver), and the number of retransmissions is 75,818. Note that the congestion control algorithm used in host h1 is *bbr* and in host h3 is *cubic*.

Step 7. In order to stop the server, press Ctrl+c in host h2's and host h4's terminals. The user can see the throughput results in the server side too.

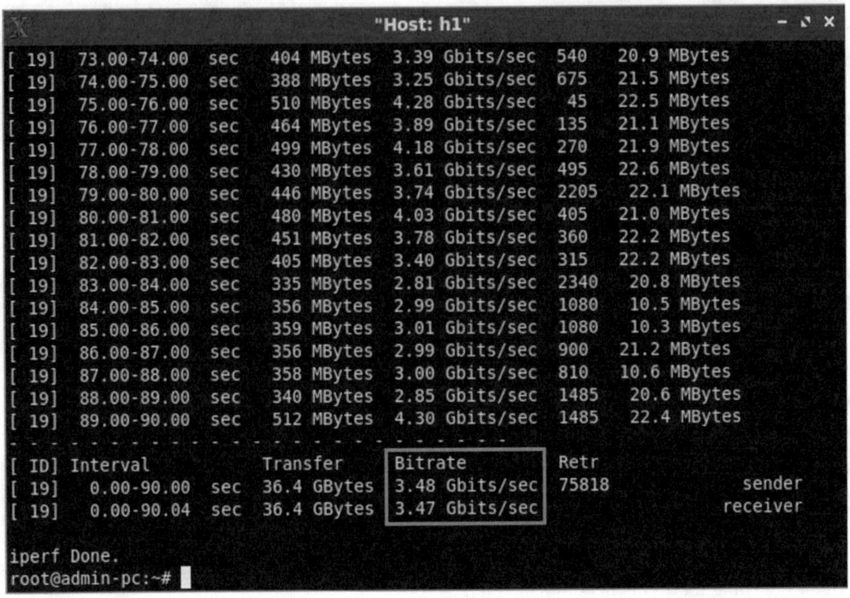

```
                            "Host: h1"                    –  ⬦ ✕
[ 19]  73.00-74.00  sec   404 MBytes  3.39 Gbits/sec  540   20.9 MBytes
[ 19]  74.00-75.00  sec   388 MBytes  3.25 Gbits/sec  675   21.5 MBytes
[ 19]  75.00-76.00  sec   510 MBytes  4.28 Gbits/sec   45   22.5 MBytes
[ 19]  76.00-77.00  sec   464 MBytes  3.89 Gbits/sec  135   21.1 MBytes
[ 19]  77.00-78.00  sec   499 MBytes  4.18 Gbits/sec  270   21.9 MBytes
[ 19]  78.00-79.00  sec   430 MBytes  3.61 Gbits/sec  495   22.6 MBytes
[ 19]  79.00-80.00  sec   446 MBytes  3.74 Gbits/sec 2205   22.1 MBytes
[ 19]  80.00-81.00  sec   480 MBytes  4.03 Gbits/sec  405   21.0 MBytes
[ 19]  81.00-82.00  sec   451 MBytes  3.78 Gbits/sec  360   22.2 MBytes
[ 19]  82.00-83.00  sec   405 MBytes  3.40 Gbits/sec  315   22.2 MBytes
[ 19]  83.00-84.00  sec   335 MBytes  2.81 Gbits/sec 2340   20.8 MBytes
[ 19]  84.00-85.00  sec   356 MBytes  2.99 Gbits/sec 1080   10.5 MBytes
[ 19]  85.00-86.00  sec   359 MBytes  3.01 Gbits/sec 1080   10.3 MBytes
[ 19]  86.00-87.00  sec   356 MBytes  2.99 Gbits/sec  900   21.2 MBytes
[ 19]  87.00-88.00  sec   358 MBytes  3.00 Gbits/sec  810   10.6 MBytes
[ 19]  88.00-89.00  sec   340 MBytes  2.85 Gbits/sec 1485   20.6 MBytes
[ 19]  89.00-90.00  sec   512 MBytes  4.30 Gbits/sec 1485   22.4 MBytes
- - - - - - - - - - - - - - - - - - - -
[ ID] Interval           Transfer    Bitrate         Retr
[ 19]   0.00-90.00  sec  36.4 GBytes 3.48 Gbits/sec  75818             sender
[ 19]   0.00-90.04  sec  36.4 GBytes 3.47 Gbits/sec                    receiver

iperf Done.
root@admin-pc:~#
```

Fig. 44 Running iPerf3 client on host h1

8   Testing Throughput with One BDP Switch's Buffer Size

In this section, you are going to change the switch S1 buffer size to one BDP
(26,214,400) using the Token Bucket Filter (tbf). Then, you will test the throughput between host h1 and host h2 while there is another TCP flow between host h3
and host h4. On each test, you will modify the congestion control algorithm in host
h1, namely *cubic*, *reno*, and *bbr*. The congestion control algorithm will still be *cubic*
in host 3 for all tests. In this section, the tbf limit value will be set to one BDP =
26,214,400 bytes.

8.1   Changing Switch S1's Buffer Size to One BDP

Step 1. Apply tbf rate limiting rule on switch S1's *s1-eth1* interface. In the client's
terminal, type the command below (Fig. 45). When prompted for a password, type
 password and hit *Enter*.

- rate : 10gbit
- burst : 5,000,000
- limit : 26,214,400

Fig. 45   Changing the buffer size to one BDP on switch S1's *s1-eth1* interface

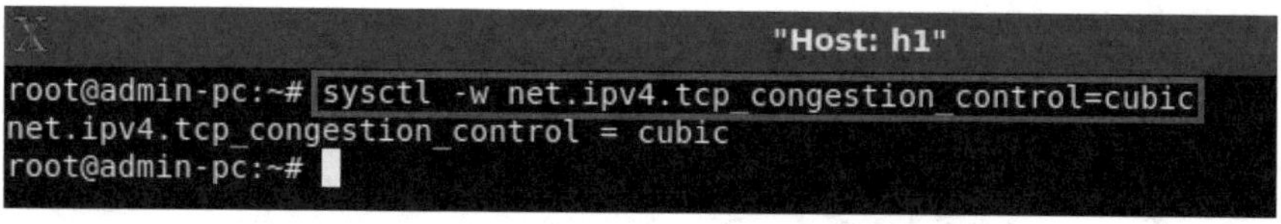

```
                                    "Host: h1"
root@admin-pc:~# sysctl -w net.ipv4.tcp_congestion_control=cubic
net.ipv4.tcp_congestion_control = cubic
root@admin-pc:~#
```

Fig. 46   Changing TCP congestion control algorithm to cubic in host h1

```
sudo tc qdisc change dev s1-eth1 parent 1: handle 2: tbf
rate 10gbit burst 5000000 limit 26214400
```

8.2   TCP Cubic

Step 1. In host h1's terminal, change the TCP congestion control algorithm to Cubic by typing the following command (Fig. 46):

```
sysctl -w net.ipv4.tcp_congestion_control=cubic
```

Note that host h3's congestion control algorithm is cubic by default.

Step 2. Launch iPerf3 in server mode on host h2's terminal (Fig. 47).

```
iperf3 -s
```

Step 3. Launch iPerf3 in server mode on host h4's terminal (Fig. 48).

```
iperf3 -s
```

Fig. 47 Starting iPerf3 server on host h2

Fig. 48 Starting iPerf3 server on host h4

Fig. 49 Typing iPerf3 client command on host h1

> The following two steps should be executed almost
> simultaneously; thus, you will type the commands displayed
> in Step 3 and Step 4, and then in Step 5 you will execute
> them.

Step 4. Type the following iPerf3 command in host h1's terminal without executing it (Fig. 49):

```
iperf3 -c 10.0.0.2 -t 90
```

Step 5. Type the following iPerf3 command in host h3's terminal without executing it (Fig. 50):

```
iperf3 -c 10.0.0.2 -t 90
```

Step 6. Press *Enter* to execute the commands, first in host h1 terminal and then in host h3 terminal (Fig. 51).

```
X                                    "Host: h3"                          — ⬚ ✕
root@admin-pc:~# iperf3 -c 10.0.0.4 -t 90▮
```

Fig. 50 Typing iPerf3 client command on host h3

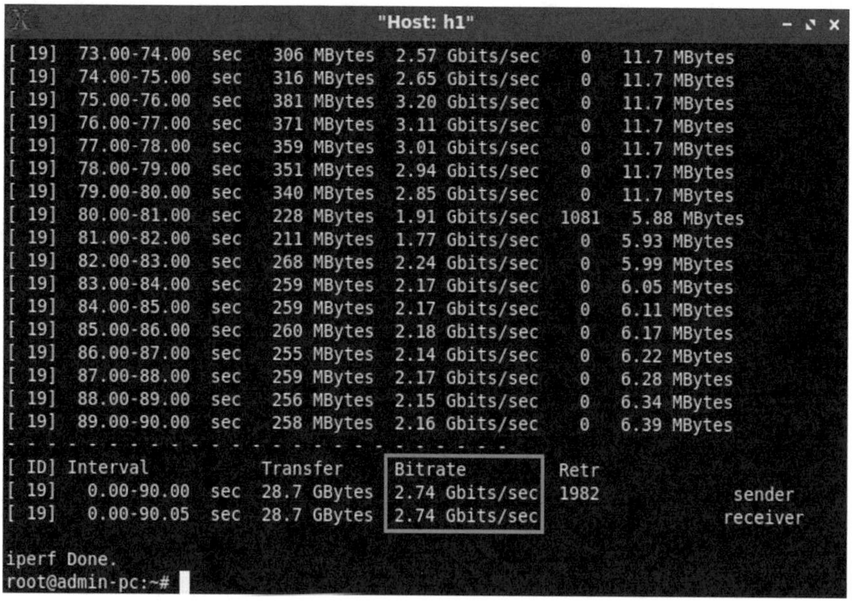

```
X                                    "Host: h1"                                     — ⬚ ✕
[ 19]  73.00-74.00  sec    306 MBytes  2.57 Gbits/sec     0    11.7 MBytes
[ 19]  74.00-75.00  sec    316 MBytes  2.65 Gbits/sec     0    11.7 MBytes
[ 19]  75.00-76.00  sec    381 MBytes  3.20 Gbits/sec     0    11.7 MBytes
[ 19]  76.00-77.00  sec    371 MBytes  3.11 Gbits/sec     0    11.7 MBytes
[ 19]  77.00-78.00  sec    359 MBytes  3.01 Gbits/sec     0    11.7 MBytes
[ 19]  78.00-79.00  sec    351 MBytes  2.94 Gbits/sec     0    11.7 MBytes
[ 19]  79.00-80.00  sec    340 MBytes  2.85 Gbits/sec     0    11.7 MBytes
[ 19]  80.00-81.00  sec    228 MBytes  1.91 Gbits/sec  1081     5.88 MBytes
[ 19]  81.00-82.00  sec    211 MBytes  1.77 Gbits/sec     0     5.93 MBytes
[ 19]  82.00-83.00  sec    268 MBytes  2.24 Gbits/sec     0     5.99 MBytes
[ 19]  83.00-84.00  sec    259 MBytes  2.17 Gbits/sec     0     6.05 MBytes
[ 19]  84.00-85.00  sec    259 MBytes  2.17 Gbits/sec     0     6.11 MBytes
[ 19]  85.00-86.00  sec    260 MBytes  2.18 Gbits/sec     0     6.17 MBytes
[ 19]  86.00-87.00  sec    255 MBytes  2.14 Gbits/sec     0     6.22 MBytes
[ 19]  87.00-88.00  sec    259 MBytes  2.17 Gbits/sec     0     6.28 MBytes
[ 19]  88.00-89.00  sec    256 MBytes  2.15 Gbits/sec     0     6.34 MBytes
[ 19]  89.00-90.00  sec    258 MBytes  2.16 Gbits/sec     0     6.39 MBytes
- - - - - - - - - - - - - - - - - - -
[ ID]  Interval           Transfer   │Bitrate       │ Retr
[ 19]   0.00-90.00  sec  28.7 GBytes │2.74 Gbits/sec│ 1982              sender
[ 19]   0.00-90.05  sec  28.7 GBytes │2.74 Gbits/sec│                   receiver

iperf Done.
root@admin-pc:~# ▮
```

Fig. 51 Running iPerf3 client on host h1

The figure above shows the iPerf3 test output report by the last 20 s. The average achieved throughput is 4.57 Gbps (sender) and 4.57 Gbps (receiver), and the number of retransmissions is 0. Note that the congestion avoidance algorithm used in host h1 and host h2 is *cubic*. Similar results are found in host h3 terminal.

Step 7. In order to stop the server, press Ctrl+c in host h2's and host h4's terminals. The user can see the throughput results in the server side too.

8.3   TCP Reno

Step 1. In host h1's terminal, change the TCP congestion control algorithm to Reno by typing the following command (Fig. 52):

```
sysctl -w net.ipv4.tcp_congestion_control=reno
```

```
                                                "Host: h1"
root@admin-pc:~# sysctl -w net.ipv4.tcp_congestion_control=reno
net.ipv4.tcp_congestion_control = reno
root@admin-pc:~#
```

Fig. 52 Changing TCP congestion control algorithm to reno   in host h1

```
                                   "Host: h2"                          –  ↘  ✕
root@admin-pc:~# iperf3 -s
----------------------------------------------------------------------------
Server listening on 5201
----------------------------------------------------------------------------
```

Fig. 53 Starting iPerf3 server on host h2

```
                                   "Host: h4"                          –  ↘  ✕
root@admin-pc:~# iperf3 -s
----------------------------------------------------------------------------
Server listening on 5201
----------------------------------------------------------------------------
```

Fig. 54 Starting iPerf3 server on host h4

Note that host h3's congestion control algorithm is cubic by default.

Step 2. Launch iPerf3 in server mode on host h2's terminal (Fig. 53).

```
iperf3 -s
```

Step 3. Launch iPerf3 in server mode on host h4's terminal (Fig. 54).

```
iperf3 -s
```

The following two steps should be executed almost simultaneously; thus, you will type the commands displayed in Step 3 and Step 4, and then in Step 5 you will execute them.

Step 4. Type the following iPerf3 command in host h1's terminal without executing it (Fig. 55):

```
iperf3 -c 10.0.0.2 -t 90
```

Fig. 55 Typing iPerf3 client command on host h1

Fig. 56 Typing iPerf3 client command on host h3

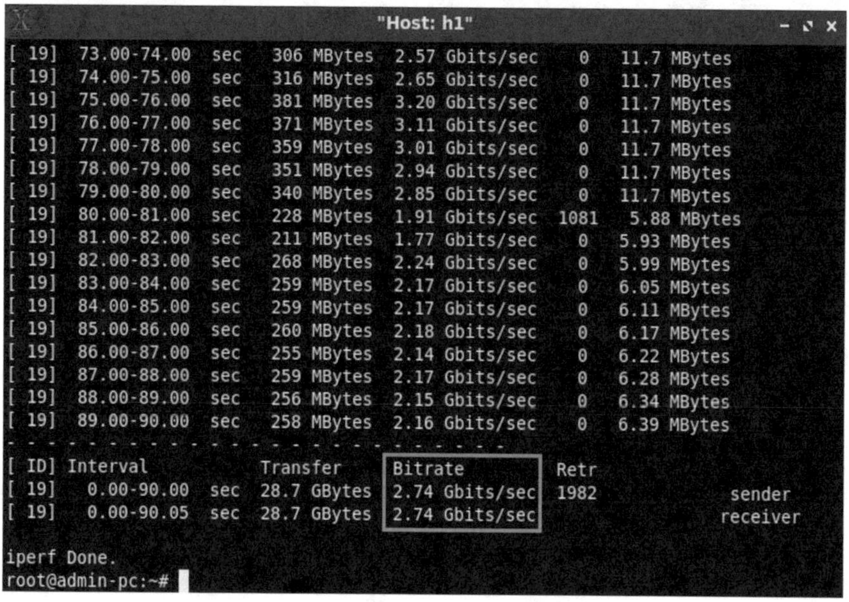

Fig. 57 Running iPerf3 client on host h1

Step 5. Type the following iPerf3 command in host h3's terminal without executing it (Fig. 56):

```
iperf3 -c 10.0.0.2 -t 90
```

Step 6. Press *Enter* to execute the commands, first in host h1 terminal and then in host h3 terminal (Fig. 57).

The figure above shows the iPerf3 test output report by the last 20 s. The average achieved throughput is 2.74 Gbps (sender) and 2.74 Gbps (receiver), and the number of retransmissions is 1982. Note that the congestion avoidance algorithm used in

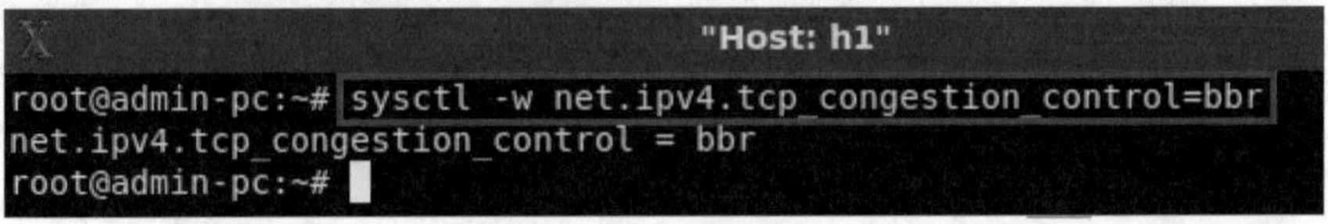

```
                                    "Host: h1"
root@admin-pc:~# sysctl -w net.ipv4.tcp_congestion_control=bbr
net.ipv4.tcp_congestion_control = bbr
root@admin-pc:~# 
```

Fig. 58 Changing TCP congestion control algorithm to bbr in host h1

```
                              "Host: h2"                          —  ⤢  ×
root@admin-pc:~# iperf3 -s
- - - - - - - - - - - - - - - - - - - - - - - - - - - - - - - - - - - - - -
Server listening on 5201
- - - - - - - - - - - - - - - - - - - - - - - - - - - - - - - - - - - - - -
```

Fig. 59 Starting iPerf3 server on host h2

host h1 is *reno* and in host h2 is *cubic*. Host h3's results are similar to the figure above; however, we are just focused on host h1's results.

Step 7. In order to stop the server, press Ctrl+c in host h2's and host h4's terminals. The user can see the throughput results in the server side too.

8.4 TCP BBR

Step 1. In host h1's terminal, change the TCP congestion control algorithm to BBR by typing the following command (Fig. 58):

```
sysctl -w net.ipv4.tcp_congestion_control=bbr
```

Note that host h3's congestion control algorithm is cubic by default.

Step 2. Launch iPerf3 in server mode on host h2's terminal (Fig. 59).

```
iperf3 -s
```

Step 3. Launch iPerf3 in server mode on host h4's terminal (Fig. 60).

```
iperf3 -s
```

Fig. 60 Starting iPerf3server on host h4

Fig. 61 Typing iPerf3 client command on host h1

Fig. 62 Typing iPerf3 client command on host h3

> The following two steps should be executed almost simultaneously; thus, you will type the commands displayed in Step 3 and Step 4, and then in Step 5 you will execute them.

Step 4. Type the following iPerf3 command in host h1's terminal without executing it (Fig. 61):

```
iperf3 -c 10.0.0.2 -t 90
```

Step 5. Type the following iPerf3 command in host h3's terminal without executing it (Fig. 62):

```
iperf3 -c 10.0.0.2 -t 90
```

Step 6. Press *Enter* to execute the commands, first in host h1 terminal and then in host h3 terminal (Fig. 63).

The figure above shows the iPerf3 test output report by the last 20 s. The average achieved throughput is 5.64 Gbps (sender) and 5.63 Gbps (receiver), and the number of retransmissions is 16,110. Note that the congestion avoidance algorithm used in host h1 is *bbr* and in host h3 is *cubic*. Host h3's results are similar to the figure above; however, we are just focused on host h1's results.

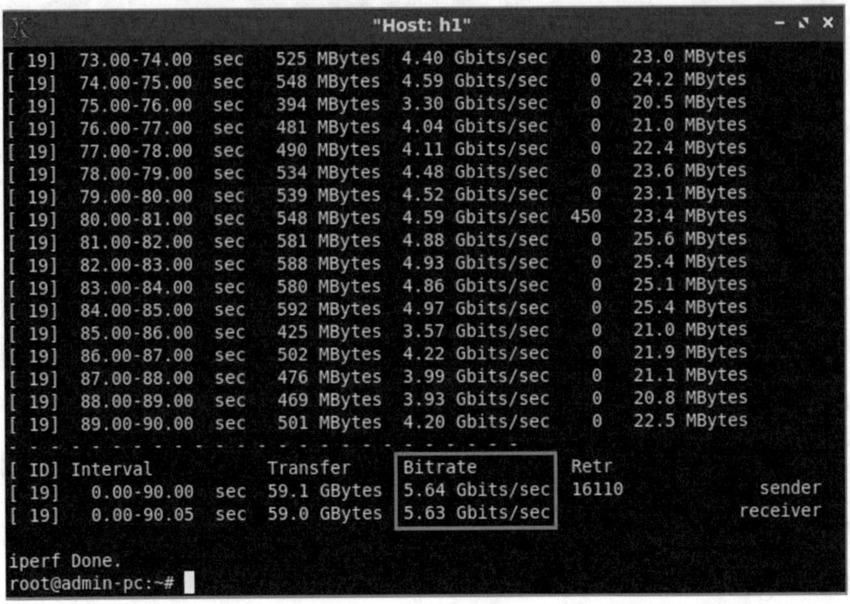

```
X                              "Host: h1"                          –  ↙ ✕
[ 19]   73.00-74.00   sec   525 MBytes  4.40 Gbits/sec    0   23.0 MBytes
[ 19]   74.00-75.00   sec   548 MBytes  4.59 Gbits/sec    0   24.2 MBytes
[ 19]   75.00-76.00   sec   394 MBytes  3.30 Gbits/sec    0   20.5 MBytes
[ 19]   76.00-77.00   sec   481 MBytes  4.04 Gbits/sec    0   21.0 MBytes
[ 19]   77.00-78.00   sec   490 MBytes  4.11 Gbits/sec    0   22.4 MBytes
[ 19]   78.00-79.00   sec   534 MBytes  4.48 Gbits/sec    0   23.6 MBytes
[ 19]   79.00-80.00   sec   539 MBytes  4.52 Gbits/sec    0   23.1 MBytes
[ 19]   80.00-81.00   sec   548 MBytes  4.59 Gbits/sec  450   23.4 MBytes
[ 19]   81.00-82.00   sec   581 MBytes  4.88 Gbits/sec    0   25.6 MBytes
[ 19]   82.00-83.00   sec   588 MBytes  4.93 Gbits/sec    0   25.4 MBytes
[ 19]   83.00-84.00   sec   580 MBytes  4.86 Gbits/sec    0   25.1 MBytes
[ 19]   84.00-85.00   sec   592 MBytes  4.97 Gbits/sec    0   25.4 MBytes
[ 19]   85.00-86.00   sec   425 MBytes  3.57 Gbits/sec    0   21.0 MBytes
[ 19]   86.00-87.00   sec   502 MBytes  4.22 Gbits/sec    0   21.9 MBytes
[ 19]   87.00-88.00   sec   476 MBytes  3.99 Gbits/sec    0   21.1 MBytes
[ 19]   88.00-89.00   sec   469 MBytes  3.93 Gbits/sec    0   20.8 MBytes
[ 19]   89.00-90.00   sec   501 MBytes  4.20 Gbits/sec    0   22.5 MBytes
- - - - - - - - - - - - - - - - - - - - - - - - -
[ ID] Interval           Transfer    Bitrate         Retr
[ 19]   0.00-90.00   sec  59.1 GBytes  5.64 Gbits/sec  16110              sender
[ 19]   0.00-90.05   sec  59.0 GBytes  5.63 Gbits/sec                   receiver

iperf Done.
root@admin-pc:~#
```

Fig. 63 Running iPerf3 client on host h1

Step 7. In order to stop the server, press Ctrl+c in host h2's and host h4's terminals. The user can see the throughput results in the server side too.

9 Emulating High-Latency WAN with Packet Loss

This section emulates a high-latency WAN with packet loss. We already have set a 20 ms RTT on the switches. Now, you will add 0.01% packet loss on the switch S1. Note that the switch S1's buffer size is set to one BDP.

Step 1. In the terminal, type the command below. When prompted for a password, type password and hit *Enter*. This command introduces 0.01% packet loss on switch S1's *s1-eth1* interface (Fig. 64).

```
sudo tc qdisc change dev s1-eth1 root handle 1: netem delay
10 ms loss 0.01%
```

Fig. 64 Adding delay of 0.01% to switch S1's *s1-eth1* interface

```
"Host: h1"
root@admin-pc:~# sysctl -w net.ipv4.tcp_congestion_control=cubic
net.ipv4.tcp_congestion_control = cubic
root@admin-pc:~#
```

Fig. 65 Changing TCP congestion control algorithm to cubic in host h1

```
"Host: h2"                                              — ⬋ ✕
root@admin-pc:~# iperf3 -s
-------------------------------------------------
Server listening on 5201
-------------------------------------------------
```

Fig. 66 Starting iPerf3 server on host h2

9.1 TCP Cubic

Step 1. In host h1's terminal, change the TCP congestion control algorithm to cubic by typing the following command (Fig. 65):

```
sysctl -w net.ipv4.tcp_congestion_control=cubic
```

Note that host h3's congestion control algorithm is cubic by default.

Step 2. Launch iPerf3 in server mode on host h2's terminal (Fig. 66).

```
iperf3 -s
```

Step 3. Launch iPerf3 in server mode on host h4's terminal (Fig. 67).

```
iperf3 -s
```

Fig. 67 Starting iPerf3 server on host h4

Fig. 68 Typing iPerf3 client command on host h1

Fig. 69 Typing iPerf3 client command on host h3

> The following two steps should be executed almost simultaneously; thus, you will type the commands displayed in Step 3 and Step 4, and then in Step 5 you will execute them.

Step 4. Type the following iPerf3 command in host h1's terminal without executing it (Fig. 68):

```
iperf3 -c 10.0.0.2 -t 90
```

Step 5. Type the following iPerf3 command in host h3's terminal without executing it (Fig. 69):

```
iperf3 -c 10.0.0.2 -t 90
```

Step 6. Press *Enter* to execute the commands, first in host h1 terminal and then in host h3 terminal (Fig. 70).

The figure above shows the iPerf3 test output report by the last 20 s. The average achieved throughput is 1.02 Gbps (sender) and 1.02 Gbps (receiver), and the number of retransmissions is 3088. Note that the congestion control algorithm used in host h1 and host h2 is *cubic*. Host h3's results are similar to the figure above; however, we are just focused on host h1's results.

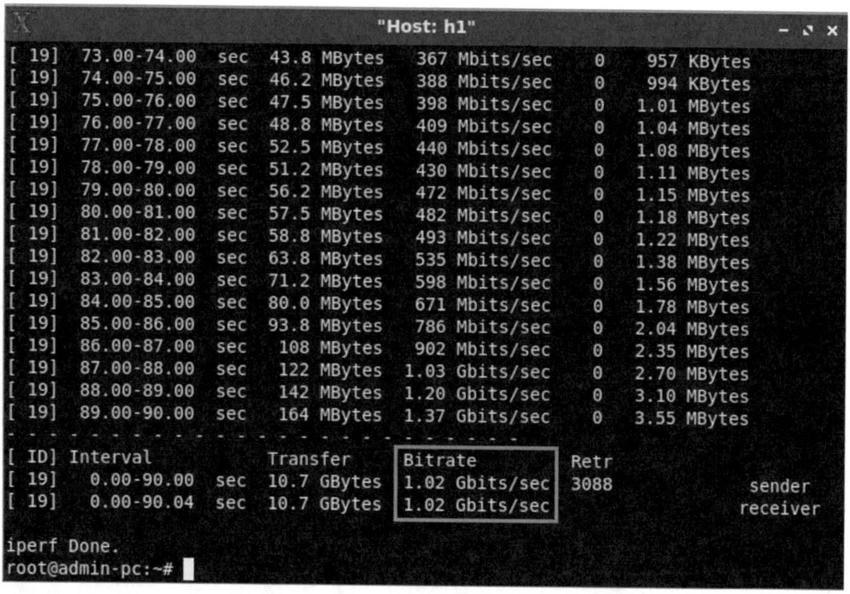

Fig. 70   Running iPerf3 client on host h1

Fig. 71   Changing TCP congestion control algorithm to reno in host h1

Step 7. In order to stop the server, press Ctrl+c in host h2's and host h4's terminals. The user can see the throughput results in the server side too.

9.2   TCP Reno

Step 1. In host h1's terminal, change the TCP congestion control algorithm to Reno by typing the following command (Fig. 71):

```
sysctl -w net.ipv4.tcp_congestion_control=reno
```

Note that host h3's congestion control algorithm is cubic by default.

Step 2. Launch iPerf3 in server mode on host h2's terminal (Fig. 72).

Fig. 72 Starting iPerf3 server on host h2

Fig. 73 Starting iPerf3 server on host h4

Fig. 74 Typing iPerf3 client command on host h1

```
iperf3 -s
```

Step 3. Launch iPerf3 in server mode on host h4's terminal (Fig. 73).

```
iperf3 -s
```

The following two steps should be executed almost simultaneously; thus, you will type the commands displayed in Step 3 and Step 4, and then in Step 5 you will execute them.

Step 4. Type the following iPerf3 command in host h1's terminal without executing it (Fig. 74):

```
iperf3 -c 10.0.0.2 -t 90
```

Step 5. Type the following iPerf3 command in host h3's terminal without executing it (Fig. 75):

```
iperf3 -c 10.0.0.2 -t 90
```

Fig. 75 Typing iPerf3 client command on host h3

```
X                              "Host: h1"                          –  ⌄  x
[ 19]  73.00-74.00  sec  63.8 MBytes   535 Mbits/sec    0   1.34 MBytes
[ 19]  74.00-75.00  sec  66.2 MBytes   556 Mbits/sec    0   1.40 MBytes
[ 19]  75.00-76.00  sec  70.0 MBytes   587 Mbits/sec    0   1.47 MBytes
[ 19]  76.00-77.00  sec  72.5 MBytes   608 Mbits/sec    0   1.54 MBytes
[ 19]  77.00-78.00  sec  77.5 MBytes   650 Mbits/sec    0   1.61 MBytes
[ 19]  78.00-79.00  sec  80.0 MBytes   671 Mbits/sec    0   1.67 MBytes
[ 19]  79.00-80.00  sec  82.5 MBytes   692 Mbits/sec    0   1.74 MBytes
[ 19]  80.00-81.00  sec  78.8 MBytes   661 Mbits/sec   45    929 KBytes
[ 19]  81.00-82.00  sec  45.0 MBytes   377 Mbits/sec    0    997 KBytes
[ 19]  82.00-83.00  sec  48.8 MBytes   409 Mbits/sec    0   1.04 MBytes
[ 19]  83.00-84.00  sec  52.5 MBytes   440 Mbits/sec    0   1.11 MBytes
[ 19]  84.00-85.00  sec  55.0 MBytes   461 Mbits/sec    0   1.17 MBytes
[ 19]  85.00-86.00  sec  57.5 MBytes   482 Mbits/sec    0   1.24 MBytes
[ 19]  86.00-87.00  sec  62.5 MBytes   524 Mbits/sec    0   1.31 MBytes
[ 19]  87.00-88.00  sec  65.0 MBytes   545 Mbits/sec    0   1.38 MBytes
[ 19]  88.00-89.00  sec  68.8 MBytes   577 Mbits/sec    0   1.44 MBytes
[ 19]  89.00-90.00  sec  71.2 MBytes   598 Mbits/sec    0   1.51 MBytes
- - - - - - - - - - - - - - - - - - - - - - - - -
[ ID] Interval           Transfer    Bitrate          Retr
[ 19]  0.00-90.00  sec  7.60 GBytes   726 Mbits/sec  19496             sender
[ 19]  0.00-90.04  sec  7.53 GBytes   718 Mbits/sec                    receiver

iperf Done.
root@admin-pc:~# █
```

Fig. 76 Running iPerf3 client on host h1

Step 6. Press *Enter* to execute the commands, first in host h1 terminal and then in host h3 terminal (Fig. 76).

The figure above shows the iPerf3 test output report by the last 20 s. The average achieved throughput is 726 Mbps (sender) and 718 Mbps (receiver), and the number of retransmissions is 19,496. Note that the congestion control algorithm used in host h1 is *reno* and in host h2 is *cubic*. Host h3's results are similar to the figure above; however, we are just focused on host h1's results.

Step 7. In order to stop the server, press $\boxed{\text{Ctrl+c}}$ in host h2's and host h4's terminals. The user can see the throughput results in the server side too.

9.3 TCP BBR

Step 1. In host h1's terminal, change the TCP congestion control algorithm to BBR by typing the following command (Fig. 77):

```
                              "Host: h1"
root@admin-pc:~# sysctl -w net.ipv4.tcp_congestion_control=bbr
net.ipv4.tcp_congestion_control = bbr
root@admin-pc:~# 
```

Fig. 77 Changing TCP congestion control algorithm to bbr in host h1

```
                              "Host: h2"                       -  ⤢  ×
root@admin-pc:~# iperf3 -s
------------------------------------------------------------
Server listening on 5201
------------------------------------------------------------
```

Fig. 78 Starting iPerf3 server on host h2

```
                              "Host: h4"                       -  ⤢  ×
root@admin-pc:~# iperf3 -s
------------------------------------------------------------
Server listening on 5201
------------------------------------------------------------
```

Fig. 79 Starting iPerf3 server on host h4

```
sysctl -w net.ipv4.tcp_congestion_control=bbr
```

Note that host h3's congestion control algorithm is cubic by default.

Step 2. Launch iPerf3 in server mode on host h2's terminal (Fig. 78).

```
iperf3 -s
```

Step 3. Launch iPerf3 in server mode on host h4's terminal (Fig. 79).

```
iperf3 -s
```

The following two steps should be executed almost simultaneously; thus, you will type the commands displayed in Step 3 and Step 4, and then in Step 5 you will execute them.

Step 4. Type the following iPerf3 command in host h1's terminal without executing it (Fig. 80):

Fig. 80 Typing iPerf3 client command on host h1

Fig. 81 Typing iPerf3 client command on host h3

```
iperf3 -c 10.0.0.2 -t 90
```

Step 5. Type the following iPerf3 command in host h3's terminal without executing it (Fig. 81):

```
iperf3 -c 10.0.0.2 -t 90
```

Step 6. Press *Enter* to execute the commands, first in host h1 terminal and then in host h3 terminal (Fig. 82).

The figure above shows the iPerf3 test output report by the last 20 s. The average achieved throughput is 8.72 Gbps (sender) and 8.71 Gbps (receiver), and the number of retransmissions is 25,740. Note that the congestion avoidance algorithm used in host h1 is *bbr* and in host h3 is *cubic*.

Step 7. In order to stop the server, press Ctrl+c in host h2's and host h4's terminals. The user can see the throughput results in the server side too.

Chapter 3—Lab 7: Router's Bufferbloat

Overview
To conduct the experiment described in this section, please login into the Academic Cloud at http://highspeednetworks.net/ and reserve a pod for Lab 7.

This lab discusses bufferbloat, a condition that occurs when a router or network device buffers too much data, leading to excessive delays. The lab describes the steps to conduct throughput tests on switched networks with different buffer sizes. Note that as the buffering process is similar in routers and switches, both terms are used interchangeably in this lab.

Objectives
By the end of this lab, students should be able to:

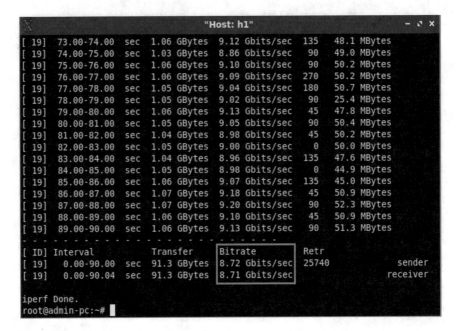

```
                              "Host: h1"                              - ⌄ x
[ 19]  73.00-74.00  sec  1.06 GBytes  9.12 Gbits/sec  135   48.1 MBytes
[ 19]  74.00-75.00  sec  1.03 GBytes  8.86 Gbits/sec   90   49.0 MBytes
[ 19]  75.00-76.00  sec  1.06 GBytes  9.10 Gbits/sec   90   50.2 MBytes
[ 19]  76.00-77.00  sec  1.06 GBytes  9.09 Gbits/sec  270   50.2 MBytes
[ 19]  77.00-78.00  sec  1.05 GBytes  9.04 Gbits/sec  180   50.7 MBytes
[ 19]  78.00-79.00  sec  1.05 GBytes  9.02 Gbits/sec   90   25.4 MBytes
[ 19]  79.00-80.00  sec  1.06 GBytes  9.13 Gbits/sec   45   47.8 MBytes
[ 19]  80.00-81.00  sec  1.05 GBytes  9.05 Gbits/sec   90   50.4 MBytes
[ 19]  81.00-82.00  sec  1.04 GBytes  8.98 Gbits/sec   45   50.2 MBytes
[ 19]  82.00-83.00  sec  1.05 GBytes  9.00 Gbits/sec    0   50.0 MBytes
[ 19]  83.00-84.00  sec  1.04 GBytes  8.96 Gbits/sec  135   47.6 MBytes
[ 19]  84.00-85.00  sec  1.05 GBytes  8.98 Gbits/sec    0   44.9 MBytes
[ 19]  85.00-86.00  sec  1.06 GBytes  9.07 Gbits/sec  135   45.0 MBytes
[ 19]  86.00-87.00  sec  1.07 GBytes  9.18 Gbits/sec   45   50.9 MBytes
[ 19]  87.00-88.00  sec  1.07 GBytes  9.20 Gbits/sec   90   52.3 MBytes
[ 19]  88.00-89.00  sec  1.06 GBytes  9.10 Gbits/sec   45   50.9 MBytes
[ 19]  89.00-90.00  sec  1.06 GBytes  9.13 Gbits/sec   90   51.3 MBytes
- - - - - - - - - - - - - - - - - - - - - - - - - -
[ ID] Interval           Transfer     Bitrate         Retr
[ 19]   0.00-90.00  sec  91.3 GBytes  8.72 Gbits/sec  25740          sender
[ 19]   0.00-90.04  sec  91.3 GBytes  8.71 Gbits/sec                 receiver

iperf Done.
root@admin-pc:~#
```

Fig. 82 Running iPerf3 client on host h1

Table 4 Credentials to
access Client1 machine

Device	Account	Password
Client1	admin	password

1. Identify and describe the components of end-to-end delay.
2. Understand the buffering process in a router.
3. Explain the concept of bufferbloat.
4. Visualize queue occupancy in a router.
5. Analyze end-to-end delay and describe how queueing delay affects end-to-end delay on networks with large routers' buffer size.
6. Modify routers' buffer size to solve the bufferbloat problem.

Lab Settings

The information in Table 4 provides the credentials of the machine containing Mininet.

Lab Roadmap

This lab is organized as follows:

1. Section 10: Introduction to bufferbloat.
2. Section 11: Lab topology.
3. Section 11.4: Testing throughput on a network with a small buffer-size switch.
4. Section 12: Testing throughput on a network with a 1·BDP buffer-size switch.
5. Section 13: Testing throughput on a network with a large buffer-size switch.

10   Introduction to Bufferbloat

10.1   1.1 Packet Delays

As a packet travels from a sender to a receiver, it experiences several types of delays at each node (router/switch) along the path. The most important of these delays are the processing delay, queuing delay, transmission delay, and propagation delay (see Fig. 83).

- Processing delay: The time required to examine the packet's header and determine where to direct the packet. For high-speed routers, this delay is on the order of microseconds or less.
- Transmission delay: The time required to put the bits on the *wire*. It is given by the packet size (in bits) divided by the bandwidth of the link (in bps). For example, for a 10 Gbps and 1500-byte packet (12,000 bits), the transmission time is $T = 12,000/10 \times 10^9 = 0.0012$ milliseconds or 1.2 microseconds.
- Queueing delay: The time a packet waits for transmission onto the link. The length of the queuing delay of a packet depends on the number of earlier-arriving packets that are queued and waiting for transmission onto the link. Queuing delays can be on the order of microseconds to milliseconds.
- Propagation delay: Once a bit is placed into the link, it needs to propagate to the other end of the link. The time required to propagate across the link is the propagation delay. In local area networks (LANs) and datacenter environments, this delay is small (microseconds to few milliseconds); however, in Wide Area Networks (WANs)/long-distance connections, the propagation delay can be on the order of hundreds of milliseconds.

10.2   Bufferbloat

In modern networks composed of high-speed routers and switches, the processing and transmission delays may be negligible. The propagation delay can be considered

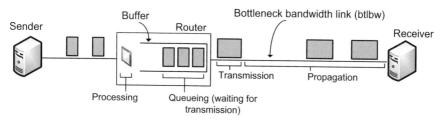

Fig. 83 Delay components: processing, queueing, transmission, and propagation delays

as a constant (i.e., it has a fixed value). Finally, the dynamics of the queues in routers results in varying queueing delays. Ideally, this delay should be minimized.

An important consideration that affects the queuing delay is the router's buffer size. While there is no consensus on how large the buffer should be, the rule of thumb has been that the amount of buffering (in bits) in a router's port should equal the average Round-Trip Time (RTT) (in seconds) multiplied by the capacity C (in bits per second) of the port:

$$\text{Router's buffer size} = C \cdot \text{RTT [bits]}$$

A large enough router's buffer size is essential for networks transporting big flows, as it absorbs transitory packet bursts and prevents losses. However, if a buffer size is excessively large, queues can be formed and substantial queueing delay be observed. This high latency produced by excess buffering of packets is referred to as bufferbloat.

The bufferbloat problem is caused by routers with large buffer size and end devices running TCP congestion control algorithms that constantly probe for additional bandwidth. Consider Fig. 7, where RT_{prop} refers to the end-to-end propagation delay from sender to receiver and then back (round-trip), and BDP refers to the bandwidth-delay product given by the product of the capacity of the bottleneck link along the path and RT_{prop}. RT_{prop} is a constant that depends on the physical distance between end devices. In the application limited region, the throughput increases as the amount of data generated by the application layer increases, while the RTT remains constant. The pipeline between sender and receiver becomes full when the inflight number of bits is equal to BDP, at the edge of the bandwidth limited region. Note that traditional TCP congestion control (e.g., Reno, Cubic, HTCP) will continue to increase the sending rate (inflight data) beyond the optimal operating point, as they probe for more bandwidth. This process is known as TCP additive increase rule. Since no packet loss is noted in the bandwidth limited region despite the increasing TCP rate (which is absorbed by the router's buffer), TCP keeps increasing the sending rate/inflight data, until eventually the router's buffer is full and a packet is dropped (the amount of bits in the network is equal to BDP plus the buffer size of the router). Beyond the application limited region, the increase in queueing delay causes the bufferbloat problem (Fig. 84).

In this lab, the reader will conduct experiments and measure the throughput and RTT under different network conditions. By modifying a router's buffer size, the bufferbloat problem will be observed.

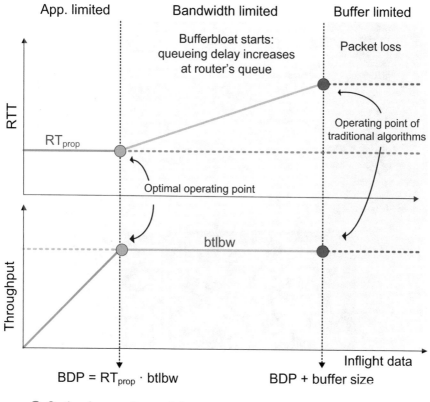

Fig. 84 Throughput and RTT as a function of inflight data

11 Lab Topology

Let's get started with creating a simple Mininet topology using MiniEdit. The topology uses 10.0.0.0/8, which is the default network assigned by Mininet (Fig. 85).

The above topology uses 10.0.0.0/8, which is the default network assigned by Mininet.

Step 1. A shortcut to MiniEdit is located on the machine's Desktop. Start MiniEdit by clicking on MiniEdit's shortcut (Fig. 86). When prompted for a password, type `password` .

Step 2. On MiniEdit's menu bar, click on *File* and then *Open* to load the lab's topology. Locate the *Lab 7.mn* topology file and click on *Open* (Fig. 87).

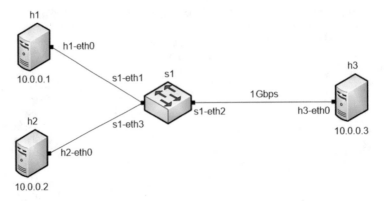

Fig. 85 Lab topology

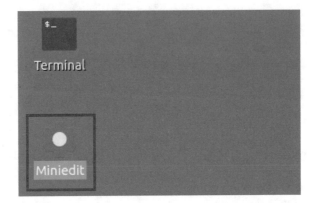

Fig. 86 MiniEdit shortcut

Step 3. Before starting the measurements between end-hosts, the network must be started. Click on the *Run* button located at the bottom left of MiniEdit's window to start the emulation (Fig. 88).

The above topology uses 10.0.0.0/8, which is the default network assigned by Mininet.

11.1 Starting Host h1, Host h2, and Host h3

Step 1. Hold right-click on host h1 and select *Terminal*. This opens the terminal of host h1 and allows the execution of commands on that host (Fig. 89).

Step 2. Apply the same steps on host h2 and host h3 and open their *Terminals*.

Step 3. Test connectivity between the end-hosts using the ⟨ping⟩ command. On host h1, type the command ⟨ping 10.0.0.3⟩. This command tests the connectivity

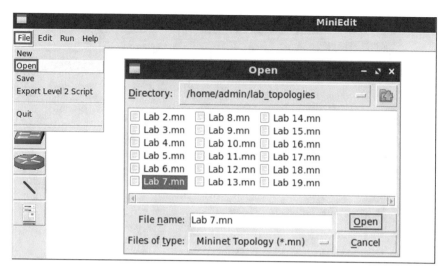

Fig. 87   MiniEdit's *Open* dialog

Fig. 88   Running the emulation

between host h1 and host h3. To stop the test, press $\boxed{\text{Ctrl+c}}$. The figure below shows a successful connectivity test (Fig. 90).

11.2   *Emulating High-Latency WAN*

This section emulates a high-latency WAN. We will emulate 20 ms delay on switch S1's *s1-eth2* interface.

Step 1.   Launch a Linux terminal by holding the $\boxed{\text{Ctrl+Alt+T}}$ keys or by clicking on the Linux terminal icon (Fig. 91).

MiniEdit
File Edit Run Help
h1
Host Options
Terminal
s1
h3
h2

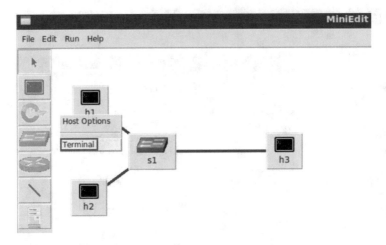

Fig. 89 Opening a terminal on host h1

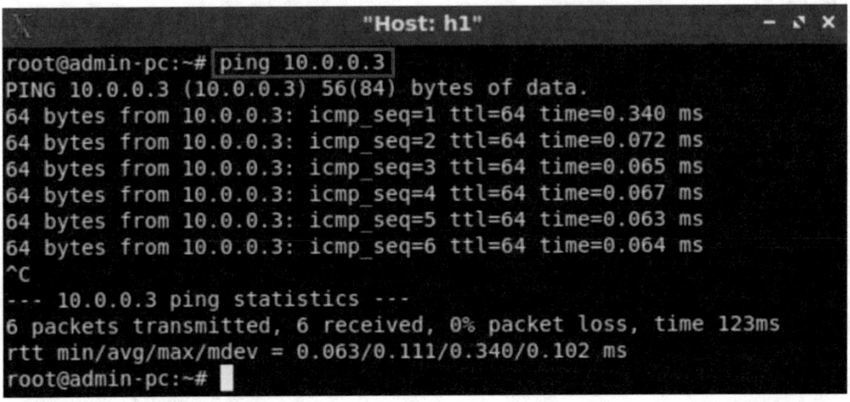

"Host: h1"
root@admin-pc:~# ping 10.0.0.3
PING 10.0.0.3 (10.0.0.3) 56(84) bytes of data.
64 bytes from 10.0.0.3: icmp_seq=1 ttl=64 time=0.340 ms
64 bytes from 10.0.0.3: icmp_seq=2 ttl=64 time=0.072 ms
64 bytes from 10.0.0.3: icmp_seq=3 ttl=64 time=0.065 ms
64 bytes from 10.0.0.3: icmp_seq=4 ttl=64 time=0.067 ms
64 bytes from 10.0.0.3: icmp_seq=5 ttl=64 time=0.063 ms
64 bytes from 10.0.0.3: icmp_seq=6 ttl=64 time=0.064 ms
^C
--- 10.0.0.3 ping statistics ---
6 packets transmitted, 6 received, 0% packet loss, time 123ms
rtt min/avg/max/mdev = 0.063/0.111/0.340/0.102 ms
root@admin-pc:~#

Fig. 90 Connectivity test using `ping` command

The Linux terminal is a program that opens a window and permits you to interact with a command-line interface (CLI). A CLI is a program that takes commands from the keyboard and sends them to the operating system to perform.

Step 2. In the terminal, type the command below. When prompted for a password, type `password` and hit *Enter*. This command introduces 10 ms delay to switch S1's *s1-eth2* interface (Fig. 92).

```
sudo tc qdisc add dev s1-eth2 root handle 1: netem delay
20ms
```

$_
Terminal
Miniedit

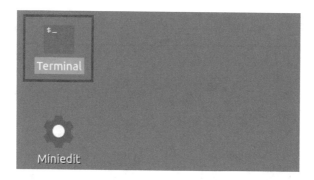

Fig. 91 Shortcut to open a Linux terminal

admin@admin-pc: ~
File Actions Edit View Help
admin@admin-pc: ~
admin@admin-pc:~$ sudo tc qdisc add dev s1-eth2 root handle 1: netem delay 20ms
[sudo] password for admin:
admin@admin-pc:~$

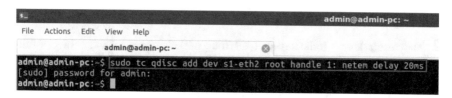

Fig. 92 Adding delay of 10 ms to switch S1's *s1-eth2* interface

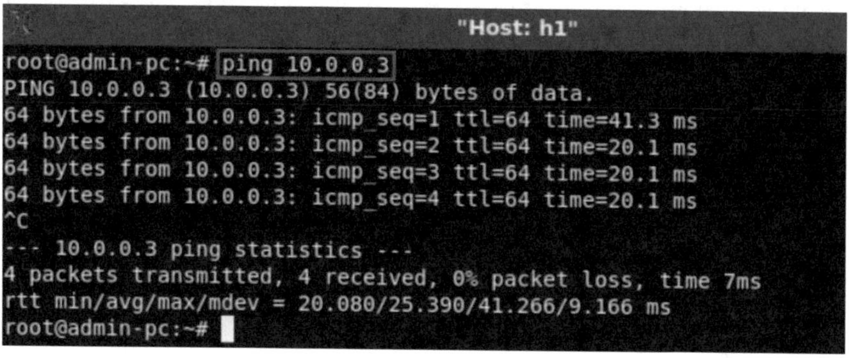

"Host: h1"
root@admin-pc:~# ping 10.0.0.3
PING 10.0.0.3 (10.0.0.3) 56(84) bytes of data.
64 bytes from 10.0.0.3: icmp_seq=1 ttl=64 time=41.3 ms
64 bytes from 10.0.0.3: icmp_seq=2 ttl=64 time=20.1 ms
64 bytes from 10.0.0.3: icmp_seq=3 ttl=64 time=20.1 ms
64 bytes from 10.0.0.3: icmp_seq=4 ttl=64 time=20.1 ms
^C
--- 10.0.0.3 ping statistics ---
4 packets transmitted, 4 received, 0% packet loss, time 7ms
rtt min/avg/max/mdev = 20.080/25.390/41.266/9.166 ms
root@admin-pc:~#

Fig. 93 Output of `ping 10.0.0.3` command

11.3 Testing Connection

To test connectivity, you can use the command `ping`.

Step 1. On the terminal of host h1, type `ping 10.0.0.3`. To stop the test, press `Ctrl+c`. The figure below shows a successful connectivity test. Host h1 (10.0.0.1) sent four packets to host h3 (10.0.0.3), successfully receiving responses back (Fig. 93).

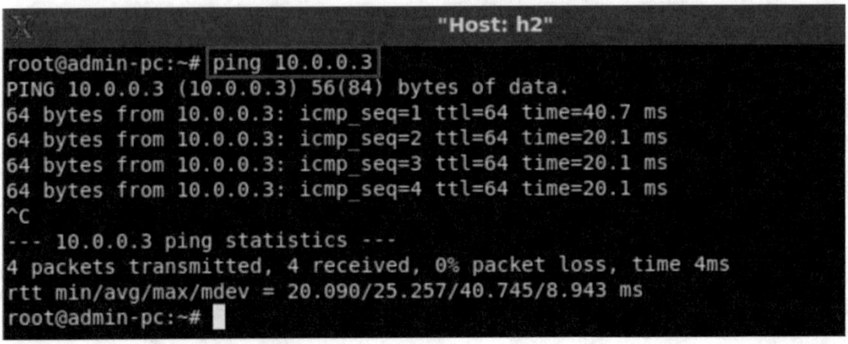

```
X                                    "Host: h2"
root@admin-pc:~# ping 10.0.0.3
PING 10.0.0.3 (10.0.0.3) 56(84) bytes of data.
64 bytes from 10.0.0.3: icmp_seq=1 ttl=64 time=40.7 ms
64 bytes from 10.0.0.3: icmp_seq=2 ttl=64 time=20.1 ms
64 bytes from 10.0.0.3: icmp_seq=3 ttl=64 time=20.1 ms
64 bytes from 10.0.0.3: icmp_seq=4 ttl=64 time=20.1 ms
^C
--- 10.0.0.3 ping statistics ---
4 packets transmitted, 4 received, 0% packet loss, time 4ms
rtt min/avg/max/mdev = 20.090/25.257/40.745/8.943 ms
root@admin-pc:~#
```

Fig. 94 Output of ping 10.0.0.3 command

The result above indicates that all four packets were received successfully (0deviation of the Round-Trip Time (RTT) were 20.080, 25.390, 41.266, and 9.166 milliseconds, respectively. The output above verifies that delay was injected successfully, as the RTT is approximately 20 ms.

Step 2. On the terminal of host h2, type ping 10.0.0.3 . The ping output in this test should be relatively similar to the results of the test initiated by host h1 in Step 1. To stop the test, press Ctrl+c (Fig. 94).

The result above indicates that all four packets were received successfully (0deviation of the Round-Trip Time (RTT) were 20.090, 25.257, 40.745, and 8.943 milliseconds, respectively. The output above verifies that delay was injected successfully, as the RTT is approximately 20 ms.

11.4 Testing Throughput on a Network with a Small Buffer-Size Switch

In this section, you are going to change the switch S1's buffer size to 100·MTU and emulate a 1 Gbps Wide Area Network (*WAN*) using the Token Bucket Filter (tbf). Then, you will test the throughput between host h1 and host h3. In this section, the MTU is 1600 bytes; thus, the tbf limit value will be set to $100 \cdot \text{MTU} = 160{,}000$ bytes.

Fig. 95 Limiting rate to 1 Gbps and setting the buffer size to 100·MTU on switch S1's interface

11.5   Setting Switch S1's Buffer Size to 100 · MTU

Step 1. Apply ⟮tbf⟯ rate limiting rule on switch S1's *s1-eth2* interface. In the client's terminal, type the command below. When prompted for a password, type ⟮password⟯ and hit *Enter* (Fig. 95).

- ⟮rate⟯: 1gbit
- ⟮burst⟯: 500,000
- ⟮limit⟯: 160,000

```
sudo tc qdisc add dev s1-eth2 parent 1: handle 2: tbf rate
1gbit burst 500000 limit 160000
```

11.6   Bandwidth-Delay Product (BDP) and Hosts' Buffer Size

In the upcoming tests, the bandwidth is limited to 1 Gbps, and the RTT (delay or latency) is 20 ms.

$$BW = 1,000,000,000 \, \text{bits/second}$$

$$RTT = 0.02 \, \text{seconds}$$

$$BDP = 1,000,000,000 \cdot 0.02 = 20,000,000 \, \text{bits}$$

$$= 2,500,000 \, \text{bytes} \approx 2.5 \, \text{Mbytes}$$

$$1 \, \text{Mbyte} = 1024^2 \, \text{bytes}$$

$$BDP = 2.5 \, \text{Mbytes} = 2.5 \cdot 1024^2 \, \text{bytes} = 2,621,440 \, \text{bytes}$$

```
                              "Host: h1"                           –  ⌄  ✕
root@admin-pc:~# sysctl -w net.ipv4.tcp_rmem='10240 87380 52428800'
net.ipv4.tcp_rmem = 10240 87380 52428800
root@admin-pc:~# ▮
```

Fig. 96 Receive window change in sysctl

```
                              "Host: h1"                           –  ⌄  ✕
root@admin-pc:~# sysctl -w net.ipv4.tcp_wmem='10240 87380 52428800'
net.ipv4.tcp_wmem = 10240 87380 52428800
root@admin-pc:~# ▮
```

Fig. 97 Send window change in sysctl

The default buffer size in Linux is 16 Mbytes, and only 8 Mbytes (half of the maximum buffer size) can be allocated. Since 8 Mbytes is greater than 2.5 Mbytes, then no need to tune the buffer sizes on end-hosts. However, in upcoming tests, we configure the buffer size on the switch to 10·BDP. To ensure that the bottleneck is not the hosts' buffers, we configure the buffers to 10·BDP (26,214,400).

Step 1. Now, we have calculated the maximum value of the TCP sending and receiving buffer size. In order to change the receiving buffer size, on host h1's terminal type the command shown below. The values set are: 10,240 (minimum), 87,380 (default), and 52,428,800 (maximum). The maximum value is doubled (2·10·BDP) as Linux only allocates half of the assigned value (Fig. 96).

```
sysctl -w net.ipv4.tcp_rmem='10240 87380 52428800'
```

The returned values are measured in bytes. 10,240 represents the minimum buffer size that is used by each TCP socket. 87,380 is the default buffer that is allocated when applications create a TCP socket. 52,428,800 is the maximum receive buffer that can be allocated for a TCP socket.

Step 2. To change the current send-window size value(s), use the following command on host h1's terminal. The values set are: 10,240 (minimum), 87,380 (default), and 52,428,800 (maximum). The maximum value is doubled as Linux allocates only half of the assigned value (Fig. 97).

```
sysctl -w net.ipv4.tcp_wmem='10240 87380 52428800'
```

Step 3. Now, we have calculated the maximum value of the TCP sending and receiving buffer size. In order to change the receiving buffer size, on host h3's terminal type the command shown below. The values set are: 10,240 (minimum), 87,380 (default), and 52,428,800 (maximum). The maximum value is doubled as Linux allocates only half of the assigned value (Fig. 98).

Fig. 98 Receive window change in $\boxed{\text{sysctl}}$

Fig. 99 Send window change in $\boxed{\text{sysctl}}$

Fig. 100 Starting iPerf3 server on host h3

```
sysctl -w net.ipv4.tcp_wmem='10240 87380 52428800'
```

Step 4. To change the current send-window size value(s), use the following command on host h1's terminal. The values set are: 10,240 (minimum), 87,380 (default), and 52,428,800 (maximum). The maximum value is doubled as Linux allocates only half of the assigned value (Fig. 99).

```
sysctl -w net.ipv4.tcp_wmem='10240 87380 52428800'
```

The returned values are measured in bytes. 10,240 represents the minimum buffer size that is used by each TCP socket. 87,380 is the default buffer that is allocated when applications create a TCP socket. 52,428,800 is the maximum receive buffer that can be allocated for a TCP socket.

11.7 Throughput Test

Step 1. Launch iPerf3 in server mode on host h3's terminal (Fig. 100).

```
iperf3 -s
```

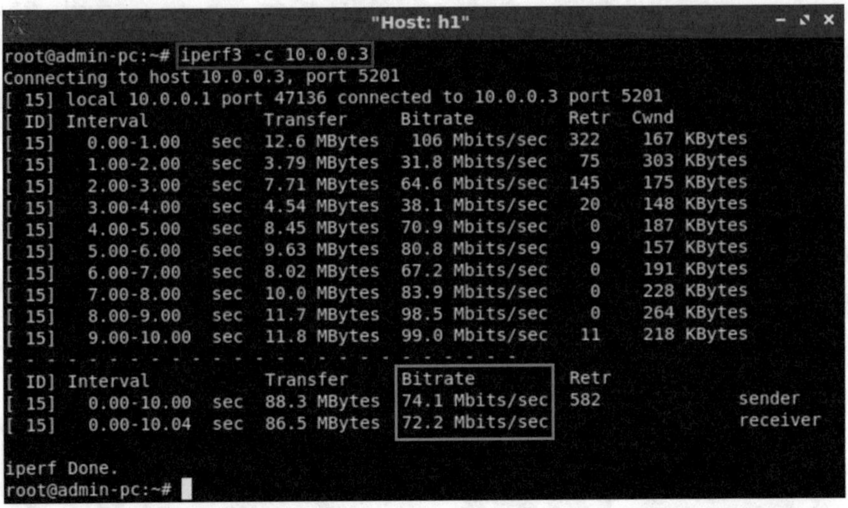

```
                          "Host: h1"                        –  ⤢ ✕
root@admin-pc:~# iperf3 -c 10.0.0.3
Connecting to host 10.0.0.3, port 5201
[ 15] local 10.0.0.1 port 47136 connected to 10.0.0.3 port 5201
[ ID] Interval           Transfer     Bitrate         Retr  Cwnd
[ 15]   0.00-1.00   sec  12.6 MBytes   106 Mbits/sec  322   167 KBytes
[ 15]   1.00-2.00   sec  3.79 MBytes  31.8 Mbits/sec   75   303 KBytes
[ 15]   2.00-3.00   sec  7.71 MBytes  64.6 Mbits/sec  145   175 KBytes
[ 15]   3.00-4.00   sec  4.54 MBytes  38.1 Mbits/sec   20   148 KBytes
[ 15]   4.00-5.00   sec  8.45 MBytes  70.9 Mbits/sec    0   187 KBytes
[ 15]   5.00-6.00   sec  9.63 MBytes  80.8 Mbits/sec    9   157 KBytes
[ 15]   6.00-7.00   sec  8.02 MBytes  67.2 Mbits/sec    0   191 KBytes
[ 15]   7.00-8.00   sec  10.0 MBytes  83.9 Mbits/sec    0   228 KBytes
[ 15]   8.00-9.00   sec  11.7 MBytes  98.5 Mbits/sec    0   264 KBytes
[ 15]   9.00-10.00  sec  11.8 MBytes  99.0 Mbits/sec   11   218 KBytes
- - - - - - - - - - - - - - - - - - - - - - - - - - -
[ ID] Interval           Transfer     Bitrate         Retr
[ 15]   0.00-10.00  sec  88.3 MBytes  74.1 Mbits/sec  582             sender
[ 15]   0.00-10.04  sec  86.5 MBytes  72.2 Mbits/sec                  receiver

iperf Done.
root@admin-pc:~# ▮
```

Fig. 101 Running iPerf3 client on host h1

Step 2. Type the following iPerf3 command in host h1's terminal (Fig. 101):

```
iperf3 -c 10.0.0.3
```

The figure above shows the iPerf3 test output report. The average achieved throughput is 74.1 Mbps (sender) and 72.2 Mbps (receiver), and the number of retransmissions is 582. Note that the maximum throughput (1 Gbps) was not achieved. This is due to having a small buffer on the switch (100 · MTU).

12 Testing Throughput on a Network with a 1 · BDP Buffer-Size Switch

In this section, you are going to change the switch S1's buffer size to 1·BDP and emulate a 1 Gbps Wide Area Network (*WAN*) using the Token Bucket Filter (tbf). Then, you will test the throughput between host h1 and host h3. The BDP is 2,621,440 bytes; thus, the tbf limit value will be set to 2,621,440.

Fig. 102 Limiting rate to 1 Gbps and setting the buffer size to 1·BDP on switch S1's interface

Fig. 103 Starting iPerf3 server on host h3

12.1   Setting Switch S1's Buffer Size to 1 · BDP

Step 1. Apply ⎡tbf⎤ rate limiting rule on switch S1's *s1-eth2* interface. In the client's terminal, type the command below. When prompted for a password, type ⎡password⎤ and hit *Enter* (Fig. 102).

- ⎡ rate ⎤: 1gbit
- ⎡ burst ⎤: 500,000
- ⎡ limit ⎤: 2,621,440

```
sudo tc qdisc change dev s1-eth2 parent 1: handle 2: tbf
rate 1gbit burst 500000 limit 2621440
```

12.2   Throughput and Latency Tests

Step 1. Launch iPerf3 in server mode on host h3's terminal (Fig. 103).

```
iperf3 -s
```

Step 2. In the client's terminal, type the command below to plot the switch's queue in real-time. When prompted for a ⎡password⎤, type password and hit *Enter* (Fig. 104).

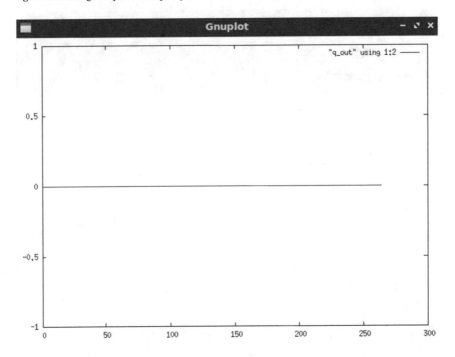

Fig. 104 Plotting the queue occupancy on switch S1's *s1-eth2* interface

Fig. 105 Queue occupancy on switch S1's *s1-eth2* interface

```
sudo plot_q.sh s1-eth2
```

A new window opens that plots the queue occupancy as shown in the figure below (Fig. 105). Since there are no active flows passing through *s1-eth2* interface on switch S1, the queue occupancy is constantly 0.

Step 3. In host h1, create a directory called *1_BDP* and navigate into it using the following command (Fig. 106):

```
mkdir 1_BDP && cd 1_BDP
```

Fig. 106 Creating and navigating into directory *1_BDP*

Fig. 107 Running iPerf3 client on host h1

Fig. 108 Typing ping command on host h2

Step 4. Type the following iPerf3 command in host h1's terminal without executing it. The -J option is used to display the output in JSON format. The redirection operator > is used to store the JSON output into a file (Fig. 107).

```
iperf3 -c 10.0.0.3 -t 90 -J >out.json
```

Step 5. Type the following ping command in host h2's terminal without executing it (Fig. 108):

```
ping 10.0.0.3 -c 90
```

Step 6. Press *Enter* to execute the commands, first in host h1 terminal and then in host h2 terminal. Then, go back to the queue plotting window and observe the queue occupancy (Fig. 109).

The graph above shows that the queue occupancy peaked at $2.5 \cdot 10^6$, which is the maximum buffer size we configure on the switch.

Step 7. In the queue plotting window, press the s key on your keyboard to stop plotting the queue.

Step 8. After the iPerf3 test finishes on host h1, enter the following command (Fig. 110):

```
plot_iperf.sh out.json && cd results
```

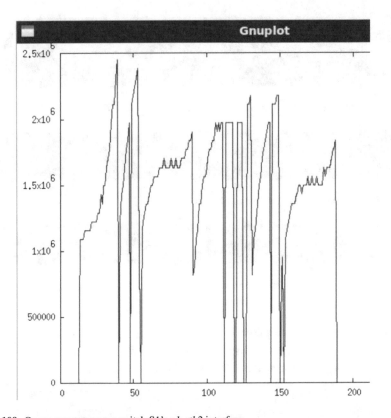

Fig. 109 Queue occupancy on switch S1's *s1-eth2* interface

```
                              "Host: h1"
root@admin-pc:~/1BDP# plot_iperf.sh out.json && cd results
root@admin-pc:~/1BDP/results# █
```

Fig. 110 Generating plotting files and entering the *results* directory

```
                              "Host: h1"
root@admin-pc:~/1BDP/results# xdg-open throughput.pdf
```

Fig. 111 Opening the *throughput.pdf* file

Step 9. Open the throughput file using the command below on host h1 (Fig. 111):

```
xdg-open throughput.pdf
```

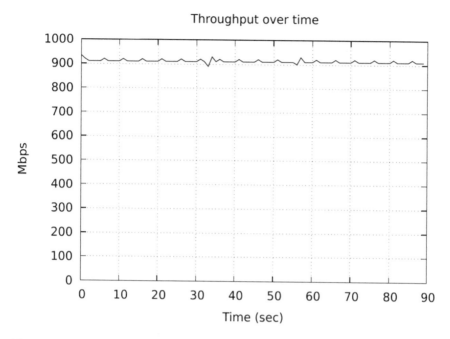

Fig. 112 Measured throughput

Fig. 113 Opening the *RTT.pdf* file

The figure above (Fig. 112) shows the iPerf3 test output report for the last 90 s. The average achieved throughput is approximately 900 Mbps. We can see now that the maximum throughput was almost achieved (1 Gbps) when we set the switch's buffer size to 1BDP.

Step 10. Close the *throughput.pdf* window and then open the Round-Trip Time (RTT) file using the command below (Fig. 113):

```
xdg-open RTT.pdf
```

The graph above (Fig. 114) shows that the RTT was between 25,000 microseconds (25 ms) and 40,000 microseconds (40 ms). The output shows that there is no bufferbloat problem as the average latency is slightly greater than the configured delay (20 ms).

Step 11. Close the *RTT.pdf* window and then open the congestion window (cwnd) file using the command below (Fig. 115):

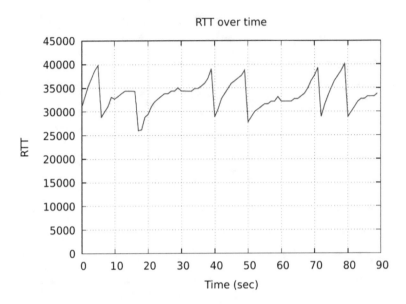

Fig. 114 Measured Round-Trip Time

Fig. 115 Opening the *cwnd.pdf* file

```
xdg-open cwnd.pdf
```

The graph above (Fig. 116) shows the evolution of the congestion window, which peaked at 4.5 Mbytes. In the next test, we see how buffer size on the switch affects the congestion window evolution.

Step 12. Close the *cwnd.pdf* window and then go back to h2's terminal to see the ping output (Fig. 117).

The result above indicates that all 90 packets were received successfully (0deviation of the Round-Trip Time (RTT) were 25.630, 32.669, 64.126, and 4.359 milliseconds, respectively. The output also verifies that there is no bufferbloat problem as the average latency (32.669) is slightly greater than the configured delay (20 ms). .

Step 13. To stop *iperf3* server in host h3, press Ctrl+c .

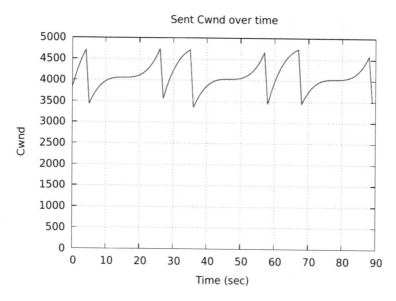

Fig. 116 Congestion window evolution

13 Testing Throughput on a Network with a Large Buffer-Size Switch

In this section, you are going to change the switch S1's buffer size to 10·BDP and emulate a 1 Gbps Wide Area Network (*WAN*) using the Token Bucket Filter (tbf). Then, you will test the throughput between host h1 and host h3. The BDP is 2,621,440 bytes; thus, the tbf limit value will be set to 26,214,400.

13.1 Setting Switch S1's Buffer Size to 10 · BDP

Step 1. Apply tbf rate limiting rule on switch S1's *s1-eth2* interface. In the client's terminal, type the command below. When prompted for a password, type password and hit *Enter* (Fig. 118).

- rate : 1gbit
- burst : 500,000
- limit : 26,214,400

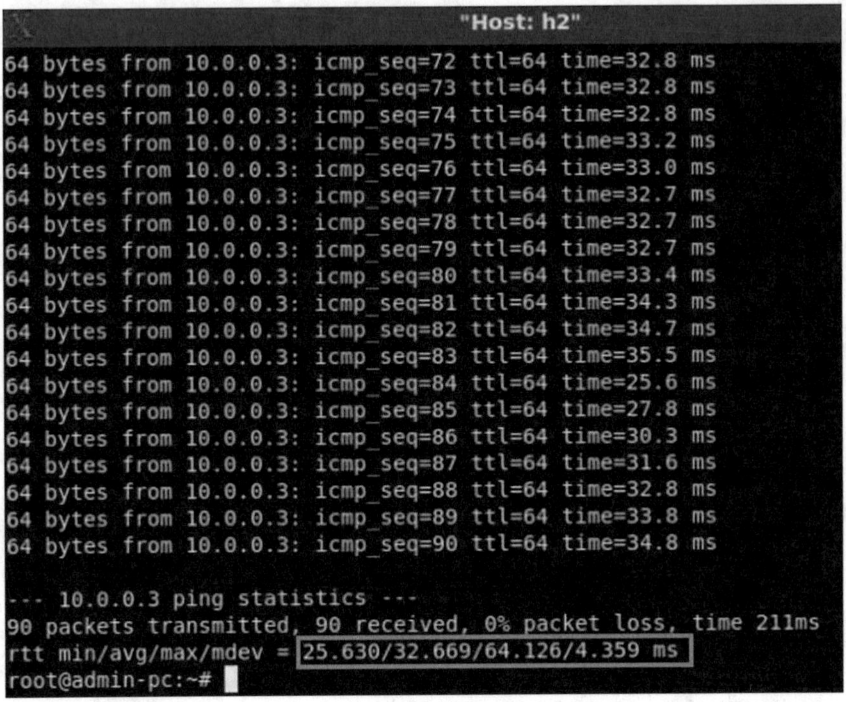

```
                              "Host: h2"
64 bytes from 10.0.0.3: icmp_seq=72 ttl=64 time=32.8 ms
64 bytes from 10.0.0.3: icmp_seq=73 ttl=64 time=32.8 ms
64 bytes from 10.0.0.3: icmp_seq=74 ttl=64 time=32.8 ms
64 bytes from 10.0.0.3: icmp_seq=75 ttl=64 time=33.2 ms
64 bytes from 10.0.0.3: icmp_seq=76 ttl=64 time=33.0 ms
64 bytes from 10.0.0.3: icmp_seq=77 ttl=64 time=32.7 ms
64 bytes from 10.0.0.3: icmp_seq=78 ttl=64 time=32.7 ms
64 bytes from 10.0.0.3: icmp_seq=79 ttl=64 time=32.7 ms
64 bytes from 10.0.0.3: icmp_seq=80 ttl=64 time=33.4 ms
64 bytes from 10.0.0.3: icmp_seq=81 ttl=64 time=34.3 ms
64 bytes from 10.0.0.3: icmp_seq=82 ttl=64 time=34.7 ms
64 bytes from 10.0.0.3: icmp_seq=83 ttl=64 time=35.5 ms
64 bytes from 10.0.0.3: icmp_seq=84 ttl=64 time=25.6 ms
64 bytes from 10.0.0.3: icmp_seq=85 ttl=64 time=27.8 ms
64 bytes from 10.0.0.3: icmp_seq=86 ttl=64 time=30.3 ms
64 bytes from 10.0.0.3: icmp_seq=87 ttl=64 time=31.6 ms
64 bytes from 10.0.0.3: icmp_seq=88 ttl=64 time=32.8 ms
64 bytes from 10.0.0.3: icmp_seq=89 ttl=64 time=33.8 ms
64 bytes from 10.0.0.3: icmp_seq=90 ttl=64 time=34.8 ms

--- 10.0.0.3 ping statistics ---
90 packets transmitted, 90 received, 0% packet loss, time 211ms
rtt min/avg/max/mdev = 25.630/32.669/64.126/4.359 ms
root@admin-pc:~#
```

Fig. 117 ping test result

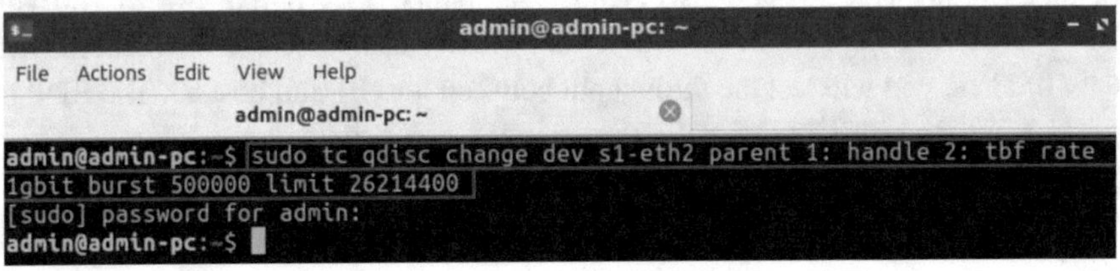

```
                          admin@admin-pc: ~                          - ⌄
 File   Actions   Edit   View   Help
                   admin@admin-pc: ~                        ⊗
admin@admin-pc:~$ sudo tc qdisc change dev s1-eth2 parent 1: handle 2: tbf rate
1gbit burst 500000 limit 26214400
[sudo] password for admin:
admin@admin-pc:~$
```

Fig. 118 Limiting rate to 1 Gbps and setting the buffer size to 10·BDP on switch S1's interface

```
sudo tc qdisc change dev s1-eth2 parent 1: handle 2: tbf
rate 1gbit burst 500000 limit 26214400
```

13.2   *Throughput and Latency Tests*

Step 1. Launch iPerf3 in server mode on host h3's terminal (Fig. 119).

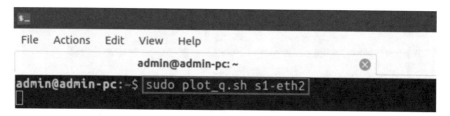

Fig. 119   Starting iPerf3 server on host h3

Fig. 120   Plotting the queue occupancy on switch S1's *s1-eth2* interface

```
iperf3 -s
```

Step 2. In the client's terminal, type the command below to plot the switch's queue in real-time. When prompted for a password, type | password | and hit *Enter* (Fig. 120).

```
sudo plot_q.sh s1-eth2
```

A new window opens that plots the queue occupancy as shown in the figure below (Fig. 121). Since there are no active flows passing through *s1-eth2* interface on switch S1, the queue occupancy is constantly 0.

Step 3. Exit from 1BDP/results directory, then create a directory *10BDP*, and navigate into it using the following command (Fig. 122):

```
cd ../../ && mkdir 10BDP && cd 10BDP
```

Step 4. Type the following iPerf3 command in host h1's terminal without executing it. The | -J | option is used to display the output in JSON format. The redirection operator | > | is used to store the JSON output into a file (Fig. 123).

```
iperf3 -c 10.0.0.3 -t 90 -J >out.json
```

Step 5. Type the following | ping | command in host h2's terminal without executing it (Fig. 124):

Gnuplot
"q_out" using 1:2
1
0.5
0
-0.5
-1
0
50
100
150
200
250
300

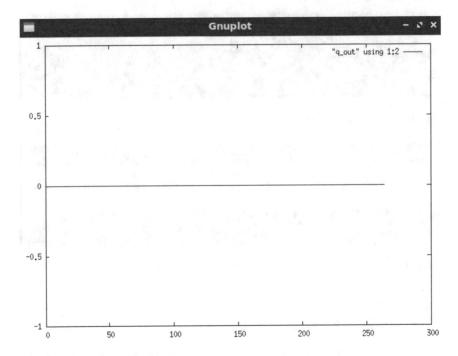

Fig. 121 Queue occupancy on switch S1's *s1-eth2* interface

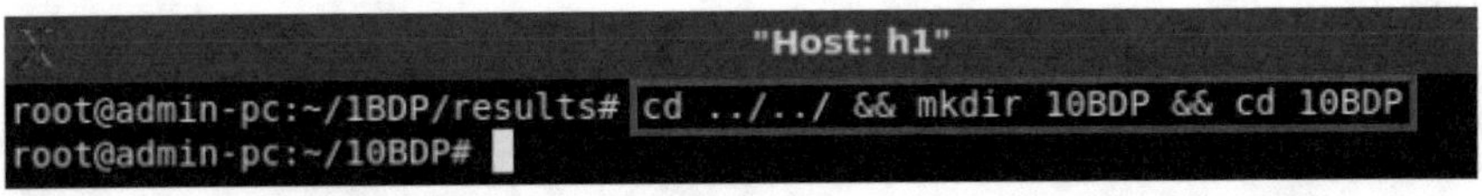

"Host: h1"
root@admin-pc:~/1BDP/results# cd ../../ && mkdir 10BDP && cd 10BDP
root@admin-pc:~/10BDP#

Fig. 122 Creating and navigating into directory *1BDP*

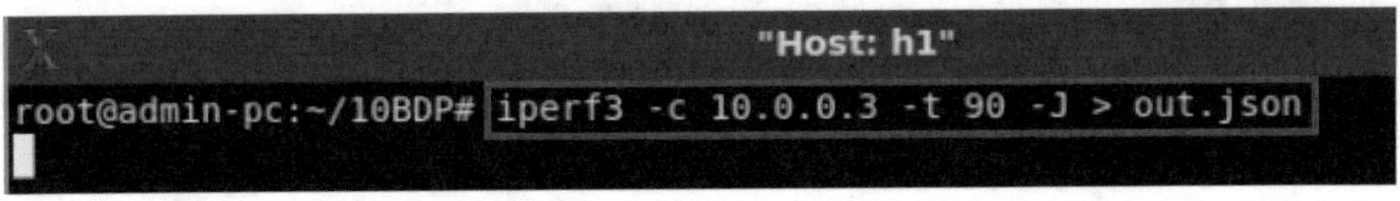

"Host: h1"
root@admin-pc:~/10BDP# iperf3 -c 10.0.0.3 -t 90 -J > out.json

Fig. 123 Running iPerf3 client on host h1

```
ping 10.0.0.3 -c 90
```

Step 6. Press *Enter* to execute the commands, first in host h1 terminal and then in host h2 terminal. Then, go back to the queue plotting window and observe the queue occupancy (Fig. 125).

Fig. 124 Typing ⟨ping⟩ command on host h2

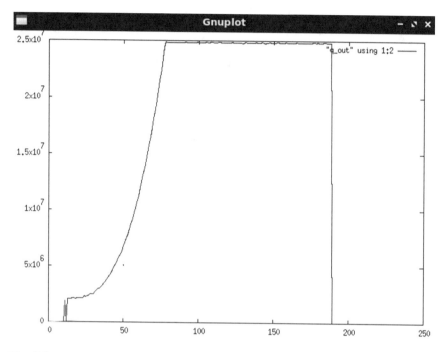

Fig. 125 Queue occupancy on switch S1's *s1-eth2* interface

The graph above shows that the queue occupancy peaked at $2.5 \cdot 10^7$, which is the maximum buffer size we configure on the switch. Note that the buffer is almost always fully occupied, which will lead to an increase in the latency as demonstrated next.

Step 7. In the queue plotting window, press the ⟨s⟩ key on your keyboard to stop plotting the queue.

Step 8. After the iPerf3 test finishes on host h1, enter the following command (Fig. 126):

```
plot_iperf.sh out.json && cd results
```

Step 9. Open the throughput file using the command below on host h1 (Fig. 127):

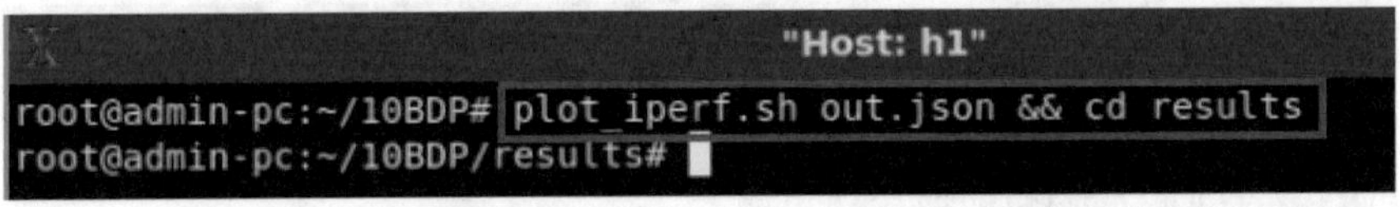

```
                                                "Host: h1"
root@admin-pc:~/10BDP# plot_iperf.sh out.json && cd results
root@admin-pc:~/10BDP/results#
```

Fig. 126 Generating plotting files and entering the *results* directory

```
                                                "Host: h1"
root@admin-pc:~/10BDP/results# xdg-open throughput.pdf
```

Fig. 127 Opening the *throughput.pdf* file

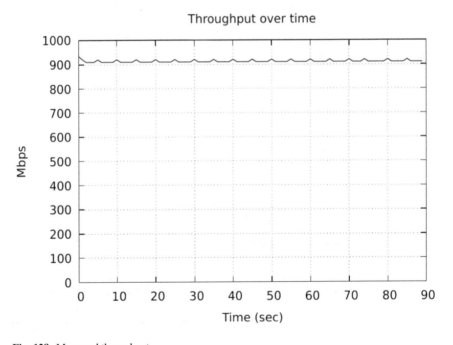

Fig. 128 Measured throughput

```
xdg-open throughput.pdf
```

The figure above (Fig. 128) shows the iPerf3 test output report for the last 90 s. The average achieved throughput is 900 Mbps. We can see now that the maximum throughput is also achieved (1 Gbps) when we set the switch's buffer size to 10·BDP.

Step 10. Close the *throughput.pdf* window and then open the Round-Trip Time (RTT) file using the command below (Fig. 129):

Fig. 129 Opening the *RTT.pdf* file

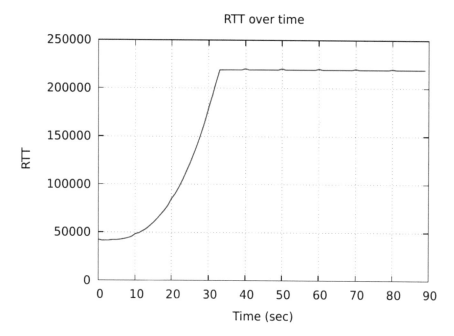

Fig. 130 Measured Round-Trip Time

```
xdg-open RTT.pdf
```

The graph above (Fig. 130) shows that the RTT increased from approximately 50,000 microseconds (50 ms) to 230,000 microseconds (230 ms). The output above shows that there is a bufferbloat problem as the average latency is significantly greater than the configured delay (20 ms). Since the buffer on the switch is accommodating a large congestion window, latency is increased as new incoming packets have to wait in the highly occupied queue.

Step 11. Close the *RTT.pdf* window and then open the congestion window (cwnd) file using the command below (Fig. 131):

```
xdg-open cwnd.pdf
```

Fig. 131 Opening the *cwnd.pdf* file

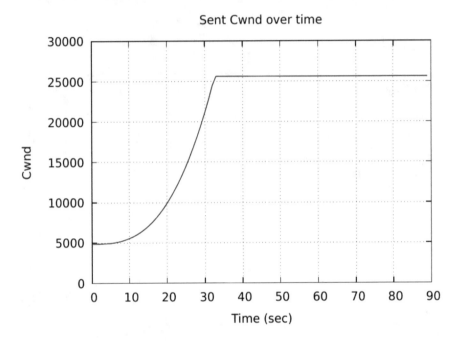

Fig. 132 Congestion window evolution

The graph above (Fig. 132) shows the evolution of the congestion window. Note how the congestion window peaked at 25.2 Mbytes compared to the previous test where it peaked at approximately 4.5 Mbytes. Since the queue size was configured with a large value, TCP continued to increase the congestion window as no packet losses were inferred.

Step 12. Close the *cwnd.pdf* window and then go back to h2's terminal to see the ping output (Fig. 133).

The result above indicates that all 90 packets were received successfully (0deviation of the Round-Trip Time (RTT) were 34.239, 167.046, 219.647, and 73.715 milliseconds, respectively. The output also verifies that there is a bufferbloat problem as the average latency (167.046) is significantly greater than the configured delay (20 ms).

Step 13. To stop *iperf3* server in host h3, press Ctrl+c.

```
                                    "Host: h2"
64 bytes from 10.0.0.3: icmp_seq=68 ttl=64 time=219 ms
64 bytes from 10.0.0.3: icmp_seq=69 ttl=64 time=219 ms
64 bytes from 10.0.0.3: icmp_seq=70 ttl=64 time=219 ms
64 bytes from 10.0.0.3: icmp_seq=71 ttl=64 time=219 ms
64 bytes from 10.0.0.3: icmp_seq=72 ttl=64 time=219 ms
64 bytes from 10.0.0.3: icmp_seq=73 ttl=64 time=220 ms
64 bytes from 10.0.0.3: icmp_seq=74 ttl=64 time=219 ms
64 bytes from 10.0.0.3: icmp_seq=75 ttl=64 time=219 ms
64 bytes from 10.0.0.3: icmp_seq=76 ttl=64 time=219 ms
64 bytes from 10.0.0.3: icmp_seq=77 ttl=64 time=219 ms
64 bytes from 10.0.0.3: icmp_seq=78 ttl=64 time=219 ms
64 bytes from 10.0.0.3: icmp_seq=79 ttl=64 time=219 ms
64 bytes from 10.0.0.3: icmp_seq=80 ttl=64 time=219 ms
64 bytes from 10.0.0.3: icmp_seq=81 ttl=64 time=219 ms
64 bytes from 10.0.0.3: icmp_seq=82 ttl=64 time=219 ms
64 bytes from 10.0.0.3: icmp_seq=83 ttl=64 time=220 ms
64 bytes from 10.0.0.3: icmp_seq=84 ttl=64 time=219 ms
64 bytes from 10.0.0.3: icmp_seq=85 ttl=64 time=219 ms
64 bytes from 10.0.0.3: icmp_seq=86 ttl=64 time=219 ms
64 bytes from 10.0.0.3: icmp_seq=87 ttl=64 time=219 ms
64 bytes from 10.0.0.3: icmp_seq=88 ttl=64 time=219 ms
64 bytes from 10.0.0.3: icmp_seq=89 ttl=64 time=219 ms
64 bytes from 10.0.0.3: icmp_seq=90 ttl=64 time=219 ms

--- 10.0.0.3 ping statistics ---
90 packets transmitted, 90 received, 0% packet loss, time 206ms
rtt min/avg/max/mdev = 34.239/167.046/219.647/73.715 ms
root@admin-pc:~#
```

Fig. 133 ping test result

Chapter 3—Lab 8: Random Early Detection (RED)

Overview
To conduct the experiment described in this section, please login into the Academic Cloud at http://highspeednetworks.net/ and reserve a pod for Lab 8.

This lab explains the Random Early Detection (RED) Active-Queue Management (AQM) algorithm. This algorithm aims at mitigating high end-to-end latency by controlling the average queue length in routers' buffers. Throughput, latency, and queue length measurements are conducted in this lab to verify the impact of the dropping policy provided RED.

Objectives
By the end of this lab, students should be able to:

1. Identify and describe the components of end-to-end latency.
2. Understand the buffering process in a router.
3. Explain the impact of RED handling the queuing policy in a router egress port.

Table 5 Credentials to
access Client1 machine

Device	Account	Password
Client1	admin	password

4. Visualize queue occupancy in a router.
5. Analyze how RED manages the queue length in order to allow end-hosts to achieve high throughput and low latency.
6. Modify the network condition in order to evaluate the performance on RED's dropping policy.

Lab Settings

The information in Table 5 provides the credentials of the machine containing Mininet.

Lab Roadmap

This lab is organized as follows:

1. Section 14: Introduction.
2. Section 15: Lab topology.
3. Section 16: Testing throughput on a network using Drop Tail AQM algorithm.
4. Section 17: Configuring RED on switch S2.

14 Introduction

End-to-end-congestion control is widely used in the current Internet to prevent congestion collapse. However, because data traffic is inherently bursty, routers are provisioned with large buffers to absorb this burstiness and maintain high link utilization. The downside of these large buffers is that if traditional drop tail buffer management is used, there will be high queuing delays at congested routers. Thus, drop tail buffer management forces network operators to choose between high utilization (requiring large buffers) and low delay (requiring small buffers).

Random Early Detection (RED) was proposed by Floyd and Van Jacobson to address network congestion responsively rather than reactively. The main goal of RED is to provide congestion avoidance by controlling the average queue size. Other goals are the avoidance of global synchronization and to introduce fairness to reduce the bias against bursty traffic. TCP global synchronization happens to a TCP flow during periods of congestion when each sender reduces and then increases their transmission rate at the same time due to packet loss.

14.1   Random Early Detection Mechanism

Figure 134a illustrates scenario where a router's buffer is managed by Random Early Detection. RED uses a low-pass filter with an exponential moving average to calculate the average queue size. Then, the average queue size is compared to two thresholds: a minimum threshold and a maximum threshold. Consequently, the packet drop probability is determined by the function shown in Fig. 134b. When the average queue size is less than the minimum threshold, no packets are dropped. When the average queue size is greater than the maximum threshold, every arriving packet is marked and therefore they are dropped. When the average queue size is between the minimum and the maximum threshold, each arriving packet is marked with drop probability. Thus, RED has two separate algorithms. First, the algorithm for computing the average queue size that determines the degree of burstiness allowed in the queue. Second, the algorithm for calculating the packet marking probability, which determines how frequently the gateway marks or drops packets, given the current level of congestion. The goal is to mark packets at evenly spaced intervals, in order to avoid biases global synchronization by marking packets to control the average queue size.

The basic red syntax used with tc is as follows:

```
tc qdisc [add |...] dev [dev_id] root red limit [BYTES] max
[BYTES] min [BYTES] burst [BYTES] avpkt [BYTES] bandwidth
[BPS] [probability [RATE]|adaptative] ecn
```

- tc : Linux traffic control tool.
- qdisc : A queue discipline (qdisc) is a set of rules that determine the order in which packets arriving from the IP protocol output are served. The queue discipline is applied to a packet queue to decide when to send each packet.
- [add |del |replace |change |show] : This is the operation on qdisc. For example, to add the token bucket algorithm on a specific interface, the operation will be add . To change or remove it, the operation will be change or del , respectively.
- dev [dev_id] : This parameter indicates the interface is to be subject to emulation.
- red : This parameter specifies the Random Early Detection algorithm.
- limit [BYTES] : Hard limit on the real (not average) queue size in bytes. Further packets are dropped. Should be set higher than max + burst .
- max [BYTES] : This parameter specifies the maximum average queue size. After this value, the dropping probability is 100%. It is recommended to set this value to limit /4.

- min [BYTES] : This parameter specifies the minimum average queue size. Below this value, no packet is dropped. Above this threshold, the dropping probability is established by probability or it increases linearly if the parameter adaptative is set.

- avpkt : Used with burst to determine the time constant for average queue size calculations. It is suggested 1000 as good value.

- burst [BYTES] : Used for determining how fast the average queue size is influenced by the real queue size. Larger values make the calculation slower, allowing longer bursts of traffic before the marking or dropping phase starts. Empirical evaluations suggest the following guideline to set this value: (2· min + max)/(3· avpkt).

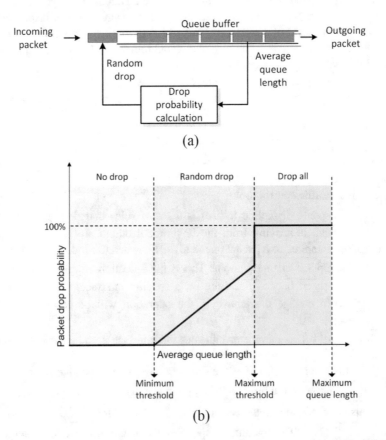

Fig. 134 Behavior of Random Early Detection AQM. (a) Buffer managed by RED AQM. (b) RED dropping function

- bandwidth [BPS] : This value is optional and used to calculate the average queue size after any idle time. It should be set to the bandwidth of the interface. This parameter does not limit the rate. The default value is 10 Mbps.
- ecn : This parameter enables RED to notify remote hosts that their rate exceeds the amount of bandwidth available. Non-ECN capable hosts can only be notified by dropping a packet.
- probability : This value specifies the dropping probability after the average queue length surpass the min threshold. It is specified as a floating point from 0.0 to 1.0. Suggested values are 0.01 or 0.02 (1% or 2%, respectively).
- adaptive : This parameter sets a dynamic value to the dropping probability. This value varies from 1% to 50%.

In this lab, we will use the red AQM algorithm to contain the queue size at the egress port of a router.

15 Lab Topology

Let's get started with creating a simple Mininet topology using MiniEdit. The topology uses 10.0.0.0/8, which is the default network assigned by Mininet (Fig. 135).

The above topology uses 10.0.0.0/8, which is the default network assigned by Mininet.

Step 1. A shortcut to MiniEdit is located on the machine's Desktop. Start MiniEdit by clicking on MiniEdit's shortcut (Fig. 136). When prompted for a password, type password .

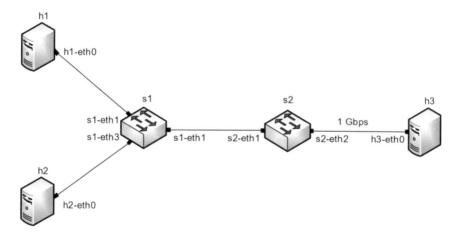

Fig. 135 Lab topology

Step 2. On MiniEdit's menu bar, click on *File* and then *Open* to load the lab's topology. Locate the *Lab 8.mn* topology file and click on *Open* (Fig. 137).

Step 3. Before starting the measurements between end-hosts, the network must be started. Click on the *Run* button located at the bottom left of MiniEdit's window to start the emulation (Fig. 138).

 The above topology uses 10.0.0.0/8, which is the default network assigned by Mininet.

15.1   Starting Host h1, Host h2, and Host h3

Step 1. Hold right-click on host h1 and select *Terminal*. This opens the terminal of host h1 and allows the execution of commands on that host (Fig. 139).

Step 2. Apply the same steps on host h2 and host h3 and open their *Terminals*.

Step 3. Test connectivity between the end-hosts using the ping command. On host h1, type the command ping 10.0.0.3 . This command tests the connectivity between host h1 and host h3 (Fig. 140). To stop the test, press Ctrl+c . The figure below shows a successful connectivity test.

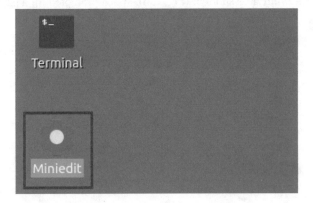

Fig. 136 MiniEdit shortcut

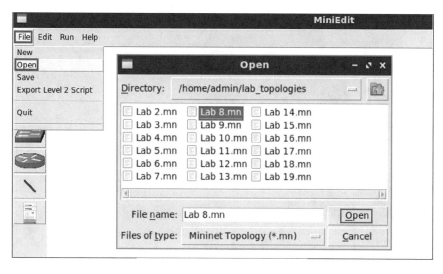

Fig. 137 MiniEdit's *Open* dialog

Fig. 138 Running the
emulation

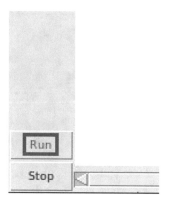

15.2 Emulating High-Latency WAN

This section emulates a high-latency WAN. We will emulate 20 ms delay on switch
S1's *s1-eth2* interface.

Step 1. Launch a Linux terminal by holding the Ctrl+Alt+T keys or by clicking
on the Linux terminal icon (Fig. 141).

The Linux terminal is a program that opens a window and permits you to interact
with a command-line interface (CLI). A CLI is a program that takes commands from
the keyboard and sends them to the operating system to perform.

MiniEdit
File Edit Run Help
Host Options
Terminal
s1
s2
h3
h2

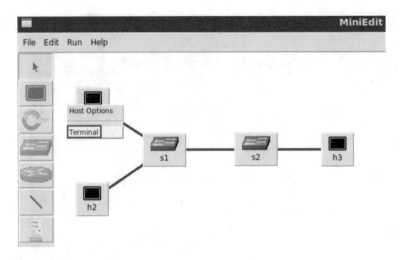

Fig. 139 Opening a terminal on host h1

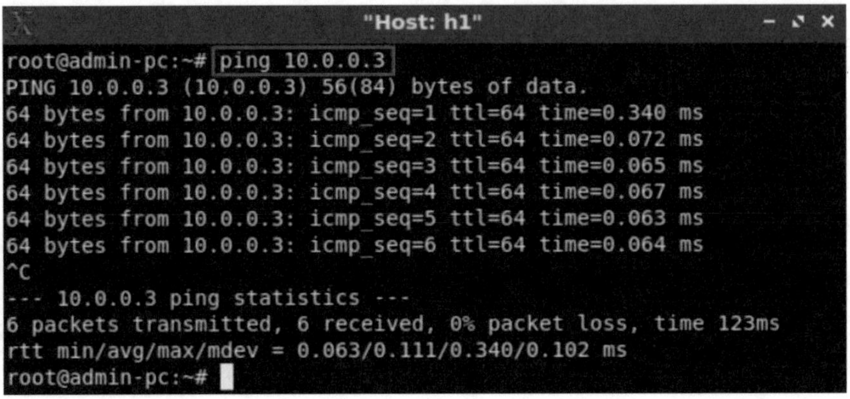

"Host: h1"
root@admin-pc:~# ping 10.0.0.3
PING 10.0.0.3 (10.0.0.3) 56(84) bytes of data.
64 bytes from 10.0.0.3: icmp_seq=1 ttl=64 time=0.340 ms
64 bytes from 10.0.0.3: icmp_seq=2 ttl=64 time=0.072 ms
64 bytes from 10.0.0.3: icmp_seq=3 ttl=64 time=0.065 ms
64 bytes from 10.0.0.3: icmp_seq=4 ttl=64 time=0.067 ms
64 bytes from 10.0.0.3: icmp_seq=5 ttl=64 time=0.063 ms
64 bytes from 10.0.0.3: icmp_seq=6 ttl=64 time=0.064 ms
^C
--- 10.0.0.3 ping statistics ---
6 packets transmitted, 6 received, 0% packet loss, time 123ms
rtt min/avg/max/mdev = 0.063/0.111/0.340/0.102 ms
root@admin-pc:~#

Fig. 140 Connectivity test using `ping` command

Step 2. In the terminal, type the command below. When prompted for a password, type `password` and hit *Enter*. This command introduces 20 ms delay to switch S1's *s1-eth1* interface (Fig. 142).

```
sudo tc qdisc add dev s1-eth1 root netem delay 20ms
```

Fig. 141 Shortcut to open a Linux terminal

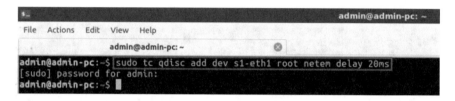

Fig. 142 Adding delay of 20 ms to switch S1's *s1-eth1* interface

15.3   *Testing Connection*

To test connectivity, you can use the command ping .

Step 1. On the terminal of host h1, type ping 10.0.0.3 . To stop the test, press
 Ctrl+c . The figure below shows a successful connectivity test. Host h1 (10.0.0.1)
sent four packets to host h3 (10.0.0.3), successfully receiving responses back
(Fig. 143).

The result above indicates that all four packets were received successfully
(0deviation of the Round-Trip Time (RTT) were 20.080, 25.390, 41.266, and
9.166 milliseconds, respectively. The output above verifies that delay was injected
successfully, as the RTT is approximately 20 ms.

Step 2. On the terminal of host h2, type ping 10.0.0.3 . The ping output in this test
should be relatively similar to the results of the test initiated by host h1 in Step 1
(Fig. 144). To stop the test, press Ctrl+c .

The result above indicates that all four packets were received successfully
(0deviation of the Round-Trip Time (RTT) were 20.090, 25.257, 40.745, and
8.943 milliseconds, respectively. The output above verifies that delay was injected
successfully, as the RTT is approximately 20 ms.

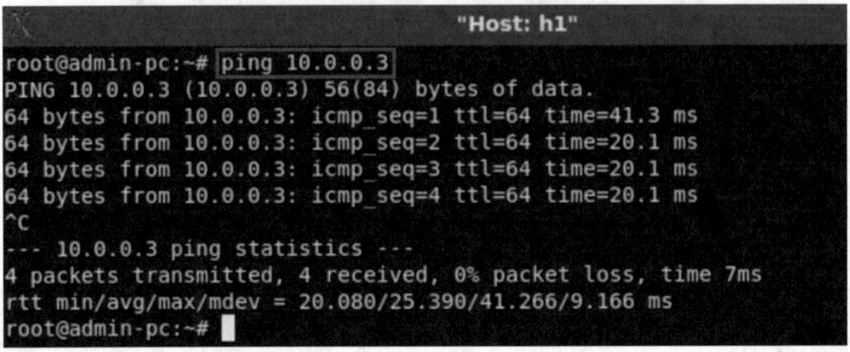

```
                                              "Host: h1"
root@admin-pc:~# ping 10.0.0.3
PING 10.0.0.3 (10.0.0.3) 56(84) bytes of data.
64 bytes from 10.0.0.3: icmp_seq=1 ttl=64 time=41.3 ms
64 bytes from 10.0.0.3: icmp_seq=2 ttl=64 time=20.1 ms
64 bytes from 10.0.0.3: icmp_seq=3 ttl=64 time=20.1 ms
64 bytes from 10.0.0.3: icmp_seq=4 ttl=64 time=20.1 ms
^C
--- 10.0.0.3 ping statistics ---
4 packets transmitted, 4 received, 0% packet loss, time 7ms
rtt min/avg/max/mdev = 20.080/25.390/41.266/9.166 ms
root@admin-pc:~# █
```

Fig. 143 Output of ping 10.0.0.3 command

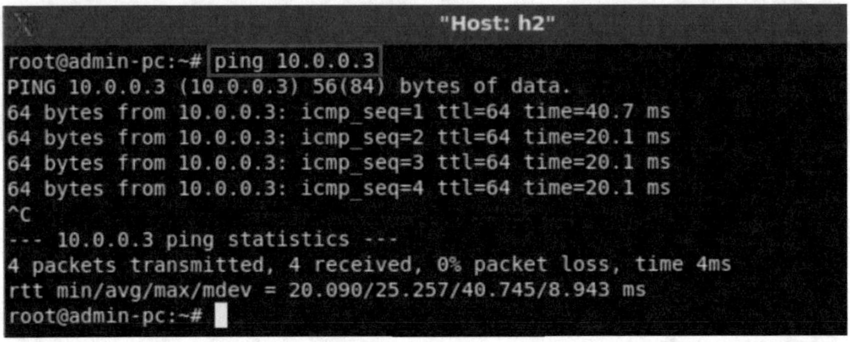

```
                                              "Host: h2"
root@admin-pc:~# ping 10.0.0.3
PING 10.0.0.3 (10.0.0.3) 56(84) bytes of data.
64 bytes from 10.0.0.3: icmp_seq=1 ttl=64 time=40.7 ms
64 bytes from 10.0.0.3: icmp_seq=2 ttl=64 time=20.1 ms
64 bytes from 10.0.0.3: icmp_seq=3 ttl=64 time=20.1 ms
64 bytes from 10.0.0.3: icmp_seq=4 ttl=64 time=20.1 ms
^C
--- 10.0.0.3 ping statistics ---
4 packets transmitted, 4 received, 0% packet loss, time 4ms
rtt min/avg/max/mdev = 20.090/25.257/40.745/8.943 ms
root@admin-pc:~# █
```

Fig. 144 Output of ping 10.0.0.3 command

16   Testing Throughput on a Network Using Drop Tail AQM Algorithm

In this section, you are going to change the switch S2's buffer size to $10 \cdot BDP$ and emulate a 1 Gbps Wide Area Network (*WAN*) using the Token Bucket Filter (tbf) as well as hosts' h1 and h3 TCP sending and receiving windows. The AQM algorithm is Drop Tail, which works by dropping newly arriving packets when the queue is full; therefore, the parameter that is configured is the queue size, which is given by the limit value set with the tbf rule. Then, you will test the throughput between host h1 and host h3. In this section, $10 \cdot BDP$ is 25 Mbytes; thus, the tbf limit value will be set to $10 \cdot BDP = 26,214,400$ bytes.

Fig. 145 Receive window change in sysctl

16.1 Bandwidth-Delay Product (BDP) and Hosts' TCP Buffer Size

In the upcoming tests, the bandwidth is limited to 1 Gbps, and the RTT (delay or latency) is 20 ms.

$$BW = 1,000,000,000 \text{ bits/second}$$

$$RTT = 0.02 \text{ seconds}$$

$$BDP = 1,000,000,000 \cdot 0.02 = 20,000,000 \text{ bits}$$

$$= 2,500,000 \text{ bytes} \approx 2.5 \text{ Mbytes}$$

$$1 \text{ Mbyte} = 1024^2 \text{ bytes}$$

$$BDP = 2.5 \text{ Mbytes} = 2.5 \cdot 1024^2 \text{ bytes} = 2,621,440 \text{ bytes}$$

The default buffer size in Linux is 16 Mbytes, and only 8 Mbytes (half of the maximum buffer size) can be allocated. Since 8 Mbytes is greater than 2.5 Mbytes, then no need to tune the buffer sizes on end-hosts. However, in upcoming tests, we configure the buffer size on the switch to 10·BDP. In addition, to ensure that the bottleneck is not the hosts' TCP buffers, we configure the buffers to 20·BDP (52,428,800).

Step 1. Now, we have calculated the maximum value of the TCP sending and receiving buffer size. In order to change the receiving buffer size, on host h1's terminal type the command shown below. The values set are: 10,240 (minimum), 87,380 (default), and 52,428,800 (maximum). The maximum value is doubled (2·10·BDP) as Linux only allocates half of the assigned value (Fig. 145).

```
sysctl -w net.ipv4.tcp_rmem='10240 87380 52428800'
```

The returned values are measured in bytes. 10,240 represents the minimum buffer size that is used by each TCP socket. 87,380 is the default buffer that is allocated when applications create a TCP socket. 52,428,800 is the maximum receive buffer that can be allocated for a TCP socket.

```
                                "Host: h1"                        –  ͏  ×
root@admin-pc:~# sysctl -w net.ipv4.tcp_wmem='10240 87380 52428800'
net.ipv4.tcp_wmem = 10240 87380 52428800
root@admin-pc:~# ▮
```

Fig. 146 Send window change in sysctl

```
                                "Host: h3"                        –  ͏  ×
root@admin-pc:~# sysctl -w net.ipv4.tcp_rmem='10240 87380 52428800'
net.ipv4.tcp_rmem = 10240 87380 52428800
root@admin-pc:~# ▮
```

Fig. 147 Receive window change in sysctl

```
                                "Host: h3"                        –  ͏  ×
root@admin-pc:~# sysctl -w net.ipv4.tcp_wmem='10240 87380 52428800'
net.ipv4.tcp_wmem = 10240 87380 52428800
root@admin-pc:~# ▮
```

Fig. 148 Send window change in sysctl

Step 2. To change the current send-window size value(s), use the following command on host h1's terminal. The values set are: 10,240 (minimum), 87,380 (default), and 52,428,800 (maximum). The maximum value is doubled as Linux allocates only half of the assigned value (Fig. 146). ·

```
sysctl -w net.ipv4.tcp_wmem='10240 87380 52428800'
```

Step 3. Now, we have calculated the maximum value of the TCP sending and receiving buffer size. In order to change the receiving buffer size, on host h3's terminal type the command shown below. The values set are: 10,240 (minimum), 87,380 (default), and 52,428,800 (maximum). The maximum value is doubled as Linux allocates only half of the assigned value (Fig. 147).

```
sysctl -w net.ipv4.tcp_rmem='10240 87380 52428800'
```

Step 4. To change the current send-window size value(s), use the following command on host h1's terminal. The values set are: 10,240 (minimum), 87,380 (default), and 52,428,800 (maximum). The maximum value is doubled as Linux allocates only half of the assigned value (Fig. 148).

```
sysctl -w net.ipv4.tcp_wmem='10240 87380 52428800'
```

The returned values are measured in bytes. 10,240 represents the minimum buffer size that is used by each TCP socket. 87,380 is the default buffer that is allocated

Fig. 149 Limiting rate to 1 Gbps and setting the buffer size to $10 \cdot$ BDP on switch S2's interface

when applications create a TCP socket. 52,428,800 is the maximum receive buffer that can be allocated for a TCP socket.

16.2 Setting Switch S2's Buffer Size to 10 · BDP

Step 1. Apply tbf rate limiting rule on switch S2's *s2-eth2* interface. In the client's terminal, type the command below. When prompted for a password, type password and hit *Enter* (Fig. 149).

- rate : 1gbit
- burst : 500,000
- limit : 26,214,400

```
sudo tc qdisc add dev s2-eth2 root handle 1: tbf rate 1gbit
burst 500000 limit 26214400
```

16.3 Throughput and Latency Tests

Step 1. Launch iPerf3 in server mode on host h3's terminal (Fig. 150).

```
iperf3 -s
```

Step 2. In the client's terminal, type the command below to plot the switch's queue in real-time. When prompted for a password, type password and hit *Enter* (Fig. 151).

```
sudo plot_q.sh s2-eth2
```

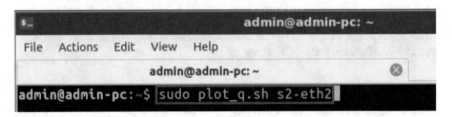

Fig. 150 Starting iPerf3 server on host h3

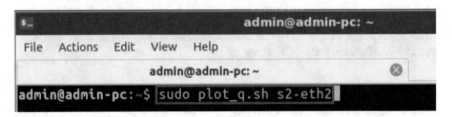

Fig. 151 Plotting the queue occupancy on switch S2's *s2-eth2* interface

A new window opens that plots the queue occupancy as shown in the figure below. Since there are no active flows passing through *s2-eth2* interface on switch S2, the queue occupancy is constantly 0 (Fig. 152).

Step 3. In host h1, create a directory called *Drop_Tail* and navigate into it using the following command (Fig. 153):

```
mkdir Drop_Tail && cd Drop_Tail
```

Step 4. Type the following iPerf3 command in host h1's terminal without executing it. The -J option is used to display the output in JSON format. The redirection operator > is used to store the JSON output into a file (Fig. 154).

```
iperf3 -c 10.0.0.3 -t 90 -J >out.json
```

Step 5. Type the following ping command in host h2's terminal without executing it (Fig. 155):

```
ping 10.0.0.3 -c 90
```

Step 6. Press *Enter* to execute the commands, first in host h1 terminal and then in host h2 terminal. Then, go back to the queue plotting window and observe the queue occupancy (Fig. 156).

The graph above shows that the queue occupancy peaked at $2.5 \cdot 10^7$, which is the maximum buffer size we configure on the switch.

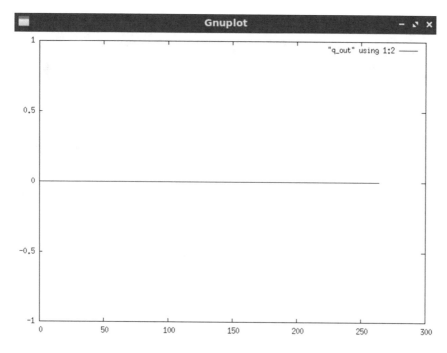

Fig. 152 Queue occupancy on switch S2's *s2-eth2* interface

```
                              "Host: h1"
root@admin-pc:~# mkdir Drop_Tail && cd Drop_Tail
root@admin-pc:~/Drop_Tail# 
```

Fig. 153 Creating and navigating into directory *Drop_Tail*

```
                              "Host: h1"
root@admin-pc:~/Drop_Tail# iperf3 -c 10.0.0.3 -t 90 -J > out.json
```

Fig. 154 Running iPerf3 client on host h1

Step 7. In the queue plotting window, press the ⎡s⎤ key on your keyboard to stop plotting the queue.

Step 8. After the iPerf3 test finishes on host h1, enter the following command (Fig. 157):

```
plot_iperf.sh out.json && cd results
```

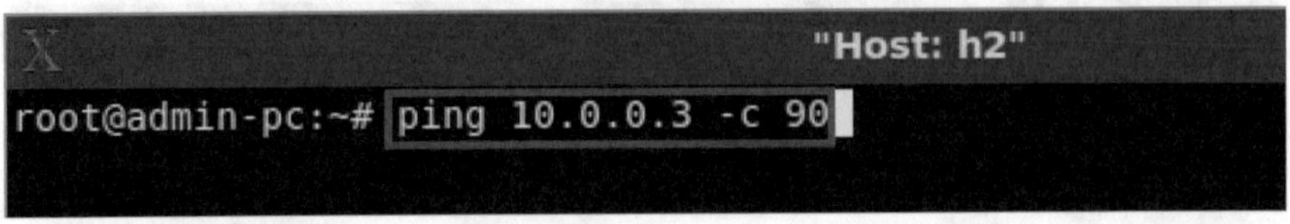

"Host: h2"
root@admin-pc:~# ping 10.0.0.3 -c 90

Fig. 155 Typing ping command on host h2

Gnuplot
"q_out" using 1:2
3x10^7
2.5x10^7
2x10^7
1.5x10^7
1x10^7
5x10^6
0
0
20
40
60
80
100
120
140
160
180
200

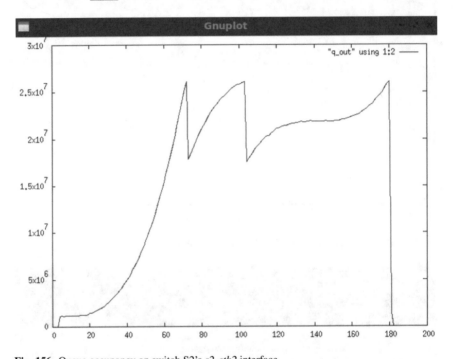

Fig. 156 Queue occupancy on switch S2's *s2-eth2* interface

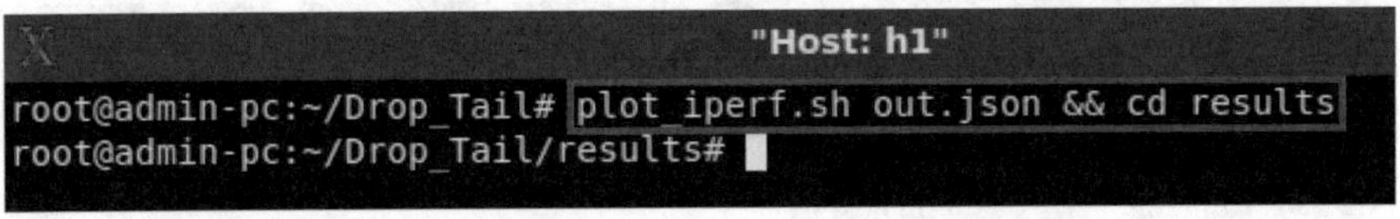

"Host: h1"
root@admin-pc:~/Drop_Tail# plot_iperf.sh out.json && cd results
root@admin-pc:~/Drop_Tail/results#

Fig. 157 Generating plotting files and entering the *results* directory

Step 9. Open the throughput file using the command below on host h1 (Fig. 158):

```
xdg-open throughput.pdf
```

The figure above (Fig. 159) shows the iPerf3 test output report for the last 90 s. The average achieved throughput is approximately 900 Mbps. We can see now that

Fig. 158 Opening the *throughput.pdf* file

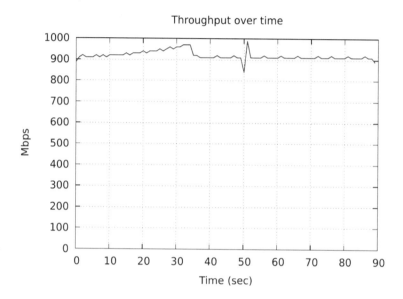

Fig. 159 Measured throughput

Fig. 160 Opening the *RTT.pdf* file

the maximum throughput was almost achieved (1 Gbps) when we set the switch's buffer size to $10 \cdot BDP$.

Step 10. Close the *throughput.pdf* window and then open the Round-Trip Time (RTT) file using the command below (Fig. 160):

```
xdg-open RTT.pdf
```

The graph above (Fig. 161) shows that the RTT was approximately 200,000 microseconds (200 ms) The output shows that there is bufferbloat as the average latency is at least ten times greater than the configured delay (20 ms).

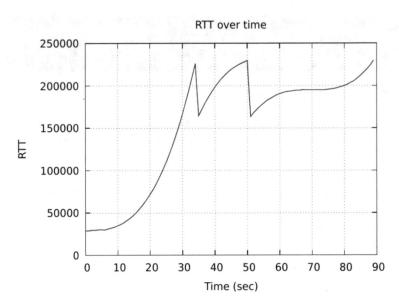

Fig. 161 Measured round-trip time

Step 11. Close the *RTT.pdf* window and then go back to h2's terminal to see the
ping output (Fig. 162).

The result above indicates that all 90 packets were received successfully (0devi-
ation of the Round-Trip Time (RTT) were 20.083, 192.823, 228.407, and 26.954
milliseconds, respectively. The output also verifies that there is bufferbloat as the
average latency (192.823) is significantly greater than the configured delay (20 ms).

Step 12. Open the congestion window (*cwnd.pdf*) file using the command below
(Fig. 163):

```
xdg-open cwnd.pdf
```

The graph above (Fig. 164) shows the evolution of the congestion window,
which peaked at 2.5 Mbytes. In the next section you will configure Random Early
Detection on switch S2 and observe how the algorithm controls the queue length.

Step 13. To stop *iperf3* server in host h3 press Ctrl+c .

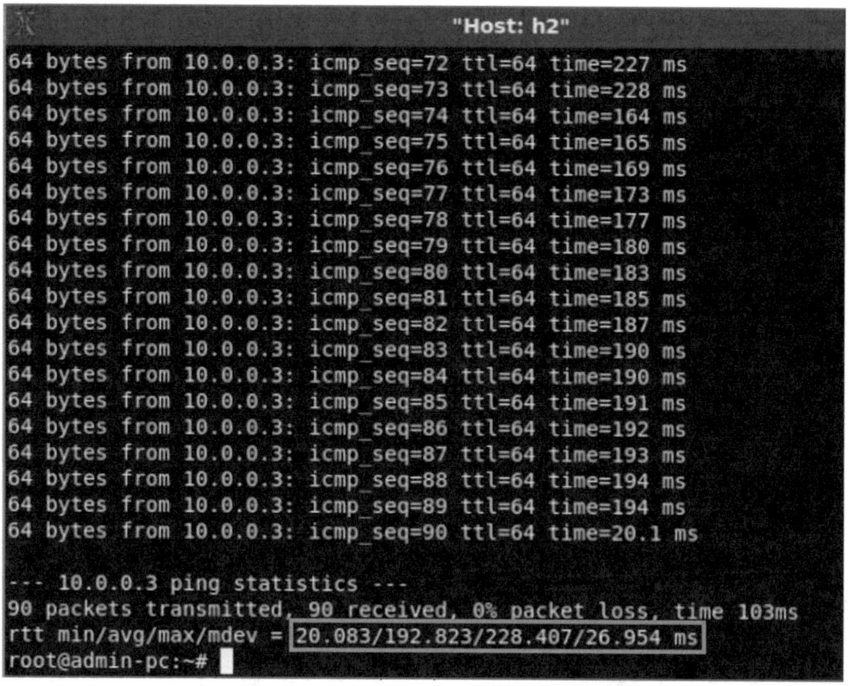

```
                              "Host: h2"
64 bytes from 10.0.0.3: icmp_seq=72 ttl=64 time=227 ms
64 bytes from 10.0.0.3: icmp_seq=73 ttl=64 time=228 ms
64 bytes from 10.0.0.3: icmp_seq=74 ttl=64 time=164 ms
64 bytes from 10.0.0.3: icmp_seq=75 ttl=64 time=165 ms
64 bytes from 10.0.0.3: icmp_seq=76 ttl=64 time=169 ms
64 bytes from 10.0.0.3: icmp_seq=77 ttl=64 time=173 ms
64 bytes from 10.0.0.3: icmp_seq=78 ttl=64 time=177 ms
64 bytes from 10.0.0.3: icmp_seq=79 ttl=64 time=180 ms
64 bytes from 10.0.0.3: icmp_seq=80 ttl=64 time=183 ms
64 bytes from 10.0.0.3: icmp_seq=81 ttl=64 time=185 ms
64 bytes from 10.0.0.3: icmp_seq=82 ttl=64 time=187 ms
64 bytes from 10.0.0.3: icmp_seq=83 ttl=64 time=190 ms
64 bytes from 10.0.0.3: icmp_seq=84 ttl=64 time=190 ms
64 bytes from 10.0.0.3: icmp_seq=85 ttl=64 time=191 ms
64 bytes from 10.0.0.3: icmp_seq=86 ttl=64 time=192 ms
64 bytes from 10.0.0.3: icmp_seq=87 ttl=64 time=193 ms
64 bytes from 10.0.0.3: icmp_seq=88 ttl=64 time=194 ms
64 bytes from 10.0.0.3: icmp_seq=89 ttl=64 time=194 ms
64 bytes from 10.0.0.3: icmp_seq=90 ttl=64 time=20.1 ms

--- 10.0.0.3 ping statistics ---
90 packets transmitted, 90 received, 0% packet loss, time 103ms
rtt min/avg/max/mdev = 20.083/192.823/228.407/26.954 ms
root@admin-pc:~#
```

Fig. 162 ping test result

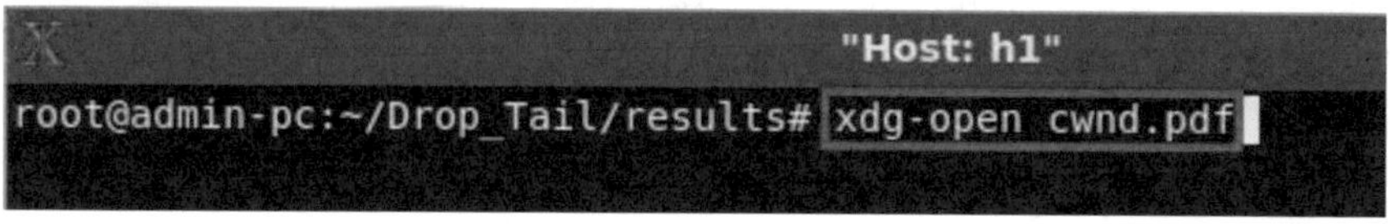

```
                              "Host: h1"
root@admin-pc:~/Drop_Tail/results# xdg-open cwnd.pdf
```

Fig. 163 Opening the *cwnd.pdf* file

17 Configuring RED on Switch S2

In this section, you are going to configure Random Early Detection in switch S2. Then, you will conduct throughput and latency measurements between host h1 and host h3. Note that the buffer size is set to 10·BDP.

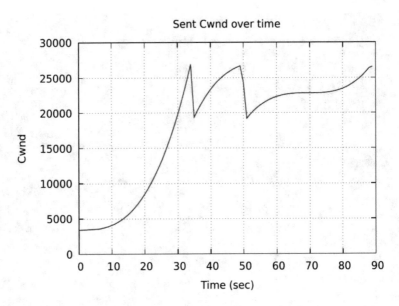

Fig. 164 Congestion window evolution

17.1   *Setting RED Parameter on Switch S2's Egress Interface*

Step 1. Apply tbf rate limiting rule on switch S2's *s2-eth2* interface. In the client's terminal, type the command below. When prompted for a password, type password and hit *Enter* (Fig. 165).

- limit : 26,214,400
- max : 8,738,133
- min : 2,184,533
- burst : 2185
- avpkt : 1000
- bandwidth : 1gbit
- adaptative

```
sudo tc qdisc add dev s2-eth2 parent 1: handle 2: red limit
26214400 max 8738133 min 2184533 burst 2185 avpkt 1000
bandwidth 1gbit adaptative
```

Fig. 165 Setting RED parameters on switch S2's *s2-eth2* interface

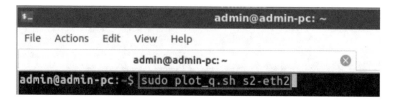

Fig. 166 Starting iPerf3 server on host h3

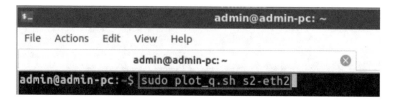

Fig. 167 Plotting the queue occupancy on switch S2's *s2-eth2* interface

17.2 Throughput and Latency Tests

Step 1. Launch iPerf3 in server mode on host h3's terminal (Fig. 166).

```
iperf3 -s
```

Step 2. In the client's terminal, type the command below to plot the switch's queue in real-time. When prompted for a password, type `password` and hit *Enter* (Fig. 167).

```
sudo plot_q.sh s2-eth2
```

A new window opens (Fig. 168) that plots the queue occupancy as shown in the figure below. Since there are no active flows passing through *s2-eth2* interface on switch S2, the queue occupancy is constantly 0.

Step 3. Exit from *Drop_Tail/results* directory, then create a directory *RED*, and navigate into it using the following command (Fig. 169):

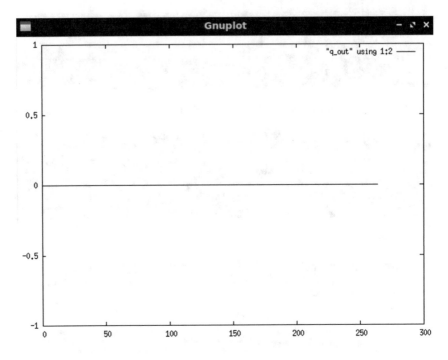

Fig. 168 Queue occupancy on switch S2's *s2-eth2* interface

```
                              "Host: h1"
root@admin-pc:~/Drop_Tail/results# cd ../.. && mkdir RED && cd RED
root@admin-pc:~/RED# 
```

Fig. 169 Creating and navigating into directory *RED*

```
cd ../../ && mkdir RED && cd RED
```

Step 4. Type the following iPerf3 command in host h1's terminal without executing it. The -J option is used to display the output in JSON format. The redirection operator > is used to store the JSON output into a file (Fig. 170).

```
iperf3 -c 10.0.0.3 -t 90 -J >out.json
```

Step 5. Type the following ping command in host h2's terminal without executing it (Fig. 171):

```
ping 10.0.0.3 -c 90
```

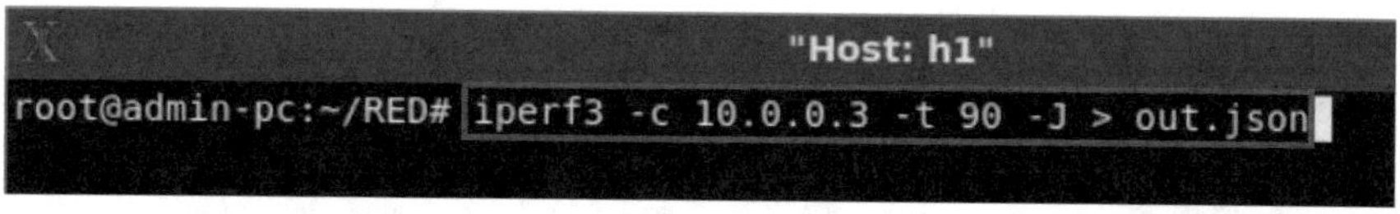

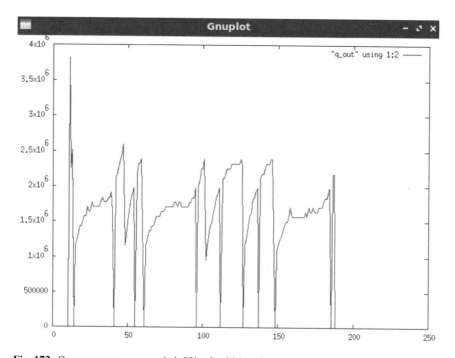

```
 X                                        "Host: h1"
root@admin-pc:~/RED# iperf3 -c 10.0.0.3 -t 90 -J > out.json
```

Fig. 170 Running iPerf3 client on host h1

```
 X                                        "Host: h2"
root@admin-pc:~# ping 10.0.0.3 -c 90
```

Fig. 171 Typing ping command on host h2

Fig. 172 Queue occupancy on switch S2's *s2-eth2* interface

Step 6. Press *Enter* to execute the commands, first in host h1 terminal and then in host h2 terminal. Then, go back to the queue plotting window and observe the queue occupancy (Fig. 172).

The graph above shows that the queue occupancy peaked around $3.5 \cdot 10^6$ bytes, which is closer to a buffer of BDP size.

Step 7. In the queue plotting window, press the $\boxed{s}$ key on your keyboard to stop plotting the queue.

Fig. 173 Generating plotting files and entering the *results* directory

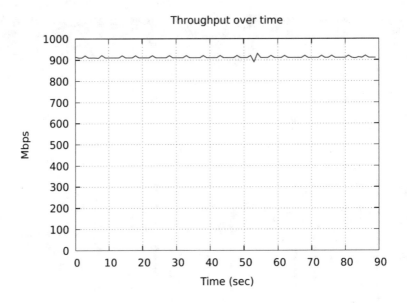

Fig. 174 Opening the *throughput.pdf* file

Throughput over time

Fig. 175 Measured throughput

Step 8. After the iPerf3 test finishes on host h1, enter the following command (Fig. 173):

```
plot_iperf.sh out.json && cd results
```

Step 9. Open the throughput file using the command below on host h1 (Fig. 174):

```
xdg-open throughput.pdf
```

The figure above (Fig. 175) shows the iPerf3 test output report for the last 90 s. The average achieved throughput is 900 Mbps. We can see now that the maximum

Fig. 176 Opening the *RTT.pdf* file

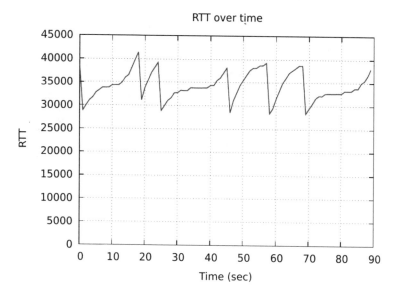

Fig. 177 Measured Round-Trip Time

throughput is also achieved (1 Gbps) when we set RED at the egress port of switch S2.

Step 10. Close the *throughput.pdf* window and then open the Round-Trip Time (*RTT.pdf*) file using the command below (Fig. 176):

```
xdg-open RTT.pdf
```

The graph above (Fig. 177) shows that the RTT was contained between 30 ms and 40 ms, which is not significantly greater that the configured delay (20 ms); thus, there is no bufferbloat. Bufferbloat is prevented because the AQM algorithm configured on the switch is applying a dropping policy to prevent unnecessary delays.

Step 11. Close the *RTT.pdf* window and then go back to h2's terminal to see the ping output (Fig. 178).

The result above indicates that all 90 packets were received successfully (0deviation of the Round-Trip Time (RTT) were 26.833, 34.048, 38.824, and 3.311 milliseconds, respectively. The output also verifies that there is not bufferbloat as

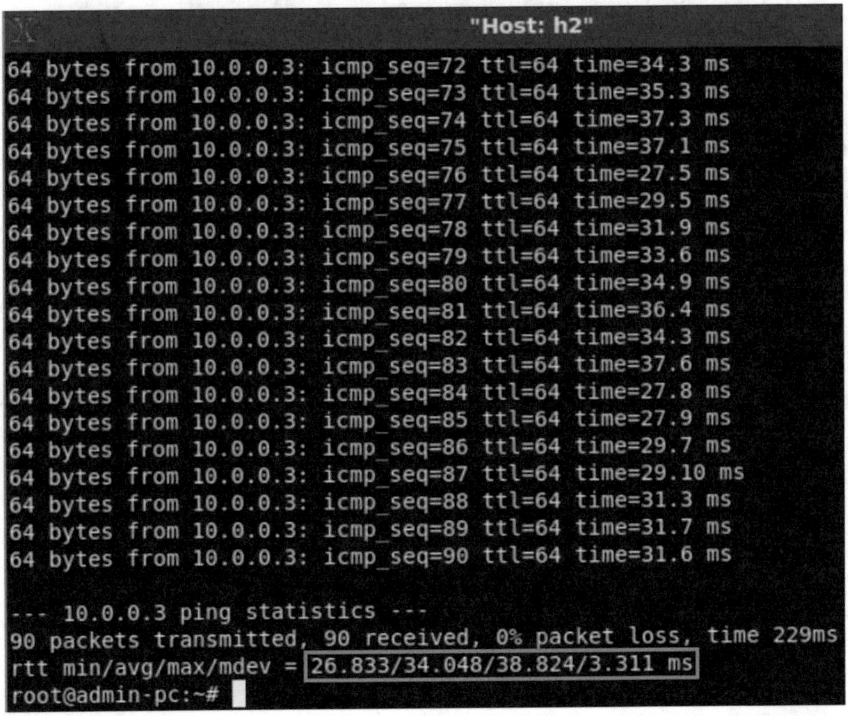

```
                                          "Host: h2"
64 bytes from 10.0.0.3: icmp_seq=72 ttl=64 time=34.3 ms
64 bytes from 10.0.0.3: icmp_seq=73 ttl=64 time=35.3 ms
64 bytes from 10.0.0.3: icmp_seq=74 ttl=64 time=37.3 ms
64 bytes from 10.0.0.3: icmp_seq=75 ttl=64 time=37.1 ms
64 bytes from 10.0.0.3: icmp_seq=76 ttl=64 time=27.5 ms
64 bytes from 10.0.0.3: icmp_seq=77 ttl=64 time=29.5 ms
64 bytes from 10.0.0.3: icmp_seq=78 ttl=64 time=31.9 ms
64 bytes from 10.0.0.3: icmp_seq=79 ttl=64 time=33.6 ms
64 bytes from 10.0.0.3: icmp_seq=80 ttl=64 time=34.9 ms
64 bytes from 10.0.0.3: icmp_seq=81 ttl=64 time=36.4 ms
64 bytes from 10.0.0.3: icmp_seq=82 ttl=64 time=34.3 ms
64 bytes from 10.0.0.3: icmp_seq=83 ttl=64 time=37.6 ms
64 bytes from 10.0.0.3: icmp_seq=84 ttl=64 time=27.8 ms
64 bytes from 10.0.0.3: icmp_seq=85 ttl=64 time=27.9 ms
64 bytes from 10.0.0.3: icmp_seq=86 ttl=64 time=29.7 ms
64 bytes from 10.0.0.3: icmp_seq=87 ttl=64 time=29.10 ms
64 bytes from 10.0.0.3: icmp_seq=88 ttl=64 time=31.3 ms
64 bytes from 10.0.0.3: icmp_seq=89 ttl=64 time=31.7 ms
64 bytes from 10.0.0.3: icmp_seq=90 ttl=64 time=31.6 ms

--- 10.0.0.3 ping statistics ---
90 packets transmitted, 90 received, 0% packet loss, time 229ms
rtt min/avg/max/mdev = 26.833/34.048/38.824/3.311 ms
root@admin-pc:~#
```

Fig. 178 ping test result

```
                                          "Host: h1"
root@admin-pc:~/RED/results# xdg-open cwnd.pdf
```

Fig. 179 Opening the *cwnd.pdf* file

the average latency (34.048) is not significantly greater than the configured delay (20 ms).

Step 12. Open the congestion window (*cwnd.pdf*) file using the command below (Fig. 179):

```
xdg-open cwnd.pdf
```

The graph above (Fig. 180) shows the evolution of the congestion window, which peaked around 5 Mbytes. In the next section you will maintain the current parameters of Random Early Detection on switch S2; however, you will change the link rate in order to verify if the algorithm performs well if the network condition changes.

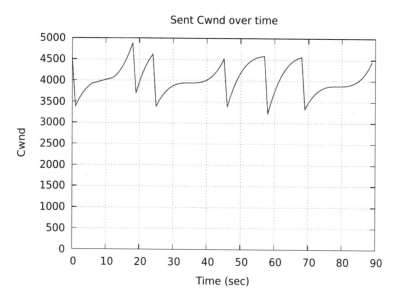

Fig. 180 Evolution of the congestion window

Step 13. To stop *iperf3* server in host h3, press $\boxed{\text{Ctrl+c}}$.

17.3 Changing the Bandwidth to 100 Mbps

This section is aimed to analyze the impact of changing the bandwidth to 100 Mbps while RED is tuned to work with the previous network condition. The results will show that RED requires a reconfiguration if the network conditions change (i.e., latency, bandwidth, loss rate). First, you will change the bandwidth to 100 Mbps, then you will observe the queue occupancy, RTT, and congestion window in order to evaluate the performance of RED when the network condition changes.

Step 1. Apply $\boxed{\text{tbf}}$ rate limiting rule on switch S2's *s2-eth2* interface. In the client's terminal, type the command below (Fig. 181). When prompted for a password, type $\boxed{\text{password}}$ and hit *Enter*.

- $\boxed{\text{rate}}$: 100mbit
- $\boxed{\text{burst}}$: 50,000
- $\boxed{\text{limit}}$: 26,214,400

admin@admin-pc: ~
File Actions Edit View Help
admin@admin-pc: ~
admin@admin-pc:~$ sudo tc qdisc change dev s2-eth2 root handle 1: tbf rate 100mbit burst 50000 limit 26214400
admin@admin-pc:~$

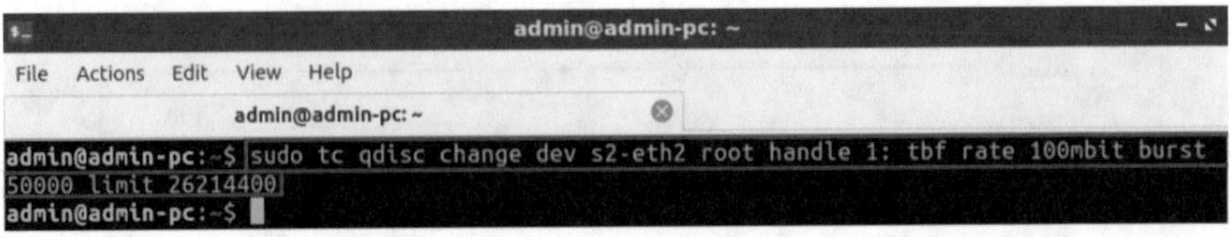

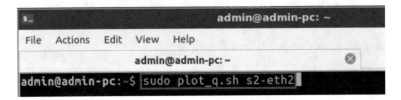

Fig. 181 Limiting rate to 100 Mbps and keeping the buffer size to 10·BDP on switch S2's interface

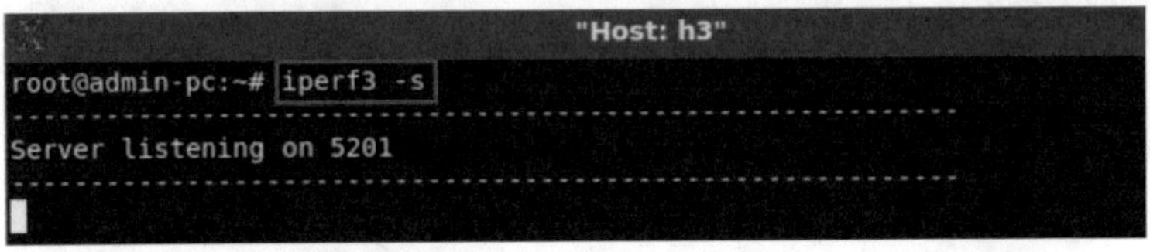

"Host: h3"
root@admin-pc:~# iperf3 -s
Server listening on 5201

Fig. 182 Starting iPerf3 server on host h3

admin@admin-pc: ~
File Actions Edit View Help
admin@admin-pc: ~
admin@admin-pc:~$ sudo plot_q.sh s2-eth2

Fig. 183 Plotting the queue occupancy on switch S2's *s2-eth2* interface

```
sudo tc qdisc change dev s2-eth2 root handle 1: tbf rate
100mbit burst 50000 limit 26214400
```

17.4 Throughput and Latency Tests

Step 1. Launch iPerf3 in server mode on host h3's terminal (Fig. 182).

```
iperf3 -s
```

Step 2. In the client's terminal, type the command below to plot the switch's queue in real-time (Fig. 183). When prompted for a password, type password and hit *Enter*.

```
sudo plot_q.sh s2-eth2
```

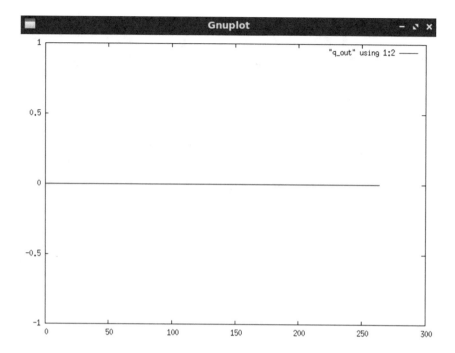

Fig. 184 Queue occupancy on switch S2's *s2-eth2* interface

Fig. 185 Creating and navigating into directory *1BDP*

A new window (Fig. 184) opens that plots the queue occupancy as shown in the figure below. Since there are no active flows passing through *s2-eth2* interface on switch S2, the queue occupancy is constantly 0.

Step 3. Exit from RED/results directory using the following command (Fig. 185):

```
cd ..
```

Step 4. Type the following iPerf3 command in host h1's terminal without executing it. The -J option is used to display the output in JSON format. The redirection operator > is used to store the JSON output into a file (Fig. 186).

```
iperf3 -c 10.0.0.3 -t 90 -J >out.json
```

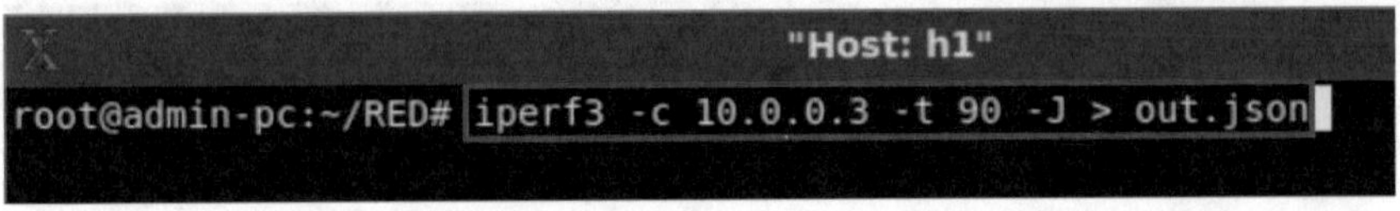

"Host: h1"
root@admin-pc:~/RED# iperf3 -c 10.0.0.3 -t 90 -J > out.json

Fig. 186 Running iPerf3 client on host h1

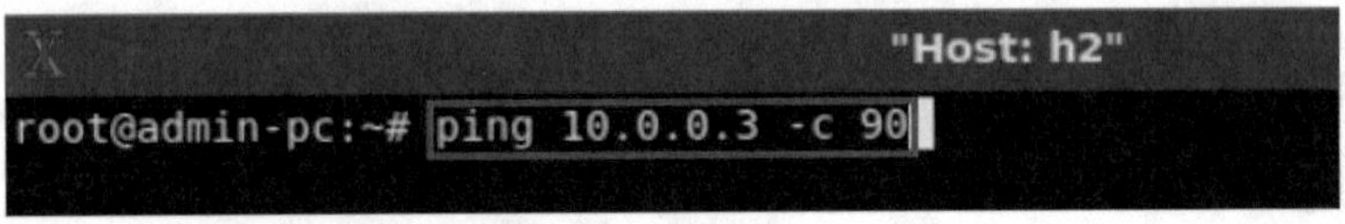

"Host: h2"
root@admin-pc:~# ping 10.0.0.3 -c 90

Fig. 187 Typing ping command on host h2

Gnuplot
"q_out" using 1:2
3x10^6
2.5x10^6
2x10^6
1.5x10^6
1x10^6
500000
0
0
50
100
150
200
250

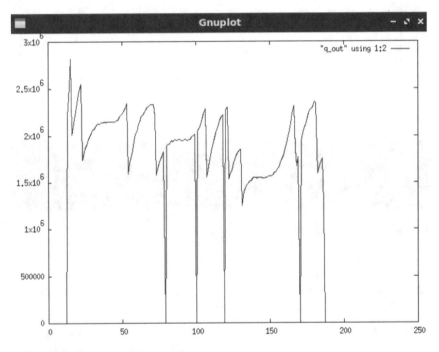

Fig. 188 Queue occupancy on switch S2's *s2-eth2* interface

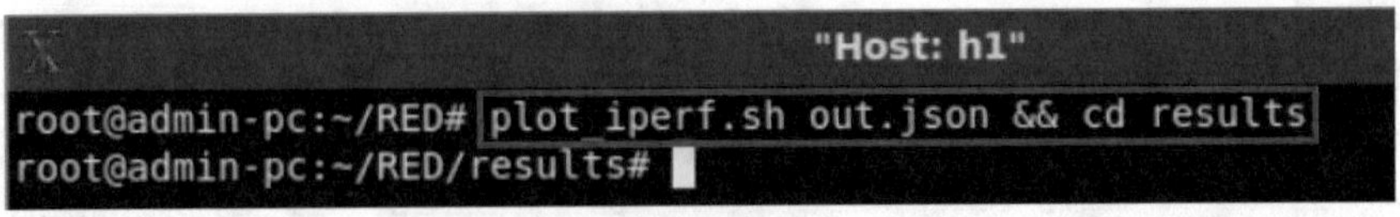

"Host: h1"
root@admin-pc:~/RED# plot_iperf.sh out.json && cd results
root@admin-pc:~/RED/results#

Fig. 189 Generating plotting files and entering the *results* directory

Fig. 190 Opening the *throughput.pdf* file

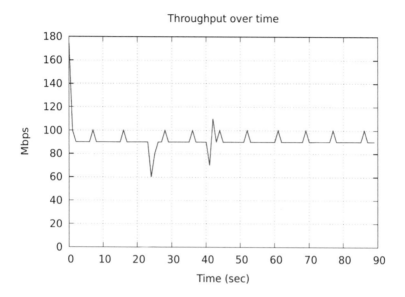

Fig. 191 Measured throughput

Fig. 192 Opening the *RTT.pdf* file

Step 5. Type the following ⎹ ping ⎸ command in host h2's terminal without executing it (Fig. 187):

```
ping 10.0.0.3 -c 90
```

Step 6. Press *Enter* to execute the commands, first in host h1 terminal and then in host h2 terminal. Then, go back to the queue plotting window and observe the queue occupancy (Fig. 188).

The graph above shows that the queue occupancy peaked over $2.5 \cdot 10^6$, which is around average queue length for a 1 Gbps link. However, in this case we set a

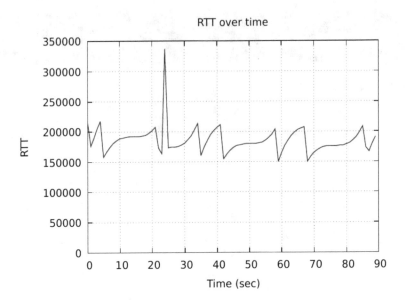

Fig. 193 Measured round-trip time

100 Mbps link when RED is configured to operate for 1 Gbps link; therefore, the point of operation changed. Consequently, bufferbloat is experienced; thus, it is necessary to reconfigure RED parameters in order to mitigate the excessive queue length.

Step 7. In the queue plotting window, press the $\boxed{\text{s}}$ key on your keyboard to stop plotting the queue.

Step 8. After the iPerf3 test finishes on host h1, enter the following command (Fig. 189):

```
plot_iperf.sh out.json && cd results
```

Step 9. Open the throughput file using the command below on host h1 (Fig. 190):

```
xdg-open throughput.pdf
```

The figure above (Fig. 191) shows the iPerf3 test output report for the last 90 s. The average achieved throughput is 100 Mbps.

Step 10. Close the *throughput.pdf* window and then open the Round-Trip Time (RTT) file using the command below (Fig. 192):

```
xdg-open RTT.pdf
```

```
                                        "Host: h2"
64 bytes from 10.0.0.3: icmp_seq=72 ttl=64 time=169 ms
64 bytes from 10.0.0.3: icmp_seq=73 ttl=64 time=171 ms
64 bytes from 10.0.0.3: icmp_seq=74 ttl=64 time=174 ms
64 bytes from 10.0.0.3: icmp_seq=75 ttl=64 time=174 ms
64 bytes from 10.0.0.3: icmp_seq=76 ttl=64 time=175 ms
64 bytes from 10.0.0.3: icmp_seq=77 ttl=64 time=174 ms
64 bytes from 10.0.0.3: icmp_seq=78 ttl=64 time=174 ms
64 bytes from 10.0.0.3: icmp_seq=79 ttl=64 time=176 ms
64 bytes from 10.0.0.3: icmp_seq=80 ttl=64 time=177 ms
64 bytes from 10.0.0.3: icmp_seq=81 ttl=64 time=178 ms
64 bytes from 10.0.0.3: icmp_seq=82 ttl=64 time=180 ms
64 bytes from 10.0.0.3: icmp_seq=83 ttl=64 time=185 ms
64 bytes from 10.0.0.3: icmp_seq=84 ttl=64 time=191 ms
64 bytes from 10.0.0.3: icmp_seq=85 ttl=64 time=198 ms
64 bytes from 10.0.0.3: icmp_seq=86 ttl=64 time=208 ms
64 bytes from 10.0.0.3: icmp_seq=87 ttl=64 time=160 ms
64 bytes from 10.0.0.3: icmp_seq=88 ttl=64 time=166 ms
64 bytes from 10.0.0.3: icmp_seq=89 ttl=64 time=180 ms
64 bytes from 10.0.0.3: icmp_seq=90 ttl=64 time=192 ms

--- 10.0.0.3 ping statistics ---
90 packets transmitted, 90 received, 0% packet loss, time 183ms
rtt min/avg/max/mdev = 148.914/186.175/468.728/33.481 ms
root@admin-pc:~#
```

Fig. 194 ping test result

```
                                        "Host: h1"
root@admin-pc:~/RED/results# xdg-open cwnd.pdf
```

Fig. 195 Opening the *cwnd.pdf* file

The graph above (Fig. 193) shows that the RTT increased from approximately ten times the default latency (20 ms). The output above shows that there is a bufferbloat problem as the average latency is significantly greater. Since RED is configured to operate on a 1 Gbps link, for this test the point of operation changed; therefore, unnecessary delay is observed.

Step 11. Close the *RTT.pdf* window and then go back to h2's terminal to see the ping output (Fig. 194).

The result above indicates that all 90 packets were received successfully (0deviation of the Round-Trip Time (RTT) were 148.914, 186.175, 468.728, and 33.481 milliseconds, respectively. The output also verifies that there is a bufferbloat problem as the average latency (186.175) is significantly greater than the configured delay (20 ms).

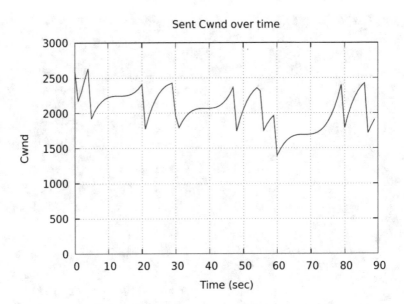

Fig. 196 Evolution of the congestion window

Step 12. Close the *RTT.pdf* window and then open the congestion window (*cwnd.pdf*) file using the command below (Fig. 195):

```
xdg-open cwnd.pdf
```

The graph above (Fig. 196) shows the evolution of the congestion window, which peaked around 2.5 Mbytes.

Step 13. To stop *iperf3* server in host h3, press Ctrl+c .

References

1. J. Moy, Open shortest path first (OSPF) Version 2, in *Internet Request for Comments, RFC Editor, RFC 2328* (1998). https://www.ietf.org/rfc/rfc2328.txt
2. Y. Rekhter, T. Li, S. Hares, Border gateway protocol 4, in *Internet Request for Comments, RFC Editor, RFC 4271* (2006). https://tools.ietf.org/html/rfc4271.
3. N. Cardwell, Y. Cheng, C. Gunn, S. Yeganeh, V. Jacobson, BBR: congestion-based congestion control. Commun. ACM **60**(2), 58–66 (2017)
4. J. Kurose, K. Ross, *Computer Networking: A Top-down Approach*, 7th edn. (Pearson, London, 2017)
5. *Router/switch Buffer Size Issues*. https://fasterdata.es.net/network-tuning/router-switch-buffer-size-issues/
6. C. Villamizar, C. Song, High performance TCP in ansnet. ACM Comput. Commun. Rev. **24**(5), 45–60 (1994)

7. R. Bush, D. Meyer, Some internet architectural guidelines and philosophy, in *Internet Request for Comments, RFC Editor, RFC 3439* (2003). https://www.ietf.org/rfc/rfc3439.txt

8. G. Appenzeller, I. Keslassy, N. McKeown, Sizing router buffers, in *Proceedings of the 2004 Conference on Applications, Technologies, Architectures, and Protocols for Computer Communications* (2004), pp. 281–292

9. J. Padhye, V. Firoiu, D. Towsley, J. Kurose, Modeling TCP throughput: a simple model and its empirical validation, in *Proceedings of the ACM SIGCOMM '98 Conference on Applications, Technologies, Architectures, and Protocols for Computer Communication* (1998), pp. 303–314

10. M. Smitasin, B. Tierney, Evaluating network buffer size requirements, in *Proceedings of the 2015 Technology Exchange Workshop* (2015). https://meetings.internet2.edu/media/medialibrary/2015/10/05/20151005-smitasin-buffersize.pdf

11. B. Tierney, Improving performance of 40G/100G data transfer nodes, in *Proceedings of the 2016 Technology Exchange Workshop* (2016). https://meetings.internet2.edu/2016-technology-exchange/detail/10004333/

12. V. Paxson, S. Floyd, Wide area traffic: the failure of poisson modeling. IEEE/ACM Trans. Networking 3(3), 226–244 (1995)

13. N. Beheshti, E. Burmeister, Y. Ganjali, J. Bowers, D. Blumenthal, N. McKeown, Optical packet buffers for backbone internet routers. IEEE/ACM Trans. Networking 18(5), 1599–1609 (2010)

14. V. Cerf, Bufferbloat and other internet challenges. IEEE Internet Comput. 18(5), 80–80 (2014)

15. H. Im, C. Joo, T. Lee, S. Bahk, Receiver-side TCP countermeasure to bufferbloat in wireless access networks. IEEE Trans. Mob. Comput. 15(8), 2080–2093 (2016)

16. K. Nichols, V. Jacobson, A. McGregor, J. Iyengar, Controlled delay active queue management, in *Internet Draft draft-ietf-aqm-codel-10* (2017). https://tools.ietf.org/html/draft-ietf-aqm-codel-10

17. Linux tuning. https://fasterdata.es.net/host-tuning/linux/

18. Cisco catalyst 6500 supervisor 2T architecture white paper, in *Cisco Systems White Paper* (2017). https://www.cisco.com/c/en/us/products/collateral/switches/catalyst-6500-series-switches/white_paper_c11-676346.html#_Toc390815326

19. N. McKeown, A. Mekkittikul, V. Anantharam, J. Walrand, Achieving 100% throughput in an input-queued switch. IEEE Trans. Commun. 47(8), 1260–1267 (1999)

20. E. Dart, L. Rotman, B. Tierney, M. Hester, J. Zurawski, The science DMZ: a network design pattern for data-intensive science, in *Proceedings of the International Conference on High Performance Computing, Networking, Storage and Analysis* (2013)

Impact of TCP on High-Speed Networks and Advances in Congestion Control Algorithms

Applications can transmit a large amount of data between end devices. Many applications require the data to be correctly delivered from one device to another (e.g., from an instrument to a DTN). This is one of the services provided by TCP and a reason why TCP is the protocol used by data transfer tools. There are several TCP attributes that should be considered when used in high-speed networks and Science DMZs, including segment size, flow control, and buffer size, selective acknowledgment, parallel connections, pacing, and congestion control. After a brief review of TCP, this chapter discusses these attributes.

1 TCP Review

TCP receives data from the application layer and places it in the TCP send buffer, as shown in Fig. 1a. Data is typically broken into MSS units. The MSS is simply the MTU minus the combined lengths of the TCP and IP headers (typically 40 bytes). Ethernet's normal MTU is 1500 bytes. Thus, the MSS's typical value is 1460. The TCP header is shown in Fig. 1b.

TCP implements flow control by requiring the receiver indicate how much spare room is available in the TCP receive buffer. For a full utilization of the path, the TCP send and receive buffers must be greater than or equal to the bandwidth-delay product. This buffer size value is the maximum number of bits that can be outstanding (inflight) if the sender continuously sends segments.

For reliability, TCP uses two fields of the TCP header: sequence number and acknowledgment (ACK) number. The sequence number is the byte-stream number of the first byte in the segment. The acknowledgment number that the receiver puts in its segment is the sequence number of the next byte the receiver is expecting from the sender. Figure 2a shows an example of the use of these two fields. If an acknowledgment for an outstanding segment is not received, TCP retransmits that

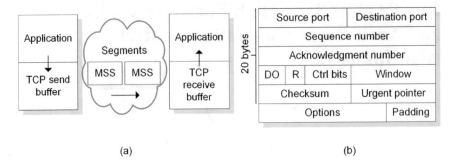

(a)　　　　　　　　　　　　　　　　　　　(b)

Fig. 1 TCP connection and header. (**a**) End points of the TCP connection. (**b**) TCP header. Ctrl, R, and DO fields stand for control, reserved, and data offset

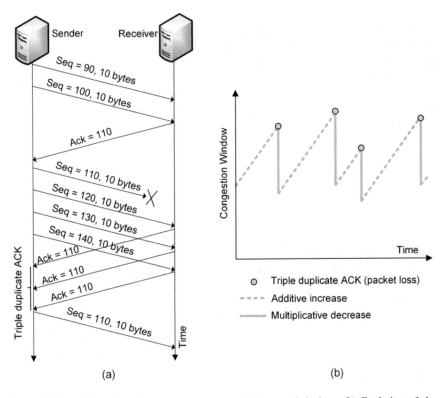

(a)　　　　　　　　　　　　　　　　　　　(b)

Fig. 2 TCP operation. (**a**) Exchange of segments between end devices. (**b**) Evolution of the congestion window

segment. Alternatively, the sender can also detect a packet loss by detecting a triple duplicate ACK.

TCP maintains a congestion window whose size is the number of bytes the sender may have in the network at any time. The connection throughput is the minimum between the flow control and the congestion window, divided by the RTT. Assuming a large TCP receive buffer, the congestion window is used to adjust the rate at which the sender sends data.

2   TCP Considerations for Science DMZs

Features such as TCP buffer size and parallel streams are usually overlooked in enterprise networks, where a slight throughput degradation is often acceptable for small flows. However, inadequate transport-layer settings may have a high negative impact for large flows. These features are discussed next.

2.1   Maximum Segment Size

One obvious advantage of using large segments is efficiency in processing because a 20-byte header overhead can be amortized over more data. Moreover, the recovery after a packet loss is proportional to the MSS. During the additive increase phase of the congestion control algorithm, TCP increases the congestion window by approximately one MSS every RTT. This means that by using a 9000-byte MSS instead of a 1500-byte MSS, the throughput increases six times faster. Even when losses are occasional, the performance improvement can be significant.

2.2   Flow Control and TCP Receive Buffer

TCP flow control imposes a limit in the utilization of the channel from the source to the destination. In order to maximize the utilization of the channel and increase throughput, the TCP buffer must be at least as large as the BDP, and preferably larger. By having a large TCP buffer, the sender can keep transmitting at full speed until the first acknowledgment comes back. Increasing the TCP buffer above BDP, for example, to a value that equals 2BDP, also adds robustness. Thus, if a sporadic loss occurs, TCP would decrease the window size to BDP. Therefore, after the sporadic loss, the sender would still fully utilize the channel.

For applications that use parallel collaborating TCP connections or streams in the transmission of a data set, the TCP buffer can be reduced. This requires an application-layer software, such as gridFTP [1–3] to orchestrate the transmission over multiple connections. Since the full bandwidth is shared by the parallel connections, the TCP buffer need not to be equal to the BDP. Instead, it can be reduced in proportion to the number of parallel connections.

2.3   Selective Acknowledgment

Much of the complexity of TCP is related to inferring which packets have arrived and which packets have been lost. The cumulative acknowledgment number does not provide this information. A selective acknowledgment (SACK) lists up to three ranges of bytes that have been received. With this information, the sender can more directly decide what segments to retransmit.

The impact of using SACK on large data transfers at 10 Gbps is not conclusive. In paths with small to medium RTT, the use of SACK is encouraged in the literature [4]. However, in paths with large RTT and bandwidth, using SACK may reduce performance. For very large BDP paths where the TCP buffer size is in the order of tens of MBs, there is a large number of inflight segments. For example, for a TCP receive buffer of 64 MB and a MSS of 1500 bytes, there could be almost 45,000 outstanding segments. When a SACK event occurs, the TCP performance may be degraded by the process of locating and resending the packets listed in the SACK lists. This in turn causes TCP to trigger a timeout and to reduce the congestion window. If such issues are observed, a solution is to disable SACK.

2.4   Parallel TCP Connections

$$\text{throughput} = \frac{MSS}{RTT \cdot \sqrt{L}}.$$ (1)

The advent of Science DMZs and the need to combat random packet losses have recently initiated new research in the use of parallel TCP connections for large flows [5–7]. Assuming that losses, RTT, and MSS are the same in each connection, the total throughput is essentially the aggregation of the K single TCP connection throughputs [8]. Since the throughput of a single TCP connection is given by Eq. (1), the aggregate throughput of K connections is given by the following equation:

$$\text{aggregate throughput} = \sum_{i=1}^{K} \frac{MSS}{RTT\sqrt{L}} = K\frac{MSS}{RTT\sqrt{L}}.$$ (2)

Thus, an application opening K parallel TCP connections essentially creates a large virtual MSS on the aggregate connection that is K times the MSS of a single connection. A larger MSS increases the rate of recovery from a loss event from one MSS per successful segment transmission to K MSSs per successful segment transmission. When the aggregate TCP connection begins to create congestion, any router or switch along the path begins dropping packets and Eq. (2) is no longer valid. Parallel TCP connections must be implemented and managed by the application layer. Its use is further discussed in Sect. V.

2.5 TCP Fair Queue Pacing

Data transmissions can be bursty, resulting in packets being buffered at routers and switches and dropped at times. End devices can contribute to the problem by sending a large number of packets in a short period of time. If those packets were transmitted at a steady pace, the formation of queues could be reduced.

TCP pacing is a technique by which a transmitter evenly spaces or paces packets at a pre-configured rate. TCP pacing has been applied for years in enterprise networks [9], with mixed results. However, its recent application to data transfers in Science DMZs suggests that its use has several advantages [10]. TCP pacing has also been applied to datacenter environments [11].

The existing TCP congestion control algorithms, with the exception of BBR [13], indicate how much data is allowed for transmission. Those algorithms do not provide a time period over which that data should be transmitted and how the data should be spread to mitigate potential bursts. The rate, however, can be enforced by a packet scheduler such as a fair queue (FQ). The packet scheduler organizes the flow of packets of each TCP connection through the network stack to meet policy objectives. Some Linux distributions such as CentOS [12] implement FQ scheduling in conjunction with TCP pacing [13, 14].

FQ is intended for locally generated traffic (e.g., a sender DTN). Figure 3 illustrates the operation of FQ pacing. Application 1 generates green packets, and application 2 generates blue packets. Each application opens a TCP connection. FQ paces each connection according to the desired rate, evenly spacing out packets within an application based on the desired rate. The periods T_1 and T_2 represent the time-space used for connections 1 and 2, respectively.

TCP pacing reduces the typical TCP sawtooth behavior [15] and is effective when there are rate mismatches along the path between the sender and the receiver. This is

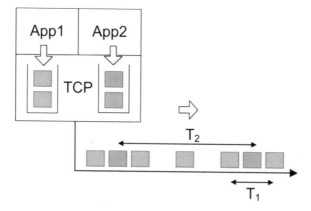

Fig. 3 TCP pacing. Packets of application 1 and application 2 are evenly spaced by T_1 and T_2 time units, respectively

the case, for example, when the ingress port of a router has a capacity of 100 Gbps, and the egress port has a capacity of 10 Gbps. Because of the TCP congestion control mechanism, the sawtooth behavior always emerges. As TCP continues to increase the size of the congestion window, eventually the bottleneck link becomes full, while the rest of the links become underutilized. These mismatches produce a continuous circle of additive increases and multiplicative decreases [15].

2.6 TCP Congestion Control Algorithms

A loss-based signal is still the main mechanism used to adjust the congestion window and thus the throughput. The key difference among loss-based congestion control algorithms is the strategy after a packet loss is detected. The rate at which the congestion window grows after the loss may follow different mathematical functions. Examples include Reno [16], Cubic [17], and HTCP [18]. Reno uses a linear rate increase, while Cubic and HTCP use cubic and quadratic functions.

Essentially, the main issue observed in high-speed networks and Science DMZs is that, after a packet loss, the additive increase is too slow to reach full speed. Consider Fig. 4a, which shows a TCP's viewpoint of a connection. At any time, the connection has exactly one slowest link or bottleneck bandwidth (btlbw) that determines the location where queues are formed. When the router's buffer is large, the loss-based congestion control keeps it full. When the router's buffer is small, the loss-based congestion control misinterprets a packet loss as a signal of congestion, leading to low throughput. The output port queue increases when the input link arrival rate exceeds btlbw. The throughput of loss-based congestion control algorithms is less than btlbw because of the frequent packet losses [13].

Figure 4b illustrates the RTT and delivery rate as functions of the amount of data inflight [13]. RTTmin is the minimum RTT, when no congestion exists. In the application limited region, the delivery rate/throughput increases as the amount of data generated by the application layer increases, while the RTT remains constant. The pipeline between the sender and the receiver becomes full when the inflight number of bits is equal to BDP, at the edge of the bandwidth limited region. The queue size starts increasing, resulting in an increase of the RTT. The delivery rate/throughput remains constant, as the bottleneck link is fully utilized. Finally, when no buffer is available at the router to store arriving packets (the amount of inflight bits is equal to BDP plus the buffer size of the router), then packets are dropped.

BBR, the recently proposed congestion control algorithm [13], is a disruption of previous algorithms in that the control is based on the rate rather than on the window. At any one time, BBR sends at a given calculated rate, instead of sending new data in response to each received acknowledgment. BBR attempts to find the optimal operating point, shown as a green dot in Fig. 4b, by estimating RTTmin and btlbw.

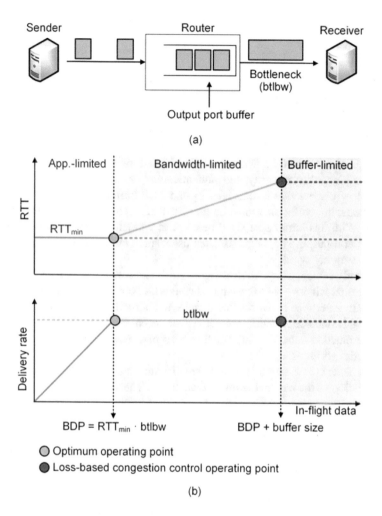

Fig. 4 TCP viewpoint of a connection and relation between throughput and RTT. (**a**) Simplified TCP interpretation of the connection. (**b**) Throughput and RTT, as a function of inflight data [13]

A natural question is how well does BBR, a rate-based congestion control algorithm, perform with respect to a Science DMZ recommended traditional loss-based congestion control? Preliminary results indicate that BBR shows better performance than traditional loss-based congestion control algorithms when packet losses occur. Of particular interest to Science DMZs is the range of corruption before the throughput completely collapses. The results, which are presented in Sect. VII, show that BBR can achieve a better performance than loss-based congestion control algorithms. Specifically, BBR can tolerate a larger rate of packet losses before the throughput collapses.

3   Transport-Layer Issues in Enterprise Networks and Science DMZs

Table 1 shows a comparison between enterprise networks and Science DMZs, regarding transport-layer features.

Reliability is required for file and data set transfers, and therefore, Science DMZ applications use TCP. While TLS [19] and SSL [20] also offer reliable service and security on top of TCP, they introduce additional overhead and a redundant service. Globus, a well-known application-layer tool for transferring large files, offers confidentiality, integrity, and authentication services.

The flow control rate is managed by the TCP buffer size. For Science DMZ applications, the buffer size must be greater than or equal to the bandwidth-delay product. With this buffer size, TCP behaves as a pipelined protocol. On the other hand, general-purpose applications often use a small buffer size, which produces a stop-and-wait behavior.

The study of congestion control algorithms is an active research area. Although the traditional window-based loss-based congestion control may not be appropriate for modern enterprise networks, there were no alternatives until recently, and thus its use can be labeled as indifferent. However, recent preliminary results, including those presented in Sect. VII, indicate that BBR performs better than window-based loss-based algorithms.

If the TCP buffer size at DTNs is smaller than the bandwidth-delay product, the utilization of the channel is lower than 100%. The sender must constantly wait for acknowledgment segments before transmitting additional data segments. On the other hand, if the buffer size is greater than or equal to the bandwidth-delay product, the path utilization approaches the maximum capacity and many data segments are allowed to be in transit, while acknowledgment segments are simultaneously received. For small and short-duration flows, this may not be essential. However, for large science flows, to achieve full performance, the buffer size must be at least equal to the bandwidth-delay product. The MSS is perhaps one of the most important features in high-throughput high-latency networks with packet losses. TCP pacing is a promising feature. The challenge for its wide adoption is the complexity of developing a mechanism to discover the bottleneck link and its capacity.

4   Academic Cloud and Virtual Laboratories

The book is accompanied by hands-on virtual laboratory experiments conducted in a cloud system, referred to as the Academic Cloud. Access to the Academic Cloud is available for a fee (six-month access) and includes all materials needed to conduct the experiments. The URL is:

http://highspeednetworks.net/

Table 1 Comparison of transport-layer features in enterprise networks and Science DMZs

Feature		Enterprise network	Science DMZ
Protocol	TCP	Used in applications requiring reliability, e.g., email, http	Used for main applications: data transfers
	UDP	Used in applications that do not require reliability, e.g., voice and video	Not used
	TLS/SSL	Used in applications requiring security, e.g., online banking	Not recommended; it adds an additional flow control layer
Flow control	Pipelined	Not necessary; BDP is typically small	Recommended; data transfers occur across high-throughput high-latency networks
	Stop-and-wait behavior	Not recommended, but performance is not dramatically impacted when RTT is small	Not recommended; throughput is severely reduced
Use of SACK	With large MSS	Indifferent	Results suggest the use of SACK may reduce throughput, especially when RTT is large
	With small MSS	Throughput is slightly improved, in particular when RTT is small	Indifferent
Congestion control	Window-based	Only alternative until recently	Only alternative until recently
	Rate-based	Performance evaluations of BBR indicate an increase in throughput in small flows. Under severe-loss scenarios, throughput can be much larger than that of window-based algorithms.	Under very low-loss scenarios, BBR's performance is between 2 and 3% lower than that of window-based. In lossy scenarios, performance is superior to that of window-based
TCP buffer size $\geq$ BDP		Not essential for small RTT.	Required for a full utilization of the end-to-end path
Large MSS		Not essential for small RTT.	Recommended; it speeds up the recovery of the congestion window
TCP pacing		Encouraging results when bottleneck bandwidth is known or can be estimated.	Encouraging results when bottleneck bandwidth is known or can be estimated
Parallel streams		Impact in small flows is not substantial.	Recommended; it minimizes the impact of packet losses

5 Chapter 4—Lab 9: Understanding Traditional TCP Congestion Control (HTCP, Cubic, Reno)

Overview

To conduct the experiment described in this section, please login into the Academic Cloud at http://highspeednetworks.net/, and reserve a pod for Lab 9.

This lab reviews key features and behavior of Transmission Control Protocol (TCP) that have a large impact on data transfers over high-throughput high-latency networks. The lab describes the behavior of TCP's congestion control algorithm, its impact on throughput, and how to modify the congestion control algorithm in a Linux machine.

Objectives

By the end of this lab, students should be able to:

1. Describe the basic operation of TCP congestion control algorithm and its impact on high-throughput networks.
2. Explain the concepts of congestion window, bandwidth probing, and Additive Increase Multiplicative Decrease (AIMD).
3. Understand TCP throughput calculation.
4. Understand the impact of packet loss on high-latency networks.
5. Deploy emulated WANs in Mininet.
6. Modify the TCP congestion control algorithm in Linux using *sysctl* tool.
7. Compare TCP Reno, HTCP, and Cubic with injected packet loss.
8. Compare TCP Reno, HTCP, and Cubic with both injected delay and packet loss.

Lab Settings

The information in Table 2 provides the credentials of the machine containing Mininet.

Lab Roadmap

This lab is organized as follows:

1. Section 6: Introduction to TCP
2. Section 7: Lab topology
3. Section 8: Introduction to *sysctl*
4. Section 9: Congestion control algorithms and *sysctl*
5. Section 10: iPerf3 throughput test

Table 2 Credentials to access Client1 machine

Device	Account	Password
Client1	admin	password

6 Introduction to TCP

6.1 TCP Review

Big data applications require the transmission of large amounts of data between end devices. Data must be correctly delivered from one device to another, e.g., from an instrument to a data transfer node (DTN). Reliability is one of the services provided by TCP and a reason why TCP is the protocol used by most data transfer tools. Thus, understanding the behavior of TCP is essential for the design and operation of networks used to transmit big data.

TCP receives data from the application layer and places it in the TCP send buffer, as shown in Fig. 5a. Data is typically broken into maximum segment size (MSS) units. Note that "segment" here refers to the protocol data unit (PDU) at the transport layer, and sometimes the terms packet and segment are interchangeably used. The MSS is simply the maximum transmission unit (MTU) minus the combined lengths of the TCP and IP headers (typically 40 bytes). Ethernet's normal MTU is 1500 bytes. Thus, MSS's typical value is 1460. The TCP header is shown in Fig. 5b.

For reliability, TCP uses two fields of the TCP header to convey information to the sender: sequence number and acknowledgment (ACK) number. The sequence number is the byte-stream number of the first byte in the segment. The acknowledgment number that the receiver puts in its segment is the sequence number of the next byte the receiver is expecting from the sender. In the example of Fig. 6a, after receiving the first two segments containing sequence number 90 (which contains bytes 90–99) and 100 (bytes 100–109), the receiver sends a segment with acknowledgment number 110. This segment is called cumulative acknowledgment.

6.2 TCP Throughput

The TCP rate limitation is defined by the receive buffer shown in Fig. 5a. If this buffer size is too small, TCP must constantly wait until an acknowledgment arrives

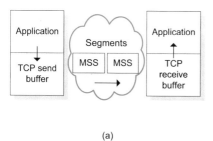

(a)

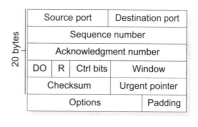

(b)

Fig. 5 (a) TCP connection. (b) TCP header

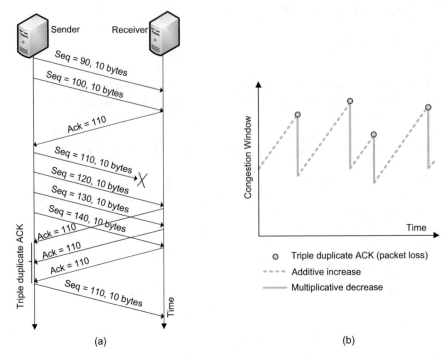

Fig. 6 (**a**) TCP operation. (**b**) Adaptation of TCP's congestion window

before sending more segments. This limitation is removed by setting a large receive buffer size.

A second limitation is imposed by the congestion control mechanism operating at the sender side, which keeps track of a variable called congestion window. The congestion window, referred to as *cwnd* (in bytes), imposes a constraint on the rate at which a TCP sender can send traffic. The *cwnd* value is the amount of unacknowledged data at the sender. To see this, note that at the beginning of every Round-Trip Time (RTT), the sender can send *cwnd* bytes of data into the connection; at the end of the RTT, the sender receives acknowledgments for the data. Thus, the sender's send rate is roughly *cwnd*/RTT bytes/sec. By adjusting the value of *cwnd*, the sender can therefore adjust the rate at which it sends data into the connection.

$$\text{TCP Throughput} \approx \frac{\text{cwnd}}{\text{RTT}} \text{ [bytes/second].}$$

6.3 TCP Packet Loss Event

TCP is a reliable transport protocol that requires each segment to be acknowledged. If an acknowledgment for an outstanding segment is not received, TCP retransmits that segment. Alternatively, instead of waiting for a timeout-triggered retransmission, the sender can also detect a packet loss before the timeout by detecting duplicate ACKs. A duplicate ACK is an ACK that reacknowledges a segment for which the sender has already received. If the TCP sender receives three duplicate ACKs for the same segment, TCP interprets this event as packet loss due to congestion and reduces the congestion window *cwnd* by half. This congestion window reduction is known as multiplicative decrease.

In steady state (ignoring the initial TCP period when a connection begins), a packet loss will be detected by a triple duplicate ACK. After decreasing *cwnd* by half, and as long as no other packet loss is detected, TCP will slowly increase *cwnd* again by 1 MSS per RTT. This congestion control phase essentially produces an additive increase in the congestion window. For this reason, TCP congestion control is referred to as an Additive Increase Multiplicative Decrease (AIMD) form of congestion control. AIMD gives rise to the "saw tooth" behavior shown in Fig. 6b, which also illustrates the idea of TCP "probing" for bandwidth—TCP linearly increases its congestion window size (and hence its transmission rate) until a triple duplicate-ACK event occurs. It then decreases its congestion window size by a factor of two but then again begins increasing it linearly, probing to see if there is additional available bandwidth.

6.4 Impact of Packet Loss in High-Latency Networks

During the additive increase phase, TCP only increases *cwnd* by 1 MSS every RTT period. This feature makes TCP very sensitive to packet loss on high-latency networks, where the RTT is large.

Consider Fig. 7, which shows the TCP throughput of a data transfer across a 10 Gbps path. The packet loss rate is 1/22,000 or 0.0046%. The purple curve is the throughput in a loss-free environment; the green curve is the theoretical throughput computed according to the equation below, where L is the packet loss rate.

$$\text{TCP Throughput} \approx \frac{\text{MSS}}{\text{RTT} \sqrt{L}} \text{ [bytes/second]}$$

The equation above indicates that the throughput of a TCP connection in steady state is directly proportional to the maximum segment size (MSS) and inversely proportional to the Round-Trip Time (RTT) and the square root of the packet loss rate (L). The red and blue curves are real throughput measurements of two

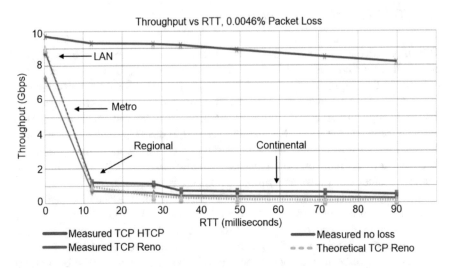

Fig. 7 Throughput vs. Round-Trip Time (RTT), for two devices connected via a 10 Gbps path. The performance of two TCP implementations is provided: Reno (blue) and Hamilton TCP (HTCP) (red). The theoretical performance with packet losses (green) and the measured throughput without packet losses (purple) are also shown

popular implementations of TCP: Reno and Hamilton TCP (HTCP). Because TCP interprets losses as network congestion, it reacts by decreasing the rate at which packets are sent. This problem is exacerbated as the latency increases between the communicating hosts. Beyond LAN transfers, the throughput decreases rapidly to less than 1 Gbps. This is often the case when research collaborators sharing data are geographically distributed.

TCP Reno is an early congestion control algorithm. TCP Cubic, HTCP, and BBR are more recent congestion control algorithms, which have demonstrated improvements with respect to TCP Reno.

7 Lab Topology

Let us get started with creating a simple Mininet topology using MiniEdit. The topology uses 10.0.0.0/8, which is the default network assigned by Mininet (Fig. 8).

Step 1. A shortcut to MiniEdit is located on the machine's Desktop. Start MiniEdit by clicking on MiniEdit's shortcut (Fig. 9). When prompted for a password, type password .

Step 2. On MiniEdit's menu bar, click on *File* and then *Open* to load the lab's topology. Locate the *Lab 9.mn* topology file and click on *Open* (Fig. 10).

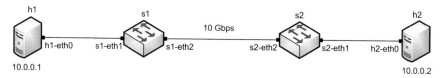

Fig. 8 Lab topology

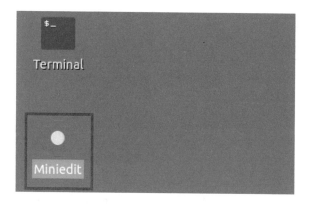

Fig. 9 MiniEdit shortcut

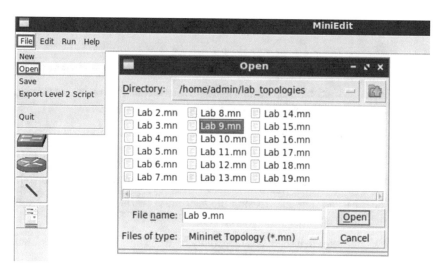

Fig. 10 MiniEdit shortcut

Step 3. Before starting the measurements between host h1 and host h2, the network must be started. Click on the *Run* button located at the bottom left of MiniEdit's window to start the emulation (Fig. 11).

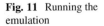

Fig. 11 Running the emulation

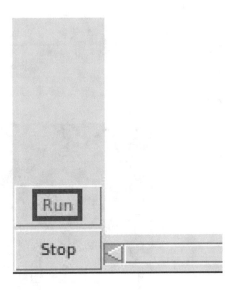

The above topology uses 10.0.0.0/8, which is the default network assigned by Mininet.

7.1   Starting Host h1 and Host h2

Step 1. Hold the right-click on host h1 and select *Terminal*. This opens the terminal of host h1 and allows the execution of commands on host h1 (Fig. 12).

Step 2. Apply the same steps on host h2 and open its *Terminal*.

Step 3. Test connectivity between the end-hosts using the ping command. On host h1, type the command ping 10.0.0.2. This command tests the connectivity between host h1 and host h2. To stop the test, press Ctrl+c. The figure below shows a successful connectivity test (Fig. 13).

Figure 13 indicates that there is connectivity between host h1 and host h2. Thus, we are ready to start the throughput measurement process.

7.2   Emulating 10 Gbps High-Latency WAN with Packet Loss

This section emulates a high-latency WAN, which is used to validate the results observed in Fig. 7. We will first set the bandwidth between host h1 and host h2 to

MiniEdit
File Edit Run Help
s1
s2
Host Options
Terminal
h2

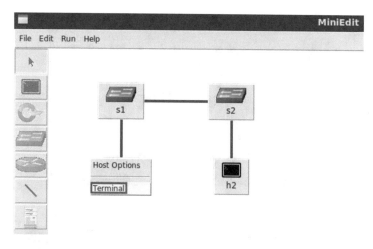

Fig. 12 Opening a terminal on host h1

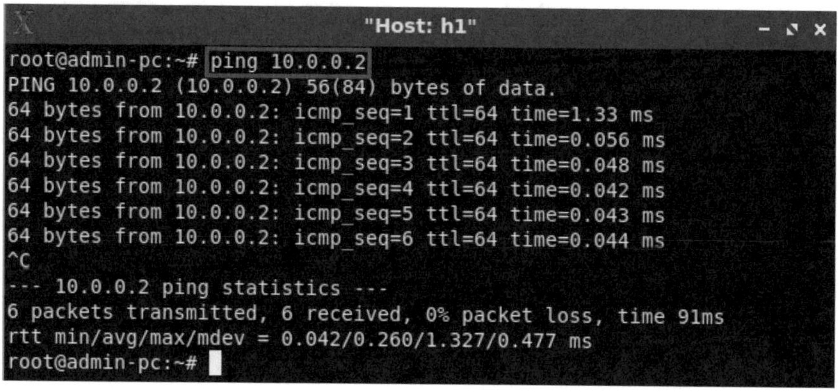

"Host: h1"
root@admin-pc:~# ping 10.0.0.2
PING 10.0.0.2 (10.0.0.2) 56(84) bytes of data.
64 bytes from 10.0.0.2: icmp_seq=1 ttl=64 time=1.33 ms
64 bytes from 10.0.0.2: icmp_seq=2 ttl=64 time=0.056 ms
64 bytes from 10.0.0.2: icmp_seq=3 ttl=64 time=0.048 ms
64 bytes from 10.0.0.2: icmp_seq=4 ttl=64 time=0.042 ms
64 bytes from 10.0.0.2: icmp_seq=5 ttl=64 time=0.043 ms
64 bytes from 10.0.0.2: icmp_seq=6 ttl=64 time=0.044 ms
^C
--- 10.0.0.2 ping statistics ---
6 packets transmitted, 6 received, 0% packet loss, time 91ms
rtt min/avg/max/mdev = 0.042/0.260/1.327/0.477 ms
root@admin-pc:~#

Fig. 13 Connectivity test using ping command

10 Gbps. Then we will emulate packet losses between switch S1 and switch S2 and measure the throughput.

Step 1. Launch a Linux terminal by holding the Ctrl+Alt+T keys or by clicking on the Linux terminal icon (Fig. 14).

The Linux terminal is a program that opens a window and permits you to Interact with a command-line interface (CLI). A CLI is a program that takes commands from the keyboard and sends them to the operating system to perform.

Step 2. In the terminal, type the command below (Fig. 15). When prompted for a password, type password and hit enter.

```
sudo tc qdisc add dev s1-eth2 root handle 1: netem loss
0.01%
```

Fig. 14 Shortcut to open a Linux terminal

Fig. 15 Adding 0.01% packet loss rate to switch S1's *s1-eth2* interface

Fig. 16 Limiting the bandwidth to 10 Gbps on switch S1's *s1-eth2* interface

Step 3. Modify the bandwidth of the link connecting the switch S1 and switch S2; on the same terminal, type the command below. This command sets the bandwidth to 10 Gbps on switch S1's *s1-eth2* interface (Fig. 16) The $\boxed{\text{tbf}}$ parameters are the following:

- $\boxed{\text{rate}}$: 10gbit
- $\boxed{\text{burst}}$: 5,000,000
- $\boxed{\text{limit}}$: 15,000,000

```
sudo tc qdisc add dev s1-eth2 parent 1: handle 2: tbf rate
10gbit burst 5000000 limit 15000000
```

```
                              "Host: h1"                              _  ↗ ×
root@admin-pc:~# ping 10.0.0.2
PING 10.0.0.2 (10.0.0.2) 56(84) bytes of data.
64 bytes from 10.0.0.2: icmp_seq=1 ttl=64 time=0.869 ms
64 bytes from 10.0.0.2: icmp_seq=2 ttl=64 time=0.075 ms
64 bytes from 10.0.0.2: icmp_seq=3 ttl=64 time=0.064 ms
64 bytes from 10.0.0.2: icmp_seq=4 ttl=64 time=0.068 ms
^C
--- 10.0.0.2 ping statistics ---
4 packets transmitted, 4 received, 0% packet loss, time 64ms
rtt min/avg/max/mdev = 0.064/0.269/0.869/0.346 ms
root@admin-pc:~# █
```

Fig. 17 Output of ping 10.0.0.2 command

7.3   Testing Connection

To test connectivity, you can use the command ping .

Step 1. On the terminal of host h1, type ping 10.0.0.2 . To stop the test, press
 Ctrl+c . The figure below shows a successful connectivity test. Host h1 (10.0.0.1)
sent four packets to host h2 (10.0.0.2), successfully receiving responses back
(Fig. 17).

The result above indicates that all four packets were received successfully (0%
packet loss) and that the minimum, average, maximum, and standard deviation of
the Round-Trip Time (RTT) were 0.064, 0.269, 0.869, and 0.346 ms, respectively.
Essentially, the standard deviation is an average of how far each ping RTT is from
the average RTT. The higher the standard deviation the more variable the RTT is.

Step 2. On the terminal of host h2, type ping 10.0.0.1 . The ping output in this test
should be relatively similar to the results of the test initiated by host h1 in Step 1.
To stop the test, press Ctrl+c .

8   Introduction to sysctl

sysctl is a tool for dynamically changing parameters in the Linux operating system.
It allows users to modify kernel parameters dynamically without rebuilding the
Linux kernel.

Step 1. Run the command below on the Client1's terminal. When prompted for a
password, type password and hit enter (Fig. 18).

```
sudo sysctl -a
```

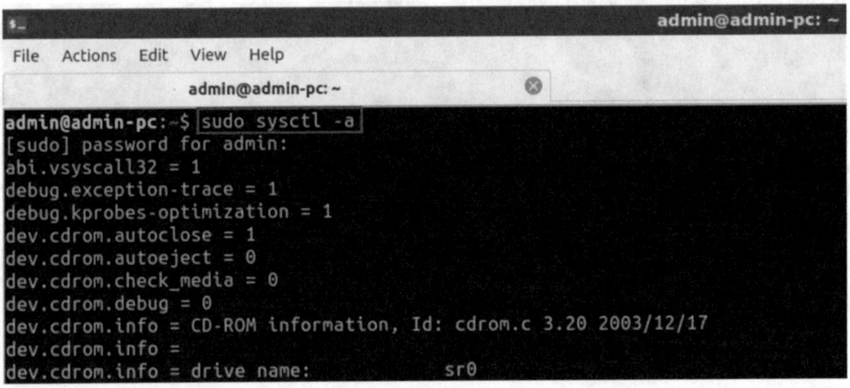

Fig. 18 Listing all system parameters in Linux

Fig. 19 Reading the value of a given key

This command produces a large output containing the kernel parameters and their values. This is represented in a key–value pair. For instance, net.ipv4.ip_forward = 0 implies that the key net.ipv4.ip_forward has the value 0.

8.1 Read sysctl Parameters

It is often useful to search for specific keys without having to manually locate the needed key. This can be achieved using the following command:

```
sysctl <key>
```

where *<key>* is replaced by the needed key. For example, the command sysctl net.ipv4.ip_forward returns net.ipv4.ip_forward = 0.

Step 1. Run the following command on the host h1's terminal (Fig. 19):

```
sysctl net.ipv4.ip_forward
```

```
                          "Host: h1"                    –  ⬡  x
root@admin-pc:~# sudo sysctl -w net.ipv4.ip_forward=1
net.ipv4.ip_forward = 1
root@admin-pc:~# █
```

Fig. 20 Modifying a system parameter

8.2 Write sysctl Parameters

It is also very useful to modify kernel parameters on the fly. The $\boxed{-w}$ switch is added to the sysctl to "write" a value for a specific key.

```
sysctl -w <key>=<value>
```

Step 1. For example, if the user decides to enable IP forwarding (i.e., to configure a device as a router), then the following command is used:

```
sudo sysctl -w net.ipv4.ip_forward=1
```

Run the above command on the host h1's terminal (Fig. 20):

The changes made to a parameter using this command are temporary. Therefore, a new boot resets the value of a key to its default value. Also, when stopping MiniEdit's emulation, the configured parameters are reset.

8.3 Configuring sysctl.conf File

If the user wishes to permanently modify the value of a specific key, then the key–value pair must be stored within the file */etc/sysctl.conf*.

Step 1. In the Linux terminal, open the */etc/sysctl.conf* file using your favorite text editor. Run the following command on the Client1's terminal. When prompted for a password, type $\boxed{\text{password}}$ and hit enter (Fig. 21).

```
sudo featherpad /etc/sysctl.conf
```

This is a text file that can be edited in any text editor ($\boxed{\text{vim}}$, $\boxed{\text{nano}}$, etc.). For simplicity, we use a Graphical User Interface (GUI)-based text editor ($\boxed{\text{featherpad}}$).

Step 2. Keys and values are appended to this file. Enable IP forwarding permanently on the system by appending $\boxed{\text{net.ipv4.ip\_forward=1}}$ to the */etc/sysctl.conf* file and saving it. Once you have saved the file, close the text editor (Fig. 22).

Fig. 21 Opening the */etc/sysctl.conf* file

```
net.ipv4.ip_ forward=1
```

Step 3. To refresh the system with the new parameters, the -p switch is passed to the sysctl command as follows (Fig. 23):

```
sudo sysctl -p
```

When prompted for a password, type password and hit enter.

Now, even after a new system boot (or reboot), the system will have IP forwarding enabled.

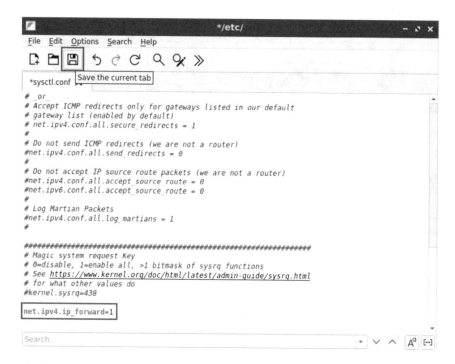

Fig. 22 Appending key+value to the */etc/sysctl.conf* file and saving

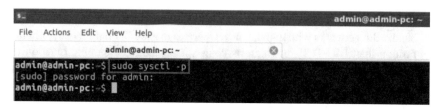

Fig. 23 Loading new *sysctl.conf* parameters

9   Congestion Control Algorithms and sysctl

Congestion control algorithms can be inspected and modified using the sysctl command and the */etc/sysctl.conf* file. Specifically, the following operations are possible:

1. Check the installed congestion control algorithms on the system.
2. Inspect the default congestion control algorithm (i.e., the current algorithm used by the system).
3. Modify the congestion control algorithm.

Fig. 24 Displaying the system's available congestion control algorithms

9.1   Inspect and Install/Load Congestion Control Algorithms

In Linux, it is possible to check the available TCP congestion control algorithms installed on the system with the command below.

Step 1. Execute the command below on the Client1's terminal (Fig. 24).

```
sysctl net.ipv4.tcp_available_congestion_control
```

Usually, the default congestion control algorithm is CUBIC or Reno, depending on the installed operating system. A list of some of the possible output is:

- reno : Traditional TCP used by almost all other operating systems characterized by slow start, congestion avoidance, and fast retransmission via triple duplicate ACKs.
- cubic : CUBIC-TCP optimized congestion control algorithm for high-bandwidth networks with high latency. Operates in a similar but more systematic fashion than BIC-TCP, in which its congestion window is a cubic function of time since the last packet loss, with the inflection point set to the window prior to the congestion event.
- bic : BIC-TCP congestion window utilizes a binary search algorithm to find the largest congestion window that will last the maximum amount of time.
- htcp : Hamilton TCP A loss-based algorithm using additive increase and multiplicative decrease to control TCP's congestion window.
- vegas : TCP Vegas emphasizes packet delay, rather than packet loss, as a signal to help determine the rate at which to send packets.
- bbr : a new algorithm, discussed in future labs. Measures bottleneck bandwidth and Round-Trip Propagation (RTP) time in its execution of congestion control.

If the above command does not return a specific congestion control algorithm, it means that it is not loaded on the distribution.

Step 2. The command used in Step 1 listed three algorithms: reno cubic bbr . To install a new algorithm, its corresponding kernel module must be loaded. This can be done using insmod or modprobe commands. For example, to load the BIC-TCP module, use the following command on the Client1's terminal (Fig. 25):

Fig. 25 Loading `tcp_bic` module into the Linux kernel

Fig. 26 Displaying the system's available congestion control algorithms after loading TCP-BIC

```
sudo modprobe tcp_bic
```

`modprobe` and `insmod` commands require high `sudo` privileges to insert kernel modules. When prompted for a password, type `password` and hit enter.

Step 3. To verify that the BIC-TCP algorithm is loaded, execute the below command on the Client1's terminal (Fig. 26).

```
sysctl net.ipv4.tcp_available_congestion_control
```

9.2 Inspect the Default (Current) Congestion Control Algorithm

To check which TCP congestion control is currently being used by the Linux kernel, the *net.ipv4.tcp_congestion_control sysctl* key is read. This key can be read on an end-host's terminal (host h1 or host h2) or on the Client1's terminal.

Step 1. Execute the following command on the Client1's terminal to determine the current congestion control algorithm (Fig. 27).

```
sysctl net.ipv4.tcp_congestion_control
```

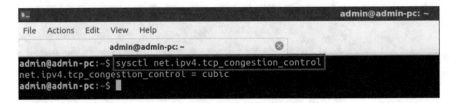

Fig. 27 Current TCP congestion control algorithm

```
admin@admin-pc:~$ sudo sysctl -w net.ipv4.tcp_congestion_control=reno
[sudo] password for admin:
net.ipv4.tcp_congestion_control = reno
admin@admin-pc:~$ 
```

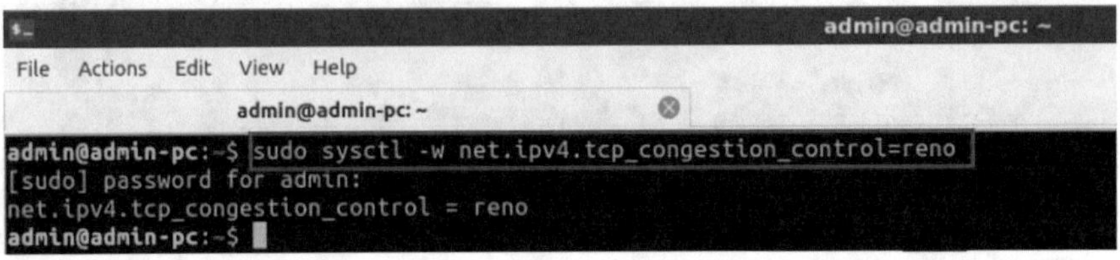

Fig. 28 Modifying the congestion control algorithm to reno

The output shows that the default congestion control algorithm is Cubic. Note that applications can set this value (congestion control algorithm) for individual connections.

9.3 Modify the Default (Current) Congestion Control Algorithm

To temporarily change the TCP congestion control algorithm, the sysctl command is used with the -w switch on the *net.ipv4.tcp_congestion_control* key.

Step 1. To modify the current algorithm to TCP Reno, the following command is used. Execute the command below on the Client1's terminal. When prompted for a password, type password and hit enter (Fig. 28).

```
sudo sysctl -w net.ipv4.tcp_congestion_control=reno
```

If no error occurred in the assignment (e.g., the module is not installed on the system), the output echoes back the new key–value pair, i.e.,

```
net.ipv4.tcp_ congestion_control=reno
```

Step 2. Execute the following command on the Client1's terminal to determine the current congestion control algorithm (Fig. 29).

Fig. 29 Current TCP congestion control algorithm after modifying to `reno`

```
sysctl net.ipv4.tcp_congestion_control
```

The output shows that the default congestion control algorithm is now Reno instead of Cubic.

10   iPerf3 Throughput Test

In this section, the throughput between host h1 and host h2 is measured using different congestion control algorithms, namely Reno, HTCP, and Cubic. Moreover, the test is repeated using various injected delays to observe the throughput variations depending on each congestion control algorithm and the selected RTT.

10.1   Throughput Test Without Delay

In this test, we measure the throughput between host h1 and host h2 without introducing delay on the switch S1's *s1-eth2* interface.

10.2   TCP Reno

Step 1. In host h1's terminal, change the TCP congestion control algorithm to Reno by typing the following command (Fig. 30):

```
sysctl -w net.ipv4.tcp_congestion_control=reno
```

Step 2. Launch iPerf3 in server mode on host h2's terminal (Fig. 31):

```
iperf3 -s
```

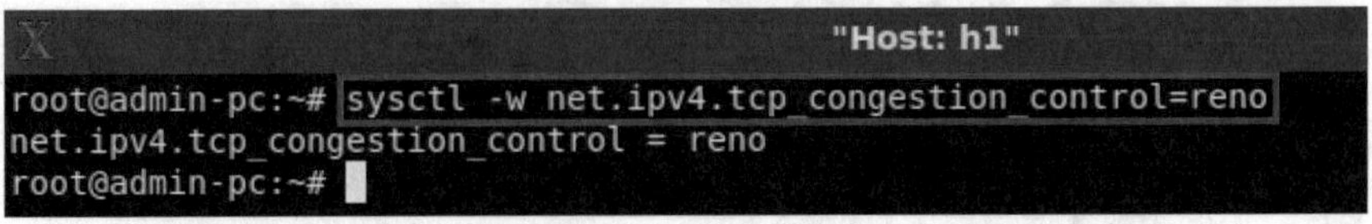

```
X                                           "Host: h1"
root@admin-pc:~# sysctl -w net.ipv4.tcp_congestion_control=reno
net.ipv4.tcp_congestion_control = reno
root@admin-pc:~# ▮
```

Fig. 30 Changing TCP congestion control algorithm to ⎡reno⎤ on host h1

```
X                                    "Host: h2"                    – ↘ ×
root@admin-pc:~# iperf3 -s
- - - - - - - - - - - - - - - - - - - - - - - - - - - - - - - - - - - - - - -
Server listening on 5201
- - - - - - - - - - - - - - - - - - - - - - - - - - - - - - - - - - - - - - -
▮
```

Fig. 31 Starting iPerf3 server on host h2

Step 3. Launch iPerf3 in client mode on host h1's terminal. The ⎡-O⎤ option is used to specify the number of seconds to omit in the resulting report. Note that this option is a capitalized "O," not a zero (Fig. 32).

```
iperf3 -c 10.0.0.2 -t 20 -O 10
```

The figure above shows the iPerf3 test output report. The average achieved throughputs are 9.56 Gbps (sender) and 9.56 Gbps (receiver), and the number of retransmissions is 1890 (due to the injected packet loss—001%).

Step 4. In order to stop the server, press ⎡Ctrl+c⎤ in host h2's terminal. The user can see the throughput results in the server side too.

10.3   Hamilton TCP (HTCP)

Step 1. In host h1's terminal, change the TCP congestion control algorithm to HTCP by typing the following command (Fig. 33):

```
sysctl -w net.ipv4.tcp_congestion_control=htcp
```

Step 2. Launch iPerf3 in server mode on host h2's terminal (Fig. 34):

```
iperf3 -s
```

Fig. 32 Running iPerf3 client on host h1

Fig. 33 Changing TCP congestion control algorithm to htcp on host h1

Fig. 34 Starting iPerf3 server on host h2

Step 3. Launch iPerf3 in client mode on host h1's terminal (Fig. 35):

```
iperf3 -c 10.0.0.2 -t 20 -O 10
```

The figure above shows the iPerf3 test output report. The average achieved throughputs are 9.56 Gbps (sender) and 9.56 Gbps (receiver), and the number of retransmissions is 1789 (due to the injected packet loss—0.01%).

Step 4. In order to stop the server, press $\boxed{\text{Ctrl+c}}$ in host h2's terminal. The user can see the throughput results in the server side too.

10.4　TCP Cubic

Step 1. In host h1's terminal, change the TCP congestion control algorithm to Cubic by typing the following command (Fig. 36):

```
sysctl -w net.ipv4.tcp_congestion_control=cubic
```

Step 2. Launch iPerf3 in server mode on host h2's terminal (Fig. 37):

```
iperf3 -s
```

Step 3. Launch iPerf3 in client mode on host h1's terminal (Fig. 38):

```
iperf3 -c 10.0.0.2 -t 20 -O 10
```

The figure above shows the iPerf3 test output report. The average achieved throughputs are 9.56 Gbps (sender) and 9.56 Gbps (receiver), and the number of retransmissions is 1845 (due to the injected packet loss—0.01%).

Step 4. In order to stop the server, press $\boxed{\text{Ctrl+c}}$ in host h2's terminal. The user can see the throughput results in the server side too.

```
X                                "Host: h1"                         - , x
root@admin-pc:~# iperf3 -c 10.0.0.2 -t 20 -O 10
Connecting to host 10.0.0.2, port 5201
[ 15] local 10.0.0.1 port 34494 connected to 10.0.0.2 port 5201
[ ID] Interval           Transfer     Bitrate         Retr  Cwnd
[ 15]   0.00-1.00   sec  1.13 GBytes  9.69 Gbits/sec  158   4.16 MBytes       (omitted)
[ 15]   1.00-2.00   sec  1.11 GBytes  9.57 Gbits/sec   45   2.49 MBytes       (omitted)
[ 15]   2.00-3.00   sec  1.11 GBytes  9.56 Gbits/sec   90   1.45 MBytes       (omitted)
[ 15]   3.00-4.00   sec  1.11 GBytes  9.56 Gbits/sec  225   956 KBytes        (omitted)
[ 15]   4.00-5.00   sec  1.11 GBytes  9.57 Gbits/sec  135   713 KBytes        (omitted)
[ 15]   5.00-6.00   sec  1.11 GBytes  9.56 Gbits/sec    0   1.85 MBytes       (omitted)
[ 15]   6.00-7.00   sec  1.11 GBytes  9.56 Gbits/sec    0   2.54 MBytes       (omitted)
[ 15]   7.00-8.00   sec  1.11 GBytes  9.57 Gbits/sec   90   1.27 MBytes       (omitted)
[ 15]   8.00-9.00   sec  1.11 GBytes  9.56 Gbits/sec   90   1.44 MBytes       (omitted)
[ 15]   9.00-10.00  sec  1.11 GBytes  9.56 Gbits/sec   45   1.68 MBytes       (omitted)
[ 15]   0.00-1.00   sec  1.11 GBytes  9.56 Gbits/sec   45   1.38 MBytes
[ 15]   1.00-2.00   sec  1.11 GBytes  9.56 Gbits/sec   90   1.61 MBytes
[ 15]   2.00-3.00   sec  1.11 GBytes  9.56 Gbits/sec   45   1.43 MBytes
[ 15]   3.00-4.00   sec  1.11 GBytes  9.56 Gbits/sec   45   1.40 MBytes
[ 15]   4.00-5.00   sec  1.11 GBytes  9.56 Gbits/sec   45   1.77 MBytes
[ 15]   5.00-6.00   sec  1.11 GBytes  9.56 Gbits/sec  135   781 KBytes
[ 15]   6.00-7.00   sec  1.11 GBytes  9.56 Gbits/sec  135   1.51 MBytes
[ 15]   7.00-8.00   sec  1.11 GBytes  9.56 Gbits/sec    0   2.30 MBytes
[ 15]   8.00-9.00   sec  1.11 GBytes  9.57 Gbits/sec    0   2.89 MBytes
[ 15]   9.00-10.00  sec  1.11 GBytes  9.56 Gbits/sec  135   1.14 MBytes
[ 15]  10.00-11.00  sec  1.11 GBytes  9.56 Gbits/sec   90   1.03 MBytes
[ 15]  11.00-12.00  sec  1.11 GBytes  9.56 Gbits/sec  214   696 KBytes
[ 15]  12.00-13.00  sec  1.11 GBytes  9.55 Gbits/sec  135   1.26 MBytes
[ 15]  13.00-14.00  sec  1.11 GBytes  9.56 Gbits/sec  270   621 KBytes
[ 15]  14.00-15.00  sec  1.11 GBytes  9.56 Gbits/sec    0   1.81 MBytes
[ 15]  15.00-16.00  sec  1.11 GBytes  9.56 Gbits/sec   45   1.90 MBytes
[ 15]  16.00-17.00  sec  1.11 GBytes  9.56 Gbits/sec  225   622 KBytes
[ 15]  17.00-18.00  sec  1.11 GBytes  9.56 Gbits/sec    0   1.81 MBytes
[ 15]  18.00-19.00  sec  1.11 GBytes  9.56 Gbits/sec   90   1.14 MBytes
[ 15]  19.00-20.00  sec  1.11 GBytes  9.56 Gbits/sec   45   1.51 MBytes
- - - - - - - - - - - - - - - - - - - - - - - - -
[ ID] Interval           Transfer     Bitrate         Retr
[ 15]   0.00-20.00  sec  22.3 GBytes  9.56 Gbits/sec  1789            sender
[ 15]   0.00-20.04  sec  22.3 GBytes  9.56 Gbits/sec                  receiver

iperf Done.
root@admin-pc:~#
```

Fig. 35 Running iPerf3 client on host h1

```
X                                "Host: h1"
root@admin-pc:~# sysctl -w net.ipv4.tcp_congestion_control=cubic
net.ipv4.tcp_congestion_control = cubic
root@admin-pc:~#
```

Fig. 36 Changing TCP congestion control algorithm to cubic on host h1

```
X                                "Host: h2"                         - , x
root@admin-pc:~# iperf3 -s
----------------------------------------------------------------
Server listening on 5201
----------------------------------------------------------------
```

Fig. 37 Starting iPerf3 server on host h2

Fig. 38 Running iPerf3 client on host h1

10.5   Throughput Test with 30 ms Delay

In this test, we measure the throughput between host h1 and host h2 while introducing 30 ms delay on the switch S1's *s1-eth2* interface. Apply the following steps:

Step 1. On the client's terminal, run the following command to modify the previous rule to include 30 ms delay. When prompted for a password, type `password` and hit enter (Fig. 39).

```
sudo tc qdisc change dev s1-eth2 root handle 1: netem loss
0.01% delay 30 ms
```

Fig. 39 Injecting 30 ms delay on switch S1's *s1-eth2* interface

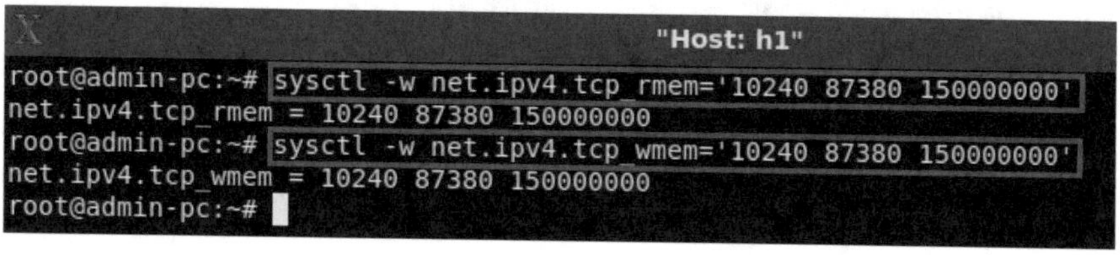

```
"Host: h1"
root@admin-pc:~# sysctl -w net.ipv4.tcp_rmem='10240 87380 150000000'
net.ipv4.tcp_rmem = 10240 87380 150000000
root@admin-pc:~# sysctl -w net.ipv4.tcp_wmem='10240 87380 150000000'
net.ipv4.tcp_wmem = 10240 87380 150000000
root@admin-pc:~#
```

Fig. 40 Modifying the TCP buffer size on host h1

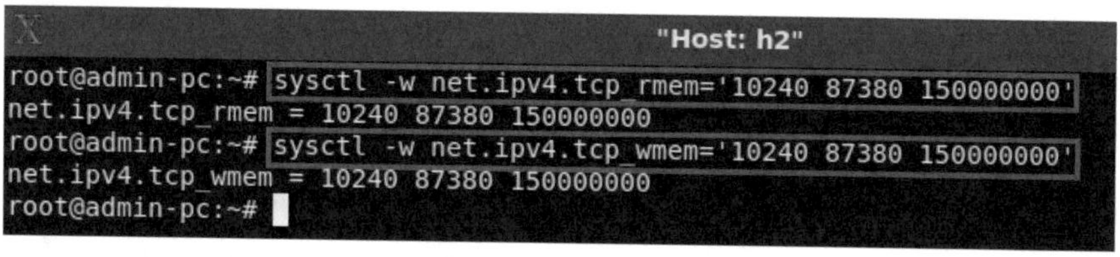

```
"Host: h2"
root@admin-pc:~# sysctl -w net.ipv4.tcp_rmem='10240 87380 150000000'
net.ipv4.tcp_rmem = 10240 87380 150000000
root@admin-pc:~# sysctl -w net.ipv4.tcp_wmem='10240 87380 150000000'
net.ipv4.tcp_wmem = 10240 87380 150000000
root@admin-pc:~#
```

Fig. 41 Modifying the TCP buffer size on host h2

Step 2. In host h1's terminal, modify the TCP buffer size by typing the following commands: *sysctl -w net.ipv4.tcp_rmem='10,240 87,380 150,000,000'* and *sysctl -w net.ipv4.tcp_wmem='10,240 87,380 150,000,000'*. This TCP buffer is explained later in future labs (Fig. 40).

```
sysctl -w net.ipv4.tcp_rmem='10240 87380 150000000'
```

```
sysctl -w net.ipv4.tcp_wmem='10240 87380 150000000'
```

Step 3. In host h2's terminal, also modify the TCP buffer size by typing the following commands: *sysctl -w net.ipv4.tcp_rmem='10,240 87,380 150,000,000'* and *sysctl -w net.ipv4.tcp_wmem='10,240 87,380 150,000,000'* (Fig. 41).

10.6   TCP Reno

Step 1. In host h1's terminal, change the TCP congestion control algorithm to Reno by typing the following command (Fig. 42):

```
X                                              "Host: h1"
root@admin-pc:~# sysctl -w net.ipv4.tcp_congestion_control=reno
net.ipv4.tcp_congestion_control = reno
root@admin-pc:~# ▊
```

Fig. 42 Changing TCP congestion control algorithm to reno on host h1

```
X                                    "Host: h2"                        — ▵ ✕
root@admin-pc:~# iperf3 -s
---------------------------------------------------------------------
Server listening on 5201
---------------------------------------------------------------------
▊
```

Fig. 43 Starting iPerf3 server on host h2

```
sysctl -w net.ipv4.tcp_congestion_control=reno
```

Step 2. Launch iPerf3 in server mode on host h2's terminal (Fig. 43):

```
iperf3 -s
```

Step 3. Launch iPerf3 in client mode on host h1's terminal. The -O option is used to specify the number of seconds to omit in the resulting report (Fig. 44).

```
iperf3 -c 10.0.0.2 -t 20 -O 10
```

The figure above shows the iPerf3 test output report. The average achieved throughputs are 472 Mbps (sender) and 472 Mbps (receiver), and the number of retransmissions is 45.

Step 4. In order to stop the server, press Ctrl+c in host h2's terminal. The user can see the throughput results in the server side too.

10.7 Hamilton TCP (HTCP)

Step 1. In host h1's terminal, change the TCP congestion control algorithm to HTCP by typing the following command (Fig. 45):

```
sysctl -w net.ipv4.tcp_congestion_control=htcp
```

```
X                               "Host: h1"                        - ↗ x
root@admin-pc:~# iperf3 -c 10.0.0.2 -t 20 -O 10
Connecting to host 10.0.0.2, port 5201
[ 15] local 10.0.0.1 port 47044 connected to 10.0.0.2 port 5201
[ ID] Interval           Transfer     Bitrate         Retr  Cwnd
[ 15]   0.00-1.00   sec   527 MBytes  4.42 Gbits/sec  5134   19.5 MBytes      (omitted)
[ 15]   1.00-2.00   sec   352 MBytes  2.96 Gbits/sec     0   10.0 MBytes      (omitted)
[ 15]   2.00-3.00   sec   335 MBytes  2.81 Gbits/sec     0   10.1 MBytes      (omitted)
[ 15]   3.00-4.00   sec   336 MBytes  2.82 Gbits/sec     0   10.1 MBytes      (omitted)
[ 15]   4.00-5.00   sec   314 MBytes  2.63 Gbits/sec    45   5.08 MBytes      (omitted)
[ 15]   5.00-6.00   sec   145 MBytes  1.22 Gbits/sec     0   5.12 MBytes      (omitted)
[ 15]   6.00-7.00   sec   134 MBytes  1.12 Gbits/sec     0   5.16 MBytes      (omitted)
[ 15]   7.00-8.00   sec  56.2 MBytes   472 Mbits/sec     0   1.74 MBytes
[ 15]   8.00-9.00   sec  57.5 MBytes   482 Mbits/sec     0   1.78 MBytes
[ 15]   9.00-10.00  sec  58.8 MBytes   493 Mbits/sec     0   1.83 MBytes
[ 15]  10.00-11.00  sec  61.2 MBytes   514 Mbits/sec     0   1.87 MBytes
[ 15]  11.00-12.00  sec  61.2 MBytes   514 Mbits/sec     0   1.92 MBytes
[ 15]  12.00-13.00  sec  63.8 MBytes   535 Mbits/sec     0   1.96 MBytes
[ 15]  13.00-14.00  sec  65.0 MBytes   545 Mbits/sec     0   2.01 MBytes
[ 15]  14.00-15.00  sec  66.2 MBytes   556 Mbits/sec     0   2.05 MBytes
[ 15]  15.00-16.00  sec  67.5 MBytes   566 Mbits/sec     0   2.10 MBytes
[ 15]  16.00-17.00  sec  70.0 MBytes   587 Mbits/sec     0   2.14 MBytes
[ 15]  17.00-18.00  sec  71.2 MBytes   598 Mbits/sec     0   2.19 MBytes
[ 15]  18.00-19.00  sec  40.0 MBytes   335 Mbits/sec    45   1.14 MBytes
[ 15]  19.00-20.00  sec  37.5 MBytes   315 Mbits/sec     0   1.18 MBytes
- - - - - - - - - - - - - - - - - - - - - - - - -
[ ID] Interval           Transfer     Bitrate         Retr
[ 15]   0.00-20.00  sec  1.10 GBytes   472 Mbits/sec    45              sender
[ 15]   0.00-20.04  sec  1.10 GBytes   472 Mbits/sec                    receiver

iperf Done.
root@admin-pc:~# ▮
```

Fig. 44   Running iPerf3 client on host h1

```
X                                                    "Host: h1"
root@admin-pc:~# sysctl -w net.ipv4.tcp_congestion_control=htcp
net.ipv4.tcp_congestion_control = htcp
```

Fig. 45   Changing TCP congestion control algorithm to htcp on host h1

```
X                               "Host: h2"                        - ↗ x
root@admin-pc:~# iperf3 -s
- - - - - - - - - - - - - - - - - - - - - - - - - - - - - - - - -
Server listening on 5201
- - - - - - - - - - - - - - - - - - - - - - - - - - - - - - - - -
```

Fig. 46   Starting iPerf3 server on host h2

Step 2. Launch iPerf3 in server mode on host h2's terminal (Fig. 46):

```
iperf3 -s
```

Step 3. Launch iPerf3 in client mode on host h1's terminal (Fig. 47):

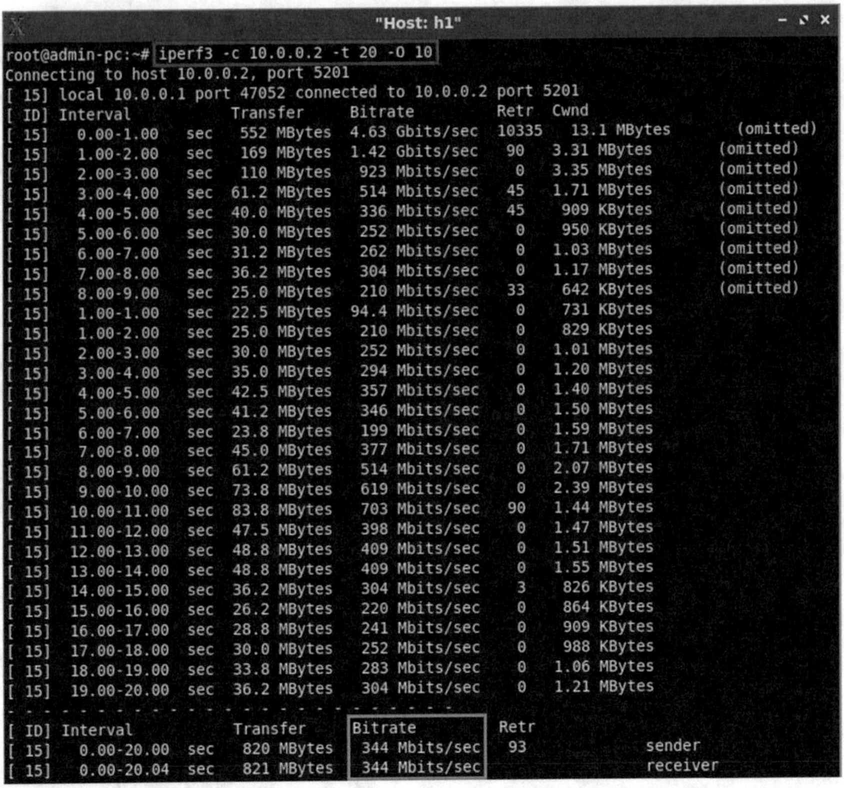

```
                                    "Host: h1"                         -  ⌕ ×
root@admin-pc:~# iperf3 -c 10.0.0.2 -t 20 -O 10
Connecting to host 10.0.0.2, port 5201
[ 15] local 10.0.0.1 port 47052 connected to 10.0.0.2 port 5201
[ ID] Interval          Transfer     Bitrate       Retr  Cwnd
[ 15]  0.00-1.00   sec   552 MBytes  4.63 Gbits/sec 10335  13.1 MBytes       (omitted)
[ 15]  1.00-2.00   sec   169 MBytes  1.42 Gbits/sec   90   3.31 MBytes       (omitted)
[ 15]  2.00-3.00   sec   110 MBytes   923 Mbits/sec    0   3.35 MBytes       (omitted)
[ 15]  3.00-4.00   sec  61.2 MBytes   514 Mbits/sec   45   1.71 MBytes       (omitted)
[ 15]  4.00-5.00   sec  40.0 MBytes   336 Mbits/sec   45    909 KBytes       (omitted)
[ 15]  5.00-6.00   sec  30.0 MBytes   252 Mbits/sec    0    950 KBytes       (omitted)
[ 15]  6.00-7.00   sec  31.2 MBytes   262 Mbits/sec    0   1.03 MBytes       (omitted)
[ 15]  7.00-8.00   sec  36.2 MBytes   304 Mbits/sec    0   1.17 MBytes       (omitted)
[ 15]  8.00-9.00   sec  25.0 MBytes   210 Mbits/sec   33    642 KBytes       (omitted)
[ 15]  1.00-1.00   sec  22.5 MBytes  94.4 Mbits/sec    0    731 KBytes
[ 15]  1.00-2.00   sec  25.0 MBytes   210 Mbits/sec    0    829 KBytes
[ 15]  2.00-3.00   sec  30.0 MBytes   252 Mbits/sec    0   1.01 MBytes
[ 15]  3.00-4.00   sec  35.0 MBytes   294 Mbits/sec    0   1.20 MBytes
[ 15]  4.00-5.00   sec  42.5 MBytes   357 Mbits/sec    0   1.40 MBytes
[ 15]  5.00-6.00   sec  41.2 MBytes   346 Mbits/sec    0   1.50 MBytes
[ 15]  6.00-7.00   sec  23.8 MBytes   199 Mbits/sec    0   1.59 MBytes
[ 15]  7.00-8.00   sec  45.0 MBytes   377 Mbits/sec    0   1.71 MBytes
[ 15]  8.00-9.00   sec  61.2 MBytes   514 Mbits/sec    0   2.07 MBytes
[ 15]  9.00-10.00  sec  73.8 MBytes   619 Mbits/sec    0   2.39 MBytes
[ 15] 10.00-11.00  sec  83.8 MBytes   703 Mbits/sec   90   1.44 MBytes
[ 15] 11.00-12.00  sec  47.5 MBytes   398 Mbits/sec    0   1.47 MBytes
[ 15] 12.00-13.00  sec  48.8 MBytes   409 Mbits/sec    0   1.51 MBytes
[ 15] 13.00-14.00  sec  48.8 MBytes   409 Mbits/sec    0   1.55 MBytes
[ 15] 14.00-15.00  sec  36.2 MBytes   304 Mbits/sec    3    826 KBytes
[ 15] 15.00-16.00  sec  26.2 MBytes   220 Mbits/sec    0    864 KBytes
[ 15] 16.00-17.00  sec  28.8 MBytes   241 Mbits/sec    0    909 KBytes
[ 15] 17.00-18.00  sec  30.0 MBytes   252 Mbits/sec    0    988 KBytes
[ 15] 18.00-19.00  sec  33.8 MBytes   283 Mbits/sec    0   1.06 MBytes
[ 15] 19.00-20.00  sec  36.2 MBytes   304 Mbits/sec    0   1.21 MBytes
- - - - - - - - - - - - - - - - - - - - - - - - -
[ ID] Interval          Transfer     Bitrate       Retr
[ 15]  0.00-20.00  sec   820 MBytes   344 Mbits/sec   93            sender
[ 15]  0.00-20.04  sec   821 MBytes   344 Mbits/sec                 receiver
```

Fig. 47 Running iPerf3 client on host h1

```
iperf3 -c 10.0.0.2 -t 20 -O 10
```

The figure above shows the iPerf3 test output report. The average achieved throughputs are 344 Mbps (sender) and 344 Mbps (receiver), and the number of retransmissions is 93.

Step 4. In order to stop the server, press Ctrl+c in host h2's terminal. The user can see the throughput results in the server side too.

10.8 TCP Cubic

Step 1. In host h1's terminal, change the TCP congestion control algorithm to Cubic by typing the following command (Fig. 48):

```
sysctl -w net.ipv4.tcp_congestion_control=cubic
```

Fig. 48 Changing TCP congestion control algorithm to cubic on host h1

Fig. 49 Starting iPerf3 server on host h2

Step 2. Launch iPerf3 in server mode on host h2's terminal (Fig. 49):

```
iperf3 -s
```

Step 3. Launch iPerf3 in client mode on host h1's terminal (Fig. 50):

```
iperf3 -c 10.0.0.2 -t 20 -O 10
```

The figure above shows the iPerf3 test output report. The average achieved throughputs are 938 Mbps (sender) and 939 Mbps (receiver), and the number of retransmissions is 180.

Step 4. In order to stop the server, press Ctrl+c in host h2's terminal. The user can see the throughput results in the server side too.

11   Chapter 4—Lab 10: Understanding Rate-Based TCP Congestion Control (BBR)

Overview

To conduct the experiment described in this section, please login into the Academic Cloud at http://highspeednetworks.net/, and reserve a pod for Lab 10.

This lab describes a new type of TCP congestion control algorithm called bottleneck bandwidth and Round-Trip Time (BBR). The lab conducts experimental results using TCP BBR and contrasts these results with those obtained using traditional congestion control algorithms such as a Reno and HTCP.

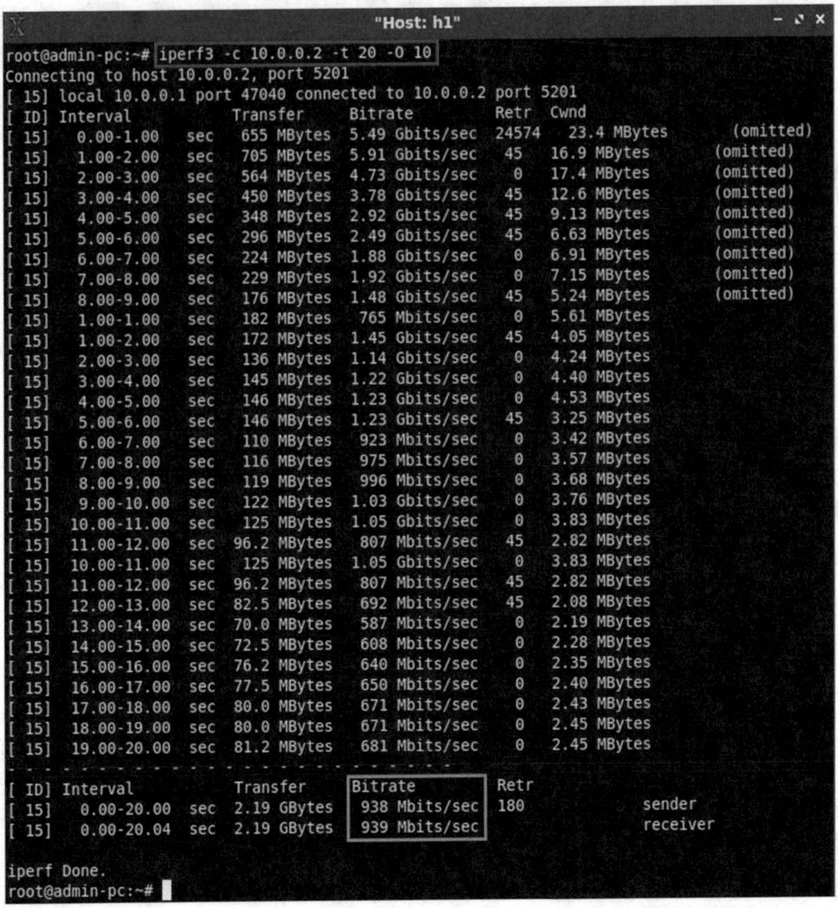

```
X                                "Host: h1"                          -  e x
root@admin-pc:~# iperf3 -c 10.0.0.2 -t 20 -O 10
Connecting to host 10.0.0.2, port 5201
[ 15] local 10.0.0.1 port 47040 connected to 10.0.0.2 port 5201
[ ID] Interval           Transfer     Bitrate       Retr  Cwnd
[ 15]   0.00-1.00   sec   655 MBytes  5.49 Gbits/sec  24574   23.4 MBytes       (omitted)
[ 15]   1.00-2.00   sec   705 MBytes  5.91 Gbits/sec   45    16.9 MBytes       (omitted)
[ 15]   2.00-3.00   sec   564 MBytes  4.73 Gbits/sec    0    17.4 MBytes       (omitted)
[ 15]   3.00-4.00   sec   450 MBytes  3.78 Gbits/sec   45    12.6 MBytes       (omitted)
[ 15]   4.00-5.00   sec   348 MBytes  2.92 Gbits/sec   45     9.13 MBytes      (omitted)
[ 15]   5.00-6.00   sec   296 MBytes  2.49 Gbits/sec   45     6.63 MBytes      (omitted)
[ 15]   6.00-7.00   sec   224 MBytes  1.88 Gbits/sec    0     6.91 MBytes      (omitted)
[ 15]   7.00-8.00   sec   229 MBytes  1.92 Gbits/sec    0     7.15 MBytes      (omitted)
[ 15]   8.00-9.00   sec   176 MBytes  1.48 Gbits/sec   45     5.24 MBytes      (omitted)
[ 15]   1.00-1.00   sec   182 MBytes   765 Mbits/sec    0     5.61 MBytes
[ 15]   1.00-2.00   sec   172 MBytes  1.45 Gbits/sec   45     4.05 MBytes
[ 15]   2.00-3.00   sec   136 MBytes  1.14 Gbits/sec    0     4.24 MBytes
[ 15]   3.00-4.00   sec   145 MBytes  1.22 Gbits/sec    0     4.40 MBytes
[ 15]   4.00-5.00   sec   146 MBytes  1.23 Gbits/sec    0     4.53 MBytes
[ 15]   5.00-6.00   sec   146 MBytes  1.23 Gbits/sec   45     3.25 MBytes
[ 15]   6.00-7.00   sec   110 MBytes   923 Mbits/sec    0     3.42 MBytes
[ 15]   7.00-8.00   sec   116 MBytes   975 Mbits/sec    0     3.57 MBytes
[ 15]   8.00-9.00   sec   119 MBytes   996 Mbits/sec    0     3.68 MBytes
[ 15]   9.00-10.00  sec   122 MBytes  1.03 Gbits/sec    0     3.76 MBytes
[ 15]  10.00-11.00  sec   125 MBytes  1.05 Gbits/sec    0     3.83 MBytes
[ 15]  11.00-12.00  sec  96.2 MBytes   807 Mbits/sec   45     2.82 MBytes
[ 15]  10.00-11.00  sec   125 MBytes  1.05 Gbits/sec    0     3.83 MBytes
[ 15]  11.00-12.00  sec  96.2 MBytes   807 Mbits/sec   45     2.82 MBytes
[ 15]  12.00-13.00  sec  82.5 MBytes   692 Mbits/sec   45     2.08 MBytes
[ 15]  13.00-14.00  sec  70.0 MBytes   587 Mbits/sec    0     2.19 MBytes
[ 15]  14.00-15.00  sec  72.5 MBytes   608 Mbits/sec    0     2.28 MBytes
[ 15]  15.00-16.00  sec  76.2 MBytes   640 Mbits/sec    0     2.35 MBytes
[ 15]  16.00-17.00  sec  77.5 MBytes   650 Mbits/sec    0     2.40 MBytes
[ 15]  17.00-18.00  sec  80.0 MBytes   671 Mbits/sec    0     2.43 MBytes
[ 15]  18.00-19.00  sec  80.0 MBytes   671 Mbits/sec    0     2.45 MBytes
[ 15]  19.00-20.00  sec  81.2 MBytes   681 Mbits/sec    0     2.45 MBytes
- - - - - - - - - - - - - - - - - - - - - - - - -
[ ID] Interval           Transfer     Bitrate       Retr
[ 15]   0.00-20.00  sec  2.19 GBytes   938 Mbits/sec  180              sender
[ 15]   0.00-20.04  sec  2.19 GBytes   939 Mbits/sec                   receiver

iperf Done.
root@admin-pc:~# █
```

Fig. 50 Running iPerf3 client on host h1

Objectives

By the end of this lab, students should be able to:

1. Describe the basic operation of TCP BBR.
2. Describe differences between rate-based congestion control and window-based loss-based congestion control.
3. Modify the TCP congestion control algorithm in Linux using sysctl tool.
4. Compare the throughput performance of TCP Reno and BBR in high-throughput high-latency networks.

Lab Settings

The information in Table 3 provides the credentials of the machine containing Mininet.

Table 3 Credentials to
access Client1 machine

Device	Account	Password
Client1	admin	password

Lab Roadmap

This lab is organized as follows:

1. Section 12: Introduction to TCP
2. Section 13: Lab topology
3. Section 14: iPerf3 throughput test

12 Introduction to TCP

12.1 Traditional TCP Congestion Control Review

TCP congestion control was introduced in the late 1980s. For many years, the main algorithm of congestion control was TCP Reno. Subsequently, multiple algorithms were proposed based on Reno's enhancements. The goal of congestion control is to determine how much capacity is available in the network, so that a source knows how many packets it can safely have in transit (inflight). Once a source has these packets in transit, it uses the arrival of an acknowledgment (ACK) as a signal that one of its packets has left the network and that it is therefore safe to insert a new packet into the network without adding to the level of congestion. By using ACKs to pace the transmission of packets, TCP is said to be self-clocking.

A major task of the congestion control algorithm is to determine the available capacity. In steady state, TCP Reno maintains an estimate of the Round-Trip Time (RTT)—the time to send a packet and receive the corresponding ACK. If the ACK stream shows that no packets are lost in transit, Reno increases the sending rate by one additional segment each RTT interval. This period is known as the additive increase. Note that "segment" here refers to the protocol data unit (PDU) at the transport layer and that sometimes the terms packet and segment are interchangeably used. Eventually, the increasing flow rate saturates the bottleneck link at a router, which drops a packet. The TCP receiver signals the missing packet by sending an ACK in response to an out-of-order received segment, as illustrated in Fig. 51a. Once the TCP sender receives three duplicate ACKs for the same out-of-order segment, it interprets this event as packet loss due to congestion and reduces the sending rate by half. This reduction is known as multiplicative decrease. Once the loss is repaired, Reno resumes the additive increase phase. This iteration of additive increase multiplicative decrease (AIMD) periods is shown in Fig. 51b.

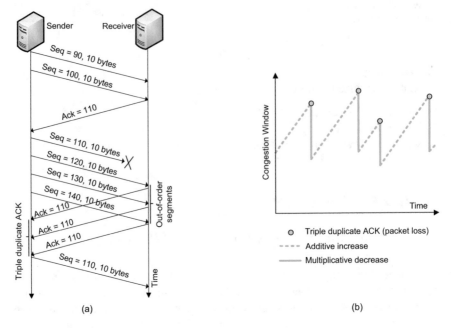

Fig. 51 (**a**) TCP operation. (**b**) Evolution of TCP's congestion window

12.2 Traditional Congestion Control Limitations

While Reno has proven to perform adequately in the past when the bulk of the TCP connections carried trivial applications such as web browsing and email, it faces severe limitations in high-throughput connections that are needed for grid computing and big science data transfers. Reno's average TCP throughput can be approximated by the following equation:

$$\text{TCP Throughput} \approx \frac{\text{MSS}}{\text{RTT}\sqrt{L}} \text{ [bytes/second]}$$

The equation above indicates that the throughput of a TCP connection in steady state is directly proportional to the maximum segment size (MSS) and inversely proportional to the product of Round-Trip Time (RTT) and the square root of the packet loss rate (L). Essentially, the equation above indicates that the TCP throughput is very sensitive to packet loss. In such environments Reno cannot achieve high throughput, especially in high-latency scenarios. Figure 52 validates the above equation. It shows the throughput as a function of RTT, for two devices connected by a 10 Gbps path. The performance of two TCP AIMD-based implementations is provided: Reno (blue) and Hamilton TCP, better known

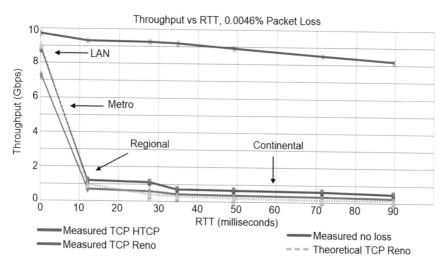

Fig. 52 Throughput vs. Round-Trip Time (RTT) for two devices connected via a 10 Gbps path. The performance of two TCP implementations are provided: Reno (blue) and HTCP (red). The theoretical performance with packet losses (green) and the measured throughput without packet losses (purple) are also shown

as HTCP (red). The theoretical performance (using the above equation) with packet losses (green) and the measured throughput without packet losses (purple) are also shown.

12.3   TCP BBR

The main issue surrounding traditional congestion control algorithms in high-speed high-latency networks is that the sender cannot recover from the packet loss and multiplicative decrease, even when the packet losses are sporadic. When the RTT is large, increasing the congestion window (and thus the sending rate) by only 1 MSS every RTT is too slow.

BBR is a new congestion control algorithm that does not adhere to the AIMD rule and the above equation. BBR is a rate-based algorithm, meaning that at any given time it sends data at a rate that is independent of current packet losses. Note that this feature is a drastic departure from traditional congestion control algorithms, which operate by reducing the sending rate by half each time a packet loss is detected.

The behavior of BBR can be described using Fig. 53, which shows a TCP's viewpoint of an end-to-end connection. At any time, the connection has exactly one slowest link or bottleneck bandwidth (*btlbw*) that determines the location where queues are formed. When router buffers are large, traditional congestion control keeps them full (i.e., they keep increasing the rate during the additive increase phase). When buffers are small, traditional congestion control misinterprets a loss as a signal of congestion, leading to low throughput. The output port queue

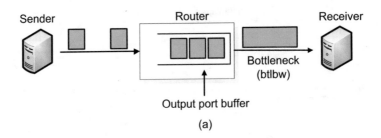

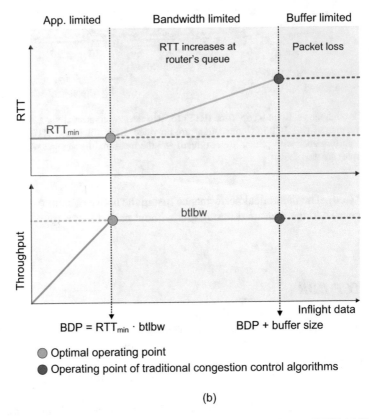

Fig. 53 TCP viewpoint of a connection and relation between throughput and RTT. (**a**) Simplified TCP interpretation of the connection. (**b**) Throughput and RTT, as a function of inflight data

increases when the input link arrival rate exceeds *btlbw*. The throughput of loss-based congestion control is less than *btlbw* because of the frequent packet losses.

Figure 53b illustrates the RTT and throughput with the amount of data inflight. RTT_{min} is the propagation time with no queueing component (the network is not congested). In the application limited region, the delivery rate/throughput increases as the amount of data generated by the application layer increases, while the RTT remains constant. The pipeline between sender and receiver becomes full when the inflight number of bits is equal to the bandwidth multiplied by the RTT. This number

Fig. 54 The rate used by the sender is the estimated bottleneck bandwidth (*btlbw*). During the probe period (1 RTT duration), the sender probes for additional bandwidth, sending at a rate of 125% of the bottleneck bandwidth. During the subsequent period, drain (1 RTT duration), the sender reduces the rate to 75% of the bottleneck bandwidth, thus allowing any bottleneck queue to drain

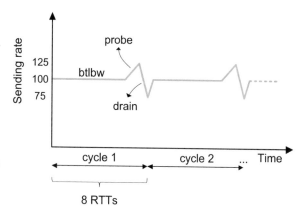

is also called bandwidth-delay product (BDP) and quantifies the number of bits that can be inflight if the sender continuously sends segments. In the bandwidth limited region, the queue size at the router of Fig. 53a starts increasing, resulting in an increase of the RTT. The throughput remains constant, as the bottleneck link is fully utilized. Finally, when no buffer is available at the router to store arriving packets (the number of inflight bits is equal to BDP plus the buffer size of the router), these are dropped.

It is important to understand that packets to be sent are paced at the estimated bottleneck rate, which is intended to avoid network queuing that would otherwise be encountered when the network performs rate adaptation at the bottleneck point. The intended operational model here is that the sender is passing packets into the network at a rate that is not anticipated to encounter queuing at any point within the entire path. This is a significant contrast to protocols such as Reno, which tends to send packet bursts at the epoch of the RTT and relies on the network's queues to perform rate adaptation in the interior of the network if the burst sending rate is higher than the bottleneck capacity.

BBR also periodically probes for additional bandwidth (Fig. 54). It spends one RTT interval deliberately sending at a rate that is higher than the current estimate bottleneck bandwidth. Specifically, it sends data at 125% of the bottleneck bandwidth. If the available bottleneck bandwidth has not changed, then the increased sending rate will cause a queue to form at the bottleneck. This will cause the ACK signaling to reveal an increased RTT, but the bottleneck bandwidth estimate will be unaltered. If this is the case, then the sender will subsequently send at a compensating reduced sending rate for an RTT interval. The reduced rate is set to 75% of the bottleneck bandwidth, allowing the bottleneck queue to drain. On the other hand, if the available bottleneck bandwidth estimate has increased because of this probe, then the sender will operate according to this new bottleneck bandwidth estimate. The entire cycle duration lasts eight RTTs and is repeated indefinitely in steady state.

13   Lab Topology

Let us get started with creating a simple Mininet topology using MiniEdit. The topology uses 10.0.0.0/8, which is the default network assigned by Mininet (Fig. 55).

Step 1. A shortcut to MiniEdit is located on the machine's Desktop. Start MiniEdit by clicking on MiniEdit's shortcut (Fig. 56). When prompted for a password, type password .

Step 2. On MiniEdit's menu bar, click on *File*, then *Open* to load the lab's topology. Locate the *Lab 10.mn* topology file and click on *Open* (Fig. 57).

Step 3. Before starting the measurements between host h1 and host h2, the network must be started. Click on the *Run* button located at the bottom left of MiniEdit's window to start the emulation (Fig. 58).

The above topology uses 10.0.0.0/8, which is the default network assigned by Mininet.

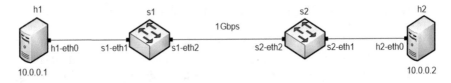

Fig. 55 Lab topology

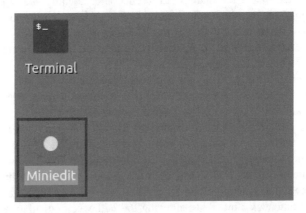

Fig. 56 MiniEdit shortcut

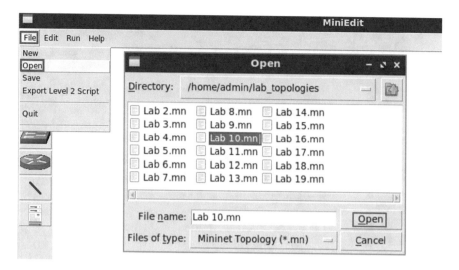

Fig. 57   MiniEdit's *Open* dialog

Fig. 58   Running the
emulation

13.1   Starting Host h1 and Host h2

Step 1. Hold the right-click on host h1 and select *Terminal*. This opens the terminal of host h1 and allows the execution of commands on that host (Fig. 59).

Step 2. Apply the same steps on host h2 and open its *Terminal*.

Step 3. Test connectivity between the end-hosts using the ⎡ping⎤ command. On host h1, type the command ⎡ping 10.0.0.2⎤. This command tests the connectivity

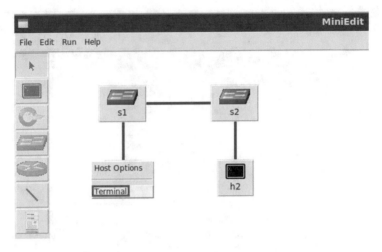

Fig. 59 Opening a terminal on host h1

```
X                           "Host: h1"                    — ↘ ✕
root@admin-pc:~# ping 10.0.0.2
PING 10.0.0.2 (10.0.0.2) 56(84) bytes of data.
64 bytes from 10.0.0.2: icmp_seq=1 ttl=64 time=1.33 ms
64 bytes from 10.0.0.2: icmp_seq=2 ttl=64 time=0.056 ms
64 bytes from 10.0.0.2: icmp_seq=3 ttl=64 time=0.048 ms
64 bytes from 10.0.0.2: icmp_seq=4 ttl=64 time=0.042 ms
64 bytes from 10.0.0.2: icmp_seq=5 ttl=64 time=0.043 ms
64 bytes from 10.0.0.2: icmp_seq=6 ttl=64 time=0.044 ms
^C
--- 10.0.0.2 ping statistics ---
6 packets transmitted, 6 received, 0% packet loss, time 91ms
rtt min/avg/max/mdev = 0.042/0.260/1.327/0.477 ms
root@admin-pc:~# █
```

Fig. 60 Connectivity test using ping command

between host h1 and host h2. To stop the test, press Ctrl+c . The figure below shows a successful connectivity test (Fig. 60).

Figure 60 indicates that there is connectivity between host h1 and host h2. Thus, we are ready to start the throughput measurement process.

13.2   Emulating 1 Gbps High-Latency WAN with Packet Loss

This section emulates a high-latency WAN, which is used to validate the results observed in Fig. 53. We will first set the bandwidth between host h1 and host h2 to

Fig. 61 Shortcut to open a Linux terminal

Fig. 62 Adding 0.01% packet loss rate to switch S1's s1-eth2 interface

1 Gbps. Then we will emulate packet losses between switch S1 and switch S2 and measure the throughput.

Step 1. Launch a Linux terminal by holding the Ctrl+Alt+T keys or by clicking on the Linux terminal icon (Fig. 61).

The Linux terminal is a program that opens a window and permits you to interact with a command-line interface (CLI). A CLI is a program that takes commands from the keyboard and sends them to the operating system for execution.

Step 2. In the terminal, type the below command. When prompted for a password, type password and hit enter. This command basically introduces a 0.01% packet loss rate on switch S1's *s1-eth2* interface (Fig. 62).

```
sudo tc qdisc add dev s1-eth2 root handle 1: netem loss
0.01%
```

Step 3. Modify the bandwidth of the link connecting the switch S1 and switch S2: on the same terminal, type the command below (Fig. 17). This command sets the bandwidth to 1 Gbps on switch S1's *s1-eth2* interface. The tbf parameters are the following (Fig. 63):

- rate : 1gbit
- burst : 500,000
- limit : 2500,000

admin@admin-pc: ~
File Actions Edit View Help
admin@admin-pc: ~
admin@admin-pc:~$ sudo tc qdisc add dev s1-eth2 parent 1: handle 2: tbf rate 1gbit
burst 500000 limit 2500000
admin@admin-pc:~$

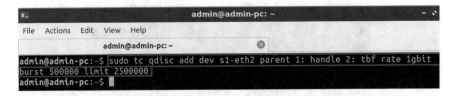

Fig. 63 Limiting the bandwidth to 1 Gbps on switch S1's s1-eth2 interface

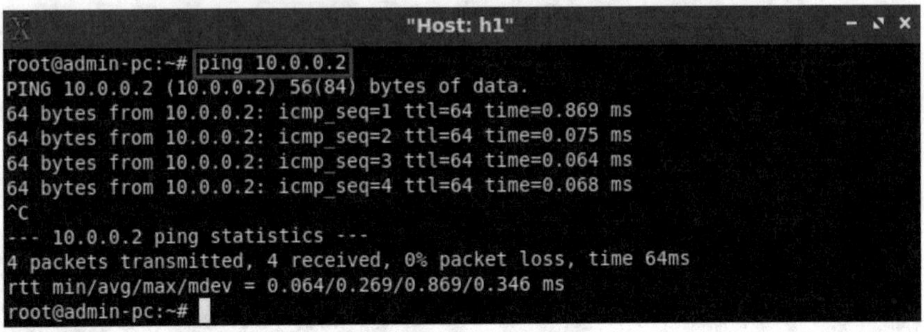
"Host: h1"
root@admin-pc:~# ping 10.0.0.2
PING 10.0.0.2 (10.0.0.2) 56(84) bytes of data.
64 bytes from 10.0.0.2: icmp_seq=1 ttl=64 time=0.869 ms
64 bytes from 10.0.0.2: icmp_seq=2 ttl=64 time=0.075 ms
64 bytes from 10.0.0.2: icmp_seq=3 ttl=64 time=0.064 ms
64 bytes from 10.0.0.2: icmp_seq=4 ttl=64 time=0.068 ms
^C
--- 10.0.0.2 ping statistics ---
4 packets transmitted, 4 received, 0% packet loss, time 64ms
rtt min/avg/max/mdev = 0.064/0.269/0.869/0.346 ms
root@admin-pc:~#

Fig. 64 Output of ping 10.0.0.2 command

```
sudo tc qdisc add dev s1-eth2 parent 1: handle 2: tbf rate
1gbit burst 500000 limit 2500000
```

13.3 Testing Connection

To test connectivity, you can use the command ping.

Step 1. On the terminal of host h1, type ping 10.0.0.2. To stop the test, press Ctrl+c. The figure below shows a successful connectivity test. Host h1 (10.0.0.1) sent four packets to host h2 (10.0.0.2), successfully receiving responses back (Fig. 64).

The result above indicates that all four packets were received successfully (0% packet loss) and that the minimum, average, maximum, and standard deviation of the Round-Trip Time (RTT) were 0.064, 0.269, 0.869, and 0.346 ms, respectively. Essentially, the standard deviation is an average of how far each ping RTT is from the average RTT. The higher the standard deviation the more variable the RTT is.

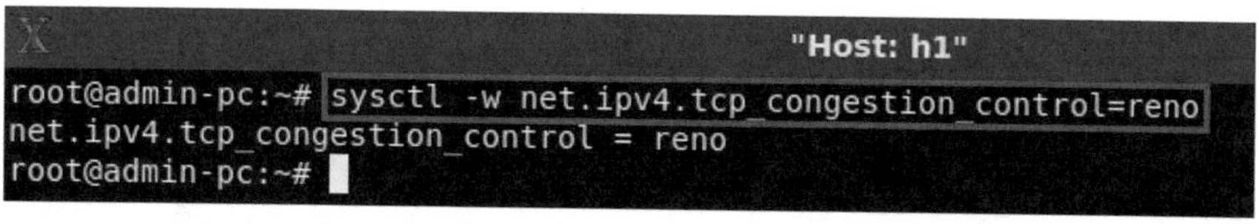

Fig. 65   Changing TCP congestion control algorithm to [reno] on host h1

Step 2.   On the terminal of host h2, type [ping 10.0.0.1]. The ping output in this test should be relatively close to the results of the test initiated by host h1 in Step 1. To stop the test, press [Ctrl+c].

14   iPerf3 Throughput Test

In this section, the throughput between host h1 and host h2 is measured using two congestion control algorithms: Reno and BBR. Moreover, the test is repeated using various injected delays to observe the throughput variations depending on each congestion control algorithm and the selected RTT.

14.1   Throughput Test Without Delay

In this test, we measure the throughput between host h1 and host h2 without introducing delay on the switch S1's *s1-eth2* interface.

14.1.1   TCP Reno

Step 1.   In host h1's terminal, change the TCP congestion control algorithm to Reno by typing the following command (Fig. 65):

```
sysctl -w net.ipv4.tcp_congestion_control=reno
```

Step 2.   Launch iPerf3 in server mode on host h2's terminal (Fig. 66):

```
iperf3 -s
```

Step 3.   Launch iPerf3 in client mode on host h1's terminal. The [-O] option is used to specify the number of seconds to omit in the resulting report (Fig. 21).

Fig. 66 Starting iPerf3 server on host h2

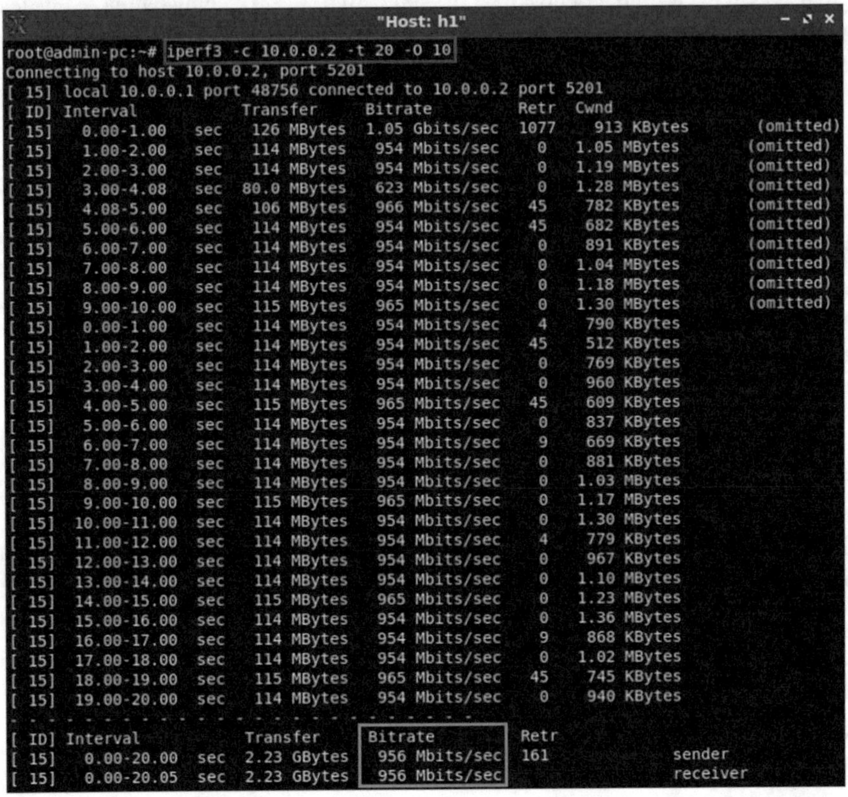

Fig. 67 Running iPerf3 client on host h1

```
iperf3 -c 10.0.0.2 -t 20 -O 10
```

The figure above shows the iPerf3 test output report. The average achieved throughputs are 956 Mbps (sender) and 956 Mbps (receiver), and the number of retransmissions is 161 (due to the injected packet loss—0.01%) (Fig. 67).

Step 4. In order to stop the server, press Ctrl+c in host h2's terminal. The user can see the throughput results in the server side too.

Fig. 68   Changing TCP congestion control algorithm to ⎡bbr⎤ on host h1

Fig. 69   Starting iPerf3 server on host h2

14.1.2   TCP BBR

Step 1.   In host h1's terminal, change the TCP congestion control algorithm to BBR by typing the following command (Fig. 68):

```
sysctl -w net.ipv4.tcp_congestion_control=bbr
```

Step 2.   Launch iPerf3 in server mode on host h2's terminal (Fig. 69):

```
iperf3 -s
```

Step 3.   Launch iPerf3 in client mode on host h1's terminal (Fig. 70):

```
iperf3 -c 10.0.0.2 -t 20 -O 10
```

Figure 70 shows the iPerf3 test output report. The average achieved throughputs are 937 Mbps (sender) and 937 Mbps (receiver), and the number of retransmissions is 92 (due to the injected packet loss—0.01%).

Step 4.   In order to stop the server, press ⎡Ctrl+c⎤ in host h2's terminal. The user can see the throughput results in the server side too.

14.2   Throughput Test with 30 ms Delay

In this test, we measure the throughput between host h1 and host h2 while introducing 30 ms delay on the switch S1's *s1-eth2* interface. Apply the following steps:

```
                              "Host: h1"                              - ↗ ×
root@admin-pc:~# iperf3 -c 10.0.0.2 -t 20 -O 10
Connecting to host 10.0.0.2, port 5201
[ 15] local 10.0.0.1 port 48760 connected to 10.0.0.2 port 5201
[ ID] Interval          Transfer     Bitrate         Retr  Cwnd
[ 15]   0.00-1.00   sec  117 MBytes   983 Mbits/sec    0    198 KBytes      (omitted)
[ 15]   1.00-2.00   sec  115 MBytes   961 Mbits/sec    0    198 KBytes      (omitted)
[ 15]   2.00-3.00   sec  114 MBytes   953 Mbits/sec    0    198 KBytes      (omitted)
[ 15]   3.00-4.00   sec  113 MBytes   952 Mbits/sec    0    198 KBytes      (omitted)
[ 15]   4.00-5.00   sec  115 MBytes   961 Mbits/sec    0    198 KBytes      (omitted)
[ 15]   5.00-6.00   sec  114 MBytes   952 Mbits/sec    0    198 KBytes      (omitted)
[ 15]   6.00-7.00   sec  115 MBytes   962 Mbits/sec    0    198 KBytes      (omitted)
[ 15]   7.00-8.00   sec  114 MBytes   952 Mbits/sec    0    198 KBytes      (omitted)
[ 15]   8.00-9.00   sec  115 MBytes   962 Mbits/sec   45    198 KBytes      (omitted)
[ 15]   9.00-10.00  sec  114 MBytes   953 Mbits/sec    0    198 KBytes      (omitted)
[ 15]   0.00-1.00   sec  114 MBytes   952 Mbits/sec    2    195 KBytes
[ 15]   1.00-2.00   sec  115 MBytes   962 Mbits/sec    0    195 KBytes
[ 15]   2.00-3.00   sec  113 MBytes   952 Mbits/sec    0    195 KBytes
[ 15]   3.00-4.00   sec  115 MBytes   962 Mbits/sec    0    195 KBytes
[ 15]   4.00-5.00   sec  113 MBytes   952 Mbits/sec    0    195 KBytes
[ 15]   5.00-6.00   sec  115 MBytes   962 Mbits/sec    0    195 KBytes
[ 15]   6.00-7.00   sec  114 MBytes   953 Mbits/sec    0    195 KBytes
[ 15]   7.00-8.00   sec  114 MBytes   954 Mbits/sec    0    195 KBytes
[ 15]   8.00-9.00   sec  115 MBytes   963 Mbits/sec    0    195 KBytes
[ 15]   9.00-10.00  sec  113 MBytes   951 Mbits/sec    0    195 KBytes
[ 15]  10.00-11.00  sec  115 MBytes   962 Mbits/sec    0    198 KBytes
[ 15]  11.00-12.00  sec  114 MBytes   952 Mbits/sec    0    198 KBytes
[ 15]  12.00-13.00  sec  113 MBytes   952 Mbits/sec   90    195 KBytes
[ 15]  13.00-14.00  sec  115 MBytes   962 Mbits/sec    0    195 KBytes
[ 15]  14.00-15.00  sec 91.2 MBytes   765 Mbits/sec    0    195 KBytes
[ 15]  15.00-16.00  sec 91.2 MBytes   765 Mbits/sec    0    195 KBytes
[ 15]  16.00-17.00  sec  113 MBytes   952 Mbits/sec    0    195 KBytes
[ 15]  17.00-18.00  sec  115 MBytes   962 Mbits/sec    0    195 KBytes
[ 15]  18.00-19.00  sec  113 MBytes   952 Mbits/sec    0    195 KBytes
[ 15]  19.00-20.00  sec  115 MBytes   962 Mbits/sec    0    195 KBytes
- - - - - - - - - - - - - - - - - - - - - - - - - - -
[ ID] Interval          Transfer     Bitrate         Retr
[ 15]   0.00-20.00  sec 2.18 GBytes   937 Mbits/sec   92            sender
[ 15]   0.00-20.04  sec 2.19 GBytes   937 Mbits/sec                 receiver
```

Fig. 70 Running iPerf3 client on host h1

Fig. 71 Injecting 30 ms delay on switch S1's *s1-eth2* interface

Step 1. In order to add delay to the switch 1 or interface s1-eth2, go back to the client's terminal, run the following command to modify the previous rule to include 30 ms delay (Fig. 71):

```
sudo tc qdisc change dev s1-eth2 root handle 1: netem loss
0.01% delay 30ms
```

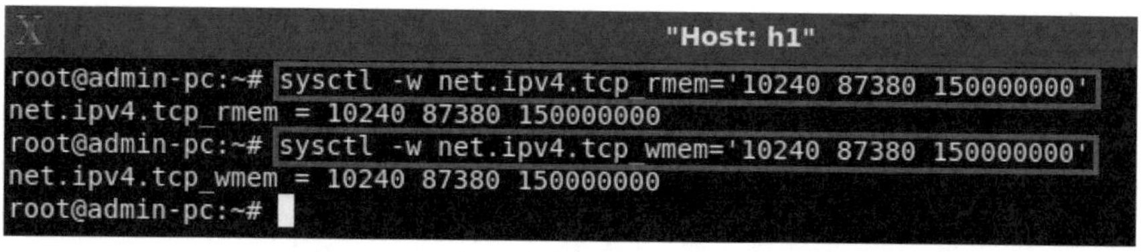

Fig. 72 Modifying the TCP buffer size on host h1

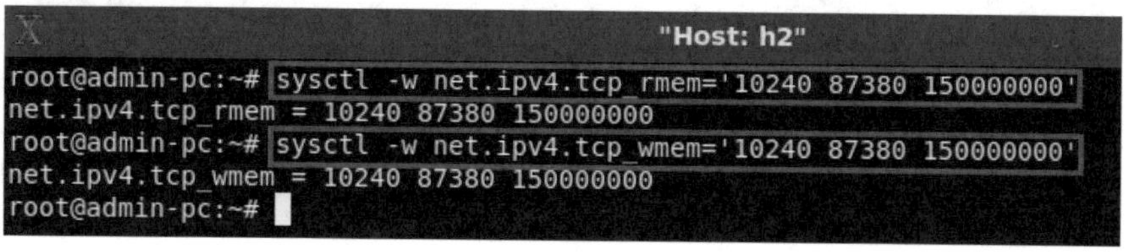

Fig. 73 Modifying the TCP buffer size on host h2

Step 2. In host h1's terminal, modify the TCP buffer size by typing the following commands: *sysctl -w net.ipv4.tcp_rmem='10,240 87,380 150,000,000'* and *sysctl -w net.ipv4.tcp_wmem='10,240 87,380 150,000,000'*. This TCP buffer is explained later in future labs (Fig. 72).

```
sysctl -w net.ipv4.tcp_rmem='10240 87380 150000000'
```

```
sysctl -w net.ipv4.tcp_wmem='10240 87380 150000000'
```

Step 3. In host h2's terminal, also modify the TCP buffer size by typing the following commands: *sysctl -w net.ipv4.tcp_rmem='10,240 87,380 150,000,000'* and *sysctl -w net.ipv4.tcp_wmem='10,240 87,380 150,000,000'* (Fig. 73).

```
sysctl -w net.ipv4.tcp_rmem='10240 87380 150000000'
```

```
sysctl -w net.ipv4.tcp_wmem='10240 87380 150000000'
```

14.2.1 TCP Reno

Step 1. In host h1's terminal, change the TCP congestion control algorithm to Reno by typing the following command (Fig. 74):

```
sysctl -w net.ipv4.tcp_congestion_control=reno
```

```
X                              "Host: h1"
root@admin-pc:~# sysctl -w net.ipv4.tcp_congestion_control=reno
net.ipv4.tcp_congestion_control = reno
root@admin-pc:~#
```

Fig. 74 Changing TCP congestion control algorithm to reno on host h1

```
X                              "Host: h2"                        —  ꙭ  ✕
root@admin-pc:~# iperf3 -s
- - - - - - - - - - - - - - - - - - - - - - - - - - - - - - - - - - - - -
Server listening on 5201
- - - - - - - - - - - - - - - - - - - - - - - - - - - - - - - - - - - - -
```

Fig. 75 Starting iPerf3 server on host h2

```
X                              "Host: h1"                        —  ꙭ  ✕
root@admin-pc:~# mkdir reno && cd reno
root@admin-pc:~/reno#
```

Fig. 76 Creating and entering a new directory *reno*

```
X                              "Host: h1"                        —  ꙭ  ✕
root@admin-pc:~/reno# iperf3 -c 10.0.0.2 -t 30 -J >reno.json
root@admin-pc:~/reno#
```

Fig. 77 Running iPerf3 client on host h1 and redirecting the output to *reno.json*

Step 2. Launch iPerf3 in server mode on host h2's terminal (Fig. 75):

```
iperf3 -s
```

Step 3. Create and enter to a new directory *reno* on host h1's terminal (Fig. 76):

```
mkdir reno && cd reno
```

Step 4. Launch iPerf3 in client mode on host h1's terminal. The -J option is used to produce a JSON output and the redirection operator > to send the standard output to a file (Fig. 77).

```
iperf3 -c 10.0.0.2 -t 30 -J >reno.json
```

Fig. 78 plot_iperf.sh script generating output results

Fig. 79 Entering the results directory using the cd command

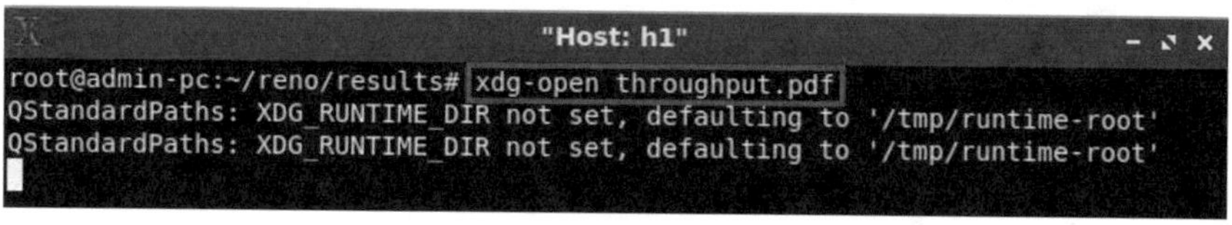

Fig. 80 Opening the *throughput.pdf* file using xdg-open

Step 5. Once the test is finished, type the following command to generate the output plots for iPerf3's JSON file (Fig. 78):

```
plot_ iperf.sh reno.json
```

This plotting script generates PDF files for the following fields: congestion window (*cwnd.pdf*), retransmits (*retransmits.pdf*), Round-Trip Time (*RTT.pdf*), Round-Trip Time variance (*RTT_Var.pdf*), throughput (*throughput.pdf*), maximum transmission unit (*MTU.pdf*), and bytes transferred (*bytes.pdf*). The plotting script also generates a CSV file (*1.dat*) to be used by applicable programs. These files are stored in a directory *results* created in the same directory where the script was executed as shown in the figure below.

Step 6. Navigate to the results folder using the cd command (Fig. 79).

```
cd results/
```

Step 7. To open any of the generated files, use the xdg-open command followed by the file name. For example, to open the *throughput.pdf* file, use the following command (Figs. 80 and 81):

```
xdg-open throughput.pdf
```

Throughput over time
1.dat
Mbps
Time (sec)
1000
900
800
700
600
500
400
300
200
100
0
0
5
10
15
20
25
30

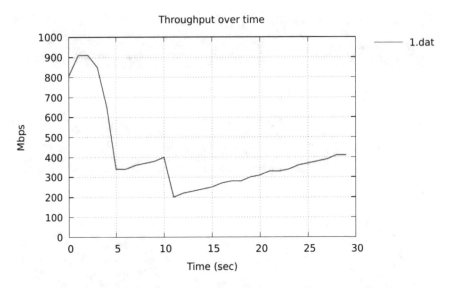

Fig. 81 Reno's throughput

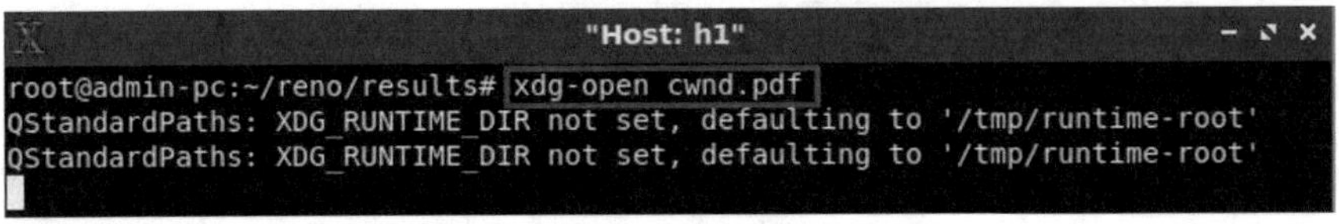

"Host: h1"
root@admin-pc:~/reno/results# xdg-open cwnd.pdf
QStandardPaths: XDG_RUNTIME_DIR not set, defaulting to '/tmp/runtime-root'
QStandardPaths: XDG_RUNTIME_DIR not set, defaulting to '/tmp/runtime-root'

Fig. 82 Opening the *throughput.pdf* file using xdg-open

Step 8. Close the *throughput.pdf* file and open the *cwnd.pdf* file using the following command (Figs. 82 and 83):

```
xdg-open cwnd.pdf
```

Step 9. In order to stop the server, press Ctrl+c in host h2's terminal. The user can see the throughput results in the server side too.

Step 10. Exit the */reno/results* directory by using the following command on host h1's terminal (Fig. 84):

```
cd ../..
```

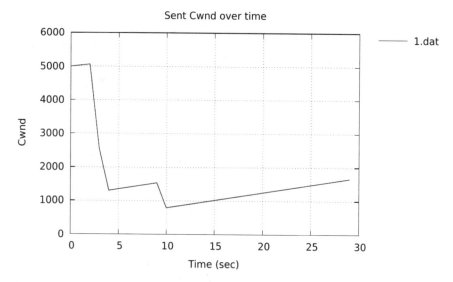

Fig. 83   Reno's congestion window

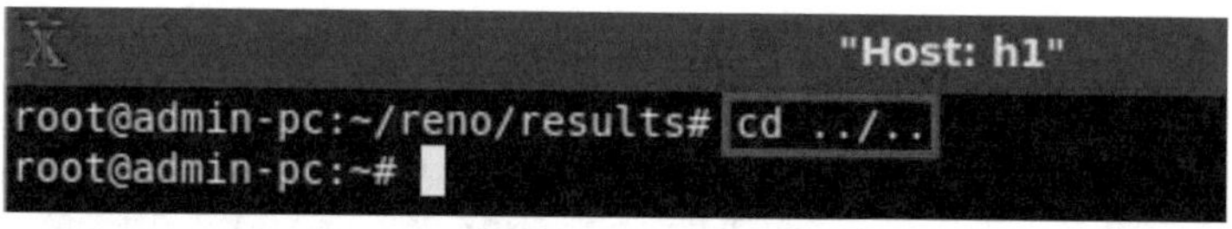

Fig. 84   Exiting the */reno/results* directory

Fig. 85   Changing TCP congestion control algorithm to bbr on host h1

14.2.2   TCP BBR

Step 1. In host h1's terminal, change the TCP congestion control algorithm to BBR by typing the following command (Fig. 85):

```
sysctl -w net.ipv4.tcp_congestion_control=bbr
```

Step 2. Launch iPerf3 in server mode on host h2's terminal (Fig. 86):

```
iperf3 -s
```

```
X                           "Host: h2"                    — ↘ ✕
root@admin-pc:~# iperf3 -s
------------------------------------------------------------
Server listening on 5201
------------------------------------------------------------
```

Fig. 86 Starting iPerf3 server on host h2

```
X                                          "Host: h1"
root@admin-pc:~# mkdir bbr && cd bbr
root@admin-pc:~/bbr#
```

Fig. 87 Creating and entering a new directory *bbr*

```
X                                    "Host: h1"
root@admin-pc:~/bbr# iperf3 -c 10.0.0.2 -t 30 -J > bbr.json
root@admin-pc:~/bbr#
```

Fig. 88 Running iPerf3 client on host h1 and redirecting the output to *bbr.json*

```
X                                        "Host: h1"
root@admin-pc:~/bbr# plot_iperf.sh bbr.json
root@admin-pc:~/bbr#
```

Fig. 89 plot_iperf.sh script generating output results

Step 3. Create and enter to a new directory *bbr* host h1's terminal (Fig. 87):

```
mkdir bbr && cd bbr
```

Step 4. Launch iPerf3 in client mode on host h1's terminal. The -J option is used to produce a JSON output and the redirection operator > to send the standard output to a file (Fig. 88).

```
iperf3 -c 10.0.0.2 -t 30 -J >bbr.json
```

Step 5. To generate the output plots for iPerf3's JSON file run the following command (Fig. 89):

```
plot_iperf.sh bbr.json
```

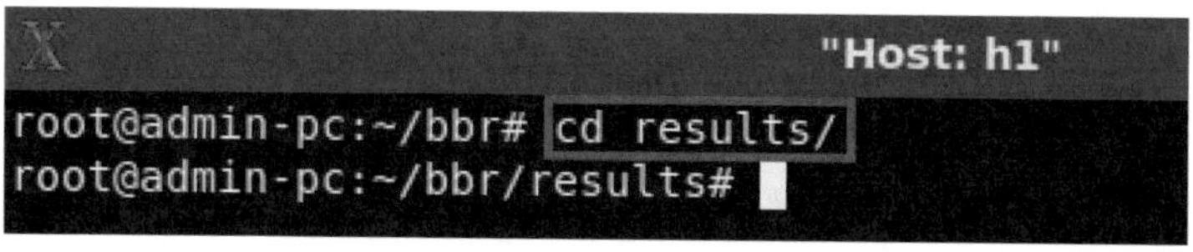

Fig. 90 Entering the results directory using the ｜cd｜ command

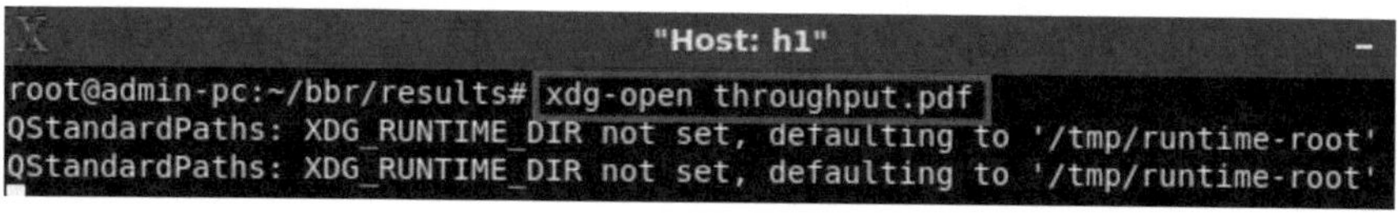

Fig. 91 Opening the *throughput.pdf* file using ｜xdg-open｜

This plotting script generates PDF files for the following fields: congestion window (*cwnd.pdf*), retransmits (*retransmits.pdf*), Round-Trip Time (*RTT.pdf*), Round-Trip Time variance (*RTT_Var.pdf*), throughput (*throughput.pdf*), maximum transmission unit (*MTU.pdf*), and bytes transferred (*bytes.pdf*). The plotting script also generates a CSV file (*1.dat*) to be used by applicable programs. These files are stored in a directory *results* created in the same directory where the script was executed as shown in the figure below.

Step 6. Navigate to the results folder using the ｜cd｜ command (Fig. 90).

```
cd results/
```

Step 7. To open any of the generated files, use the ｜xdg-open｜ command followed by the file name. For example, to open the *throughput.pdf* file, use the following command (Fig. 91):

```
xdg-open throughput.pdf
```

Step 8. Figure 92 shows that in steady state, BBR has already attained the maximum throughput, which is over 900 Mbps (the bottleneck bandwidth is 1 Gbps, with an observed effective bandwidth of ∼937 Gbps). Note also the periodic (short) drain intervals, where the throughput decreases to ∼75% of the maximum throughput, as discussed in Sect. 12.3. To proceed, close the *throughput.pdf* file and open the *cwnd.pdf* file using the following command (Figs. 93 and 94):

```
xdg-open cwnd.pdf
```

Throughput over time
1000
900
800
700
600
500
400
300
200
100
0
Mbps
0
5
10
15
20
25
30
Time (sec)

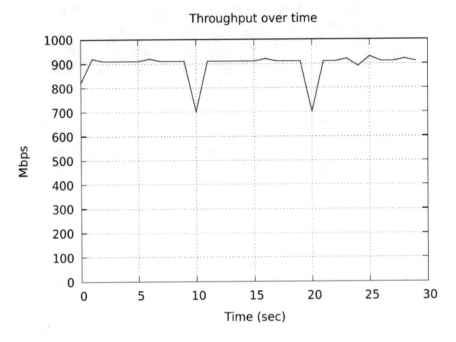

Fig. 92 BBR's throughput

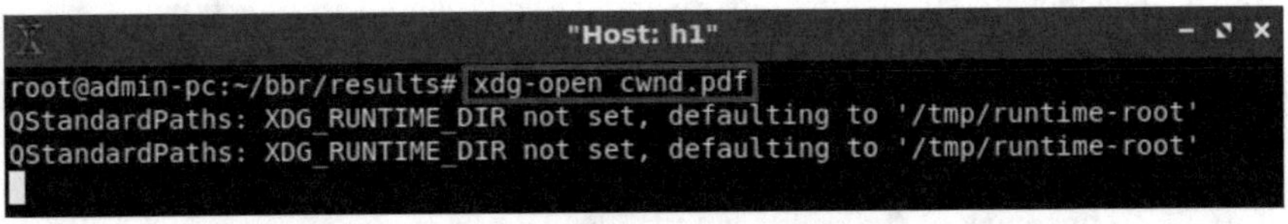

"Host: h1"
root@admin-pc:~/bbr/results# xdg-open cwnd.pdf
QStandardPaths: XDG_RUNTIME_DIR not set, defaulting to '/tmp/runtime-root'
QStandardPaths: XDG_RUNTIME_DIR not set, defaulting to '/tmp/runtime-root'

Fig. 93 Opening the *cwnd.pdf* file using xdg-open

Step 9. In order to stop the server, press Ctrl+c in host h2's terminal. The user can see the throughput results in the server side too.

Step 10. Exit the */bbr/results* directory by using the following command on host h1's terminal (Fig. 95):

```
cd ../..
```

It is clear from the above test that when introducing delay, BBR preforms significantly better than Reno.

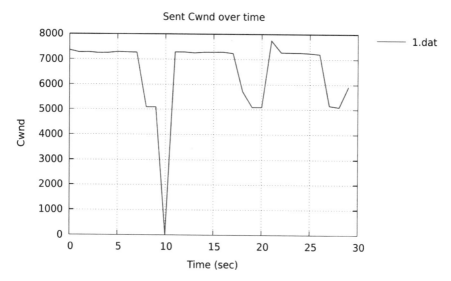

Fig. 94 BBR's congestion window

Fig. 95 Exiting the /bbr/results directory

15   Chapter 4—Lab 11: Bandwidth-Delay Product and TCP Buffer Size

Overview

To conduct the experiment described in this section, please login into the Academic Cloud at http://highspeednetworks.net/, and reserve a pod for Lab 11.

This lab explains the bandwidth-delay product (BDP) and how to modify the TCP buffer size in Linux systems. Throughput measurements are also conducted to test and verify TCP buffer configurations based on the BDP.

Objectives

By the end of this lab, students should be able to:

1. Understand BDP.
2. Define and calculate TCP window size.
3. Modify the TCP buffer size with sysctl, based on BDP calculations.
4. Emulate WAN properties in Mininet.
5. Achieve full throughput in WANs by modifying the size of TCP buffers.

	Device	Account	Password
Table 4 Credentials to access Client1 machine	Client1	admin	password

Lab Settings

The information in Table 4 provides the credentials of the machine containing Mininet.

Lab Roadmap

This lab is organized as follows:

1. Section 16: Introduction to TCP buffers, BDP, and TCP window
2. Section 17: Lab topology
3. Section 18: BDP and buffer size experiments
4. Section 19: Modifying buffer size and throughput test

16 Introduction to TCP buffers, BDP, and TCP Window

16.1 TCP Buffers

The TCP send and receive buffers may impact the performance of Wide Area Networks (WAN) data transfers. Consider Fig. 96. At the sender side, TCP receives data from the application layer and places it in the TCP send buffer. Typically, TCP fragments the data in the buffer into maximum segment size (MSS) units. In this example, the MSS is 100 bytes. Each segment carries a sequence number, which is the byte-stream number of the first byte in the segment. The corresponding acknowledgment (Ack) carries the number of the next expected byte (e.g., Ack-101 acknowledges bytes 1–100, carried by the first segment). At the receiver, TCP receives data from the network layer and places it into the TCP receive buffer. TCP delivers the data *in order* to the application, e.g., bytes contained in a segment, say segment 2 (bytes 101–200), cannot be delivered to the application layer before the bytes contained in segment 1 (bytes 1–100) are delivered to the application. At any given time, the TCP receiver indicates the TCP sender how many bytes the latter can send, based on how much free buffer space is available at the receiver.

16.2 Bandwidth-Delay Product

In many WANs, the Round-Trip Time (RTT) is dominated by the propagation delay. Long RTTs along and TCP buffer size have throughput implications. Consider a 10 Gbps WAN with a 50-ms RTT. Assume that the TCP send and receive buffer sizes are set to 1 Mbyte (1 Mbyte $= 1024^2$ bytes $= 1,048576$ bytes $= 1,048,576 \cdot$

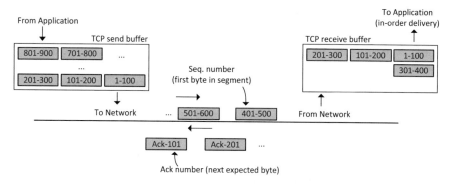

Fig. 96 TCP send and receive buffers

8 bits $= 8{,}388{,}608$ bits). With a bandwidth (Bw) of 10 Gbps, this number of bits is approximately transmitted in

$$T_{tx} = \frac{\#\,bits}{Bw} = \frac{8,388,608}{10 \cdot 10^9} = 0.84 \text{ milliseconds.}$$

That is, after 0.84 ms the content of the TCP send buffer will be completely sent. At this point, TCP must wait for the corresponding acknowledgments, which will only start arriving at $t = 50$ ms. This means that the sender only uses 0.84/50 or 1.68% of the available bandwidth.

The solution to that above problem lies in allowing the sender to continuously transmit segments until the corresponding acknowledgments arrive back. Note that the first acknowledgment arrives after an RTT. The number of bits that can be transmitted in a RTT period is given by the bandwidth of the channel in bits per second multiplied by the RTT. This quantity is referred to as the bandwidth-delay product (BDP). For the above example, the buffer size must be greater than or equal to the BDP:

$$TCP\,buffer\,size \geq BDP = \left(10 \cdot 10^9\right)\left(50 \cdot 10^{-3}\right) = 500{,}000{,}000 \text{ bits}$$

$$= 62{,}500{,}000 \text{ bytes.}$$

The first factor $(10 \cdot 10^9)$ is the bandwidth; the second factor $(50 \cdot 10^{-3})$ is the RTT. For practical purposes, the TCP buffer can also be expressed in Mbytes (1 Mbyte $= 1024^2$ bytes) or Gbytes (1 Gbyte $= 1024^3$ bytes). The above expression in Mbytes is

$$TCP\,buffer\,size \geq 62{,}500{,}000 \text{ bytes} = 59.6 \text{ Mbytes} \approx 60 \text{ Mbytes.}$$

16.3   Practical Observations on Setting TCP Buffer Size

Linux Systems Configuration Linux assumes that half of the send/receive TCP buffers are used for internal structures. Thus, only half of the buffer size is used to store segments. This implies that if a TCP connection requires certain buffer size, then the administrator must configure the buffer size equals to twice that size. For the previous example, the TCP buffer size must be

$$\text{TCP buffer size} \geq 2 \cdot 60\,\text{Mbytes} = 120\,\text{Mbytes}.$$

Packet Loss Scenarios and TCP BBR TCP provides a reliable, in-order delivery service. Reliability means that bytes successfully received must be acknowledged. In-order delivery means that the receiver only delivers bytes to the application layer in sequential order. The memory occupied by those bytes will be deallocated from the receive buffer after passing the bytes to the application layer. This process has some performance implications, as illustrated next. Consider Fig. 97, which shows a TCP receive buffer. Assume that the segment carrying bytes 101–200 is lost. Although the receiver has successfully received bytes 201–900, it cannot deliver to the application layer until bytes 101–200 are received. Note that the receive buffer may become full, which would block the sender from utilizing the channel.

   While setting the buffer size equal to the BDP is acceptable when traditional congestion control algorithms are used (e.g., Reno, Cubic, HTCP), this size may not allow the full utilization of the channel with BBR. In contrast to other algorithms, BBR does not reduce the transmission rate after a packet loss. For example, suppose that a packet sent at t = 0 is lost, as shown in Fig. 98. At t = RTT, the acknowledgment identifying the packet to retransmit is received. By then, the sender

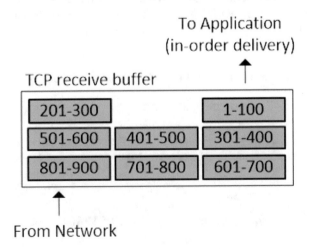

Fig. 97 TCP receive buffer. Although bytes 301–900 have been received, they cannot be delivered to the application until the segment carrying bytes 201–300 are received

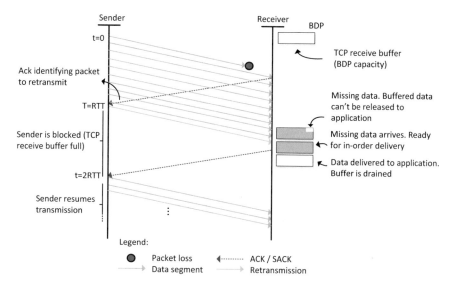

Fig. 98 A scenario where a TCP receive buffer size of BDP cannot prevent throughput degradation

has sent BDP bits, which will be stored in the receive buffer. This data cannot be delivered yet to the application because of the in-order delivery requirement. Since the receive buffer has a capacity of BDP only, the sender is temporarily blocked until the acknowledgment for the retransmitted data is received at t = 2·RTT. Thus, the throughput over the period t = 0 to t = 2·RTT is reduced by half:

$$\text{throughput} = \frac{\#\,\text{bits transmitted}}{\text{period}} = \frac{Bw \cdot RTT}{2 \cdot RTT} = \frac{Bw}{2}.$$

With BBR, to fully utilize the available bandwidth, the TCP send and receive buffers must be large enough to prevent such situation. Figure 99 shows an example on how a TCP buffer size of 2·BDP mitigates packet loss.

High to moderate packet loss scenarios, using TCP BBR:

$$\text{TCP send/receive buffer} \geq \text{several BDPs (e.g., } 4 \cdot \text{BDP).}$$

Continuing with the example of Sect. 16.2, in a Linux system using TCP BBR, the send/receive buffers for a BDP of 60 Mbytes in a high to moderate packet loss scenario should be

$$\text{TCP buffer size} \geq (2 \cdot 60\,\text{Mbytes}) \cdot 4 = 480\,\text{Mbytes}.$$

The factor 2 is because of the Linux systems configuration, and the factor 4 is because of the use of TCP BBR in a high to moderate packet loss scenario.

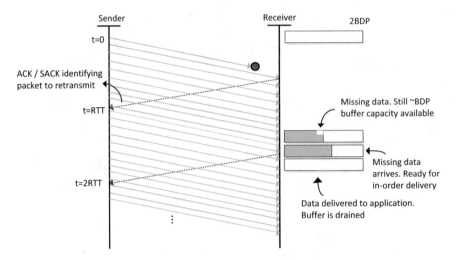

Fig. 99 A scenario where a TCP buffer size of 2·BDP mitigates packet loss

```
▸ Flags: 0x010 (ACK)
  Window size value: 2049
  [Calculated window size: 1049088]
  [Window size scaling factor: 512]
```

Fig. 100 Window scaling in Wireshark

16.4   TCP Window Size Calculated Value

The receiver must constantly communicate with the sender to indicate how much free buffer space is available in the TCP receive buffer. This information is carried in a TCP header field called window size. The window size has a maximum value of 65,535 bytes, as the header value allocated for the window size is two bytes long (16 bits; $2^{16} - 1 = 65{,}535$). However, this value is not large enough for high-bandwidth high-latency networks. Therefore, *TCP window scale option* was standardized in RFC 1323. By using this option, the calculated window size may be increased up to a maximum value of 1,073,725,440 bytes. When advertising its window, a device also advertises the *scale factor* (multiplier) that will be used throughout the session. The TCP window size is calculated as follows:

$$\text{Scaled TCP}_{\text{Win}} = \text{TCP}_{\text{Win}} \cdot \text{Scaling Factor}$$

As an example, consider the following example. For an advertised TCP window of 2049 and a scale factor of 512, then the final window size is 1,049,088 bytes. Figure 100 displays a packet inspected in Wireshark protocol analyzer for this numerical example

16.5 Zero Window

When the TCP buffer is full, a window size of zero is advertised to inform the other end that it cannot receive any more data. When a client sends a TCP window of zero, the server will pause its data transmission and wait for the client to recover. Once the client is recovered, it digests data and informs the server to resume the transmission again by setting again the TCP window.

17 Lab Topology

Let us get started with creating a simple Mininet topology using Miniedit. The topology uses 10.0.0.0/8, which is the default network assigned by Mininet (Fig. 101).

Step 1. A shortcut to Miniedit is located on the machine's Desktop. Start Miniedit by clicking on Miniedit's shortcut (Fig. 102). When prompted for a password, type ⎢ password ⎥.

Step 2. On Miniedit's menu bar, click on *File* and then *Open* to load the lab's topology. Locate the *Lab 11.mn* topology file and click on *Open* (Fig. 103).

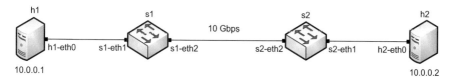

Fig. 101 Lab topology

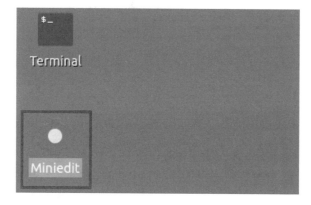

Fig. 102 Miniedit shortcut

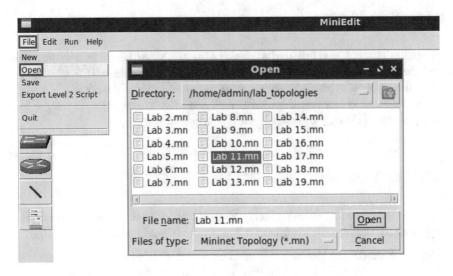

Fig. 103 Miniedit's *Open* dialog

Fig. 104 Running the
emulation

Step 3. Before starting the measurements between host h1 and host h2, the network must be started. Click on the *Run* button located at the bottom left of Miniedit's window to start the emulation (Fig. 104).

The above topology uses 10.0.0.0/8, which is the default network assigned by Mininet.

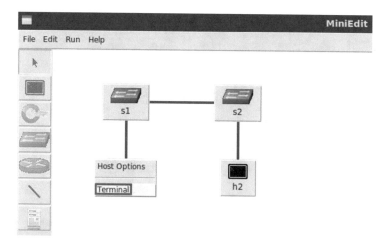

Fig. 105 Opening a terminal on host h1

```
                          "Host: h1"                              - ʌ x
root@admin-pc:~# ping 10.0.0.2
PING 10.0.0.2 (10.0.0.2) 56(84) bytes of data.
64 bytes from 10.0.0.2: icmp_seq=1 ttl=64 time=1.33 ms
64 bytes from 10.0.0.2: icmp_seq=2 ttl=64 time=0.056 ms
64 bytes from 10.0.0.2: icmp_seq=3 ttl=64 time=0.048 ms
64 bytes from 10.0.0.2: icmp_seq=4 ttl=64 time=0.042 ms
64 bytes from 10.0.0.2: icmp_seq=5 ttl=64 time=0.043 ms
64 bytes from 10.0.0.2: icmp_seq=6 ttl=64 time=0.044 ms
^C
--- 10.0.0.2 ping statistics ---
6 packets transmitted, 6 received, 0% packet loss, time 91ms
rtt min/avg/max/mdev = 0.042/0.260/1.327/0.477 ms
root@admin-pc:~# ▮
```

Fig. 106 Connectivity test using `ping` command

17.1   Starting Host h1 and Host h2

Step 1. Hold the right-click on host h1 and select *Terminal*. This opens the terminal of host h1 and allows the execution of commands on that host (Fig. 105).

Step 2. Apply the same steps on host h2 and open its *Terminal*.

Step 3. Test connectivity between the end-hosts using the `ping` command. On host h1, type the command `ping 10.0.0.2` (Fig. 106). This command tests the connectivity between host h1 and host h2. To stop the test, press `Ctrl+c`. The figure below shows a successful connectivity test.

Fig. 107 Shortcut to open a Linux terminal

Fig. 108 Adding 20 ms delay to switch S1's *s1-eth2* interface

Figure 106 indicates that there is connectivity between host h1 and host h2.

17.2 Emulating 10 Gbps High-Latency WAN

This section emulates a high-latency WAN by introducing delays to the network. We will first set the bandwidth between hosts 1 and 2 to 10 Gbps. Then, we will emulate a 20 ms delay and measure the throughput.

Step 1. Launch a Linux terminal by holding the Ctrl+Alt+T keys or by clicking on the Linux terminal icon (Fig. 107).

   The Linux terminal is a program that opens a window and permits you to interact with a command-line interface (CLI). A CLI is a program that takes commands from the keyboard and sends them to the operating system to perform.

Step 2. In the terminal, type the command below. When prompted for a password, type password and hit enter. This command introduces 20 ms delay on S1's *s1-eth2* interface (Fig. 108).

```
sudo tc qdisc add dev s1-eth2 root handle 1: netem delay
20 ms
```

admin@admin-pc: ~
File Actions Edit View Help
admin@admin-pc: ~
admin@admin-pc:~$ sudo tc qdisc add dev s1-eth2 parent 1: handle 2: tbf rate 10gbit burst 5000000 limit 25000000
admin@admin-pc:~$

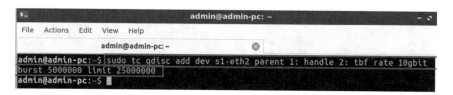

Fig. 109 Limiting the bandwidth to 10 Gbps on switch S1's *s1-eth2* interface

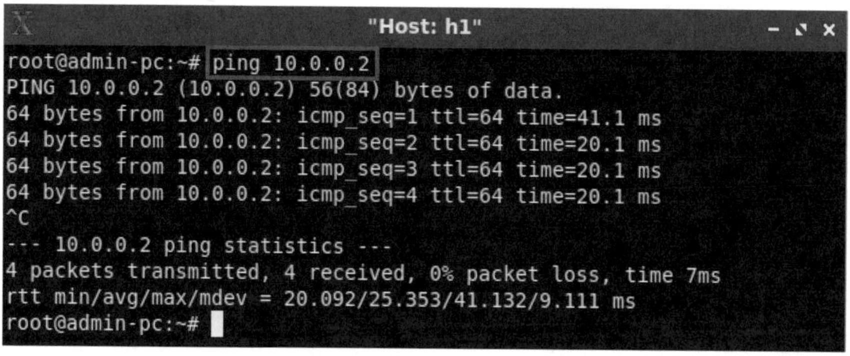

"Host: h1"
root@admin-pc:~# ping 10.0.0.2
PING 10.0.0.2 (10.0.0.2) 56(84) bytes of data.
64 bytes from 10.0.0.2: icmp_seq=1 ttl=64 time=41.1 ms
64 bytes from 10.0.0.2: icmp_seq=2 ttl=64 time=20.1 ms
64 bytes from 10.0.0.2: icmp_seq=3 ttl=64 time=20.1 ms
64 bytes from 10.0.0.2: icmp_seq=4 ttl=64 time=20.1 ms
^C
--- 10.0.0.2 ping statistics ---
4 packets transmitted, 4 received, 0% packet loss, time 7ms
rtt min/avg/max/mdev = 20.092/25.353/41.132/9.111 ms
root@admin-pc:~#

Fig. 110 Output of `ping 10.0.0.2` command

Step 3. Modify the bandwidth of the link connecting the switches S1 and S2: on the same terminal, type the command below. This command sets the bandwidth to 10 Gbps on S1's *s1-eth2* interface (Fig. 109). The `tbf` parameters are the following:

- `rate`: 10gbit
- `burst`: 5,000,000
- `limit`: 25,000,000

```
sudo tc qdisc add dev s1-eth2 parent 1: handle 2: tbf rate
10gbit burst 5000000 limit 25000000
```

Step 3. On h1's terminal, type `ping 10.0.0.2`. This command tests the connectivity between host h1 and host h2. The test was initiated by h1 as the command is executed on h1's terminal.

To stop the test, press `Ctrl+c`. The figure below shows a successful connectivity test. Host h1 (10.0.0.1) sent four packets to host h2 (10.0.0.2), successfully receiving responses back (Fig. 110).

The result above indicates that all four packets were received successfully (0% packet loss) and that the minimum, average, maximum, and standard deviation of the Round-Trip Time (RTT) were 20.092, 25.353, 41.132, and 9.111 ms, respectively.

```
X                              "Host: h2"                           - ⌕ ×
root@admin-pc:~# iperf3 -s
- - - - - - - - - - - - - - - - - - - - - - - - - - - - - - - - - - - - - - - -
Server listening on 5201
- - - - - - - - - - - - - - - - - - - - - - - - - - - - - - - - - - - - - - - -
```

Fig. 111 Host h2 running iperf3 as server

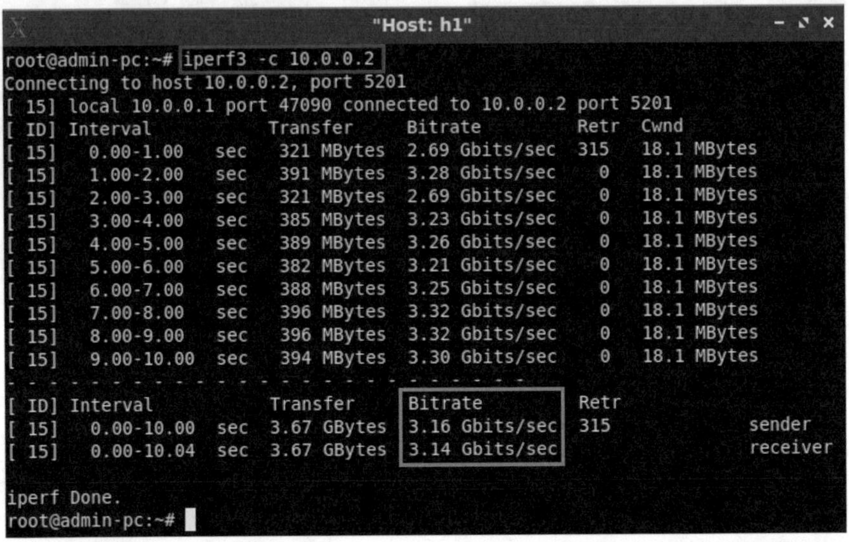

```
X                              "Host: h1"                           - ⌕ ×
root@admin-pc:~# iperf3 -c 10.0.0.2
Connecting to host 10.0.0.2, port 5201
[ 15] local 10.0.0.1 port 47090 connected to 10.0.0.2 port 5201
[ ID] Interval           Transfer     Bitrate         Retr  Cwnd
[ 15]   0.00-1.00   sec   321 MBytes  2.69 Gbits/sec  315   18.1 MBytes
[ 15]   1.00-2.00   sec   391 MBytes  3.28 Gbits/sec    0   18.1 MBytes
[ 15]   2.00-3.00   sec   321 MBytes  2.69 Gbits/sec    0   18.1 MBytes
[ 15]   3.00-4.00   sec   385 MBytes  3.23 Gbits/sec    0   18.1 MBytes
[ 15]   4.00-5.00   sec   389 MBytes  3.26 Gbits/sec    0   18.1 MBytes
[ 15]   5.00-6.00   sec   382 MBytes  3.21 Gbits/sec    0   18.1 MBytes
[ 15]   6.00-7.00   sec   388 MBytes  3.25 Gbits/sec    0   18.1 MBytes
[ 15]   7.00-8.00   sec   396 MBytes  3.32 Gbits/sec    0   18.1 MBytes
[ 15]   8.00-9.00   sec   396 MBytes  3.32 Gbits/sec    0   18.1 MBytes
[ 15]   9.00-10.00  sec   394 MBytes  3.30 Gbits/sec    0   18.1 MBytes
- - - - - - - - - - - - - - - - - - - - - - - - - - - - - - - - - - -
[ ID] Interval           Transfer     Bitrate         Retr
[ 15]   0.00-10.00  sec  3.67 GBytes  3.16 Gbits/sec  315           sender
[ 15]   0.00-10.04  sec  3.67 GBytes  3.14 Gbits/sec                receiver

iperf Done.
root@admin-pc:~#
```

Fig. 112 iPerf3 throughput test

The output above verifies that delay was injected successfully, as the RTT is approximately 20 ms.

Step 4. The user can now verify the rate limit configuration by using the iperf3 tool to measure throughput. To launch iPerf3 in server mode, run the command iperf3 -s in H2's terminal (Fig. 111):

```
iperf3 -s
```

Step 5. Now to launch iPerf3 in client mode again by running the command iperf3 -c 10.0.0.2 in h1's terminal (Fig. 112):

```
iperf3 -c 10.0.0.2
```

Notice the measured throughput now is approximately 3 Gbps, which is different than the value assigned in our $\boxed{\text{tbf}}$ rule. Next, we explain why the 10 Gbps maximum theoretical bandwidth is not achieved.

Step 6. In order to stop the server, press $\boxed{\text{Ctrl+c}}$ in host h2's terminal. The user can see the throughput results in the server side too.

18 BDP and Buffer Size

In connections that have a small BDP (either because the link has a low bandwidth or because the latency is small), buffers are usually small. However, in high-bandwidth high-latency networks, where the BDP is large, a larger buffer is required to achieve the maximum theoretical bandwidth.

18.1 Window Size in sysctl

The tool *sysctl* is used for dynamically changing parameters in the Linux operating system. It allows users to modify kernel parameters dynamically without rebuilding the Linux kernel.

The *sysctl key* for the receive window size is $\boxed{\text{net.ipv4.tcp\_rmem}}$, and the send window size is $\boxed{\text{net.ipv4.tcp\_wmem}}$.

Step 1. To read the current receiver window size value of host h1, use the following command on h1's terminal (Fig. 113):

```
sysctl net.ipv4.tcp_rmem
```

Step 2. To read the current send window size value of host h1, use the following command on h1's terminal (Fig. 114):

```
sysctl net.ipv4.tcp_wmem
```

Fig. 113 Receive window read in $\boxed{\text{sysctl}}$

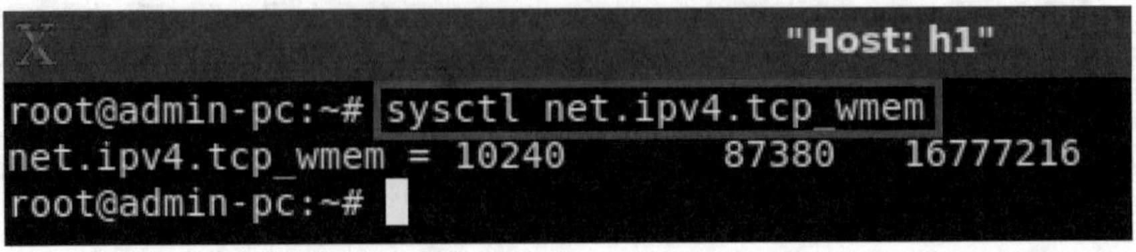

```
X                                              "Host: h1"

root@admin-pc:~# sysctl net.ipv4.tcp_wmem
net.ipv4.tcp_wmem = 10240          87380     16777216
root@admin-pc:~# 
```

Fig. 114 Send window read in sysctl

Fig. 115 Sample window
size from the previous test

```
Window size value: 16129
[Calculated window size: 8258048]
[Window size scaling factor: 512]
```

The returned values of both keys (net.ipv4.tcp_rmem and net.ipv4.tcp_wmem) are measured in bytes. The first number represents the minimum buffer size that is used by each TCP socket. The middle one is the default buffer that is allocated when applications create a TCP socket. The last one is the maximum receive buffer that can be allocated for a TCP socket.

The default values used by Linux are:

- Minimum: 10,240
- Default: 87,380
- Maximum: 16,777,216

In the previous test (10 Gbps, 20 ms delay), the buffer size was not modified on end-hosts. The BDP for the above test is

$$\text{BDP} = (10 \cdot 10^9) \cdot (20 \cdot 10^{-3}) = 200{,}000{,}000 \, \text{bits} = 25{,}000{,}000 \, \text{bytes}$$
$$\approx 25 \, \text{Mbytes.}$$

Note that this value is significantly greater than the maximum buffer size (16 Mbytes), and therefore, the pipe is not getting filled, which leads to network resources underutilization. Moreover, since Linux systems by default use half of the send/receive TCP buffers for internal kernel structures (see Sect. 16.3 Linux systems configuration), only half of the buffer size is used to store TCP segments. Figure 115 shows the calculated window size of a sample packet of the previous test—approximately 8 Mbytes. This is 50% of the default buffer size used by Linux (16 Mbytes).

Note that the observation in Fig. 115 reinforces the best practice described in Sect. 16.3: in Linux systems, the TCP buffer size must be at least twice the BDP.

19 Modifying Buffer Size and Throughput Test

This section repeats the throughput test of Sect. 19 after modifying the buffer size according to the formula described above. This test assumes the same network parameters introduced in the previous test; therefore, the bandwidth is limited to 1 Gbps, and the RTT (delay or latency) is 20 ms. The send and receive buffer sizes should be set to at least 2 •BDP (if BBR is used as the congestion control algorithm, this should be set to even a larger value, as described in Sect. 19). We will use 25 Mbytes value for the BDP instead of 25,000,000 bytes (1 Mbyte = 1024^2 bytes)

$$BDP = 25\,Mbytes = 25 \cdot 1024^2\,bytes = 26{,}214{,}400\,bytes$$

$$TCP\,buffer\,size = 2 \cdot BDP = 2 \cdot 26{,}214{,}400\,bytes = 52{,}428{,}800\,bytes.$$

Step 1. To change the TCP receive window size value(s), use the following command on h1's terminal. The values set are: 10,240 (minimum), 87,380 (default), and 52,428,800 (maximum, calculated by doubling the BDP) (Fig. 116).

```
sysctl -w net.ipv4.tcp_rmem='10240 87380 52428800'
```

The returned values are measured in bytes. 10,240 represents the minimum buffer size that is used by each TCP socket. 87,380 is the default buffer that is allocated when applications create a TCP socket. 52428800 is the maximum receive buffer that can be allocated for a TCP socket.

Step 2. To change the current send window size value(s), use the following command on h1's terminal. The values set are: 10,240 (minimum), 87,380 (default), and 52,428,800 (maximum, calculated by doubling the BDP) (Fig. 117).

Fig. 116 Receive window change in sysctl

Fig. 117 Send window change in sysctl

```
"Host: h2"
root@admin-pc:~# sysctl -w net.ipv4.tcp_wmem='10240 87380 52428800'
net.ipv4.tcp_wmem = 10240 87380 52428800
root@admin-pc:~#
```

Fig. 118 Receive window change in sysctl

```
"Host: h2"                                          –  ↙ ×
root@admin-pc:~# iperf3 -s
- - - - - - - - - - - - - - - - - - - - - - - - - - - - - - - - - - -
Server listening on 5201
- - - - - - - - - - - - - - - - - - - - - - - - - - - - - - - - - - -
```

Fig. 119 Send window change in sysctl

```
sysctl -w net.ipv4.tcp_wmem='10240 87380 52428800'
```

Next, the same commands must be configured on host h2.

Step 3. To change the current receiver window size value(s), use the following command on h2's terminal. The values set are: 10,240 (minimum), 87,380 (default), and 52,428,800 (maximum, calculated by doubling the BDP) (Fig. 118).

```
sysctl -w net.ipv4.tcp_rmem='10240 87380 52428800'
```

Step 4. To change the current send window size value(s), use the following command on h2's terminal. The values set are: 10,240 (minimum), 87,380 (default), and 52,428,800 (maximum, calculated by doubling the BDP) (Fig. 119).

```
sysctl -w net.ipv4.tcp_wmem='10240 87380 52428800'
```

Step 5. The user can now verify the rate limit configuration by using the iperf3 tool to measure throughput. To launch iPerf3 in server mode, run the command iperf3 -s in h2's terminal (Fig. 120):

```
iperf3 -s
```

Step 6. Now to launch iPerf3 in client mode again by running the command iperf3 -c 10.0.0.2 in h1's terminal (Fig. 121):

```
                                    "Host: h1"                          -  ⌄ x
root@admin-pc:~# iperf3 -c 10.0.0.2
Connecting to host 10.0.0.2, port 5201
[ 15] local 10.0.0.1 port 47094 connected to 10.0.0.2 port 5201
[ ID] Interval           Transfer     Bitrate         Retr  Cwnd
[ 15]   0.00-1.00   sec   925 MBytes  7.76 Gbits/sec    45   39.8 MBytes
[ 15]   1.00-2.00   sec  1.11 GBytes  9.57 Gbits/sec     0   39.8 MBytes
[ 15]   2.00-3.00   sec  1.11 GBytes  9.56 Gbits/sec     0   39.8 MBytes
[ 15]   3.00-4.00   sec  1.11 GBytes  9.56 Gbits/sec     0   39.8 MBytes
[ 15]   4.00-5.00   sec  1.11 GBytes  9.56 Gbits/sec     0   39.8 MBytes
[ 15]   5.00-6.00   sec  1.11 GBytes  9.55 Gbits/sec     0   39.8 MBytes
[ 15]   6.00-7.00   sec  1.11 GBytes  9.56 Gbits/sec     0   39.8 MBytes
[ 15]   7.00-8.00   sec  1.11 GBytes  9.56 Gbits/sec     0   39.8 MBytes
[ 15]   8.00-9.00   sec  1.11 GBytes  9.56 Gbits/sec     0   39.8 MBytes
[ 15]   9.00-10.00  sec  1.11 GBytes  9.56 Gbits/sec     0   39.8 MBytes
- - - - - - - - - - - - - - - - - - - - - - - - - - -
[ ID] Interval           Transfer     Bitrate         Retr
[ 15]   0.00-10.00  sec  10.9 GBytes  9.38 Gbits/sec    45              sender
[ 15]   0.00-10.04  sec  10.9 GBytes  9.34 Gbits/sec                    receiver

iperf Done.
root@admin-pc:~# █
```

Fig. 120 Host h2 running iPerf3 as server

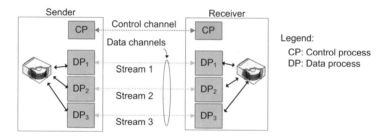

Fig. 121 iPerf3 throughput test

```
iperf3 -c 10.0.0.2
```

Note the measured throughput now is approximately 10 Gbps, which is close to the value assigned in our tbf rule (10 Gbps).

20   Chapter 4—Lab 12: Enhancing TCP Throughput with Parallel Streams

Overview
To conduct the experiment described in this section, please login into the Academic Cloud at http://highspeednetworks.net/, and reserve a pod for Lab 12.

Table 5 Credentials to access Client1 machine. Average throughput $\approx \frac{MSS}{RTT} \sqrt{L}$ bytes per second

Device	Account	Password
Client1	admin	password

This lab introduces TCP parallel streams in Wide Area Networks (WANs) and explains how they are used to achieve higher throughput. Then, throughput tests using parallel streams are conducted.

Objectives

By the end of this lab, students should be able to:

1. Understand TCP parallel streams.
2. Describe the advantages of TCP parallel streams.
3. Specify the number of parallel streams in an iPerf3 test.
4. Conduct tests and measure performance of parallel streams on an emulated WAN.

Lab Settings

The information in Table 5 provides the credentials of the machine containing Mininet.

Lab Roadmap

This lab is organized as follows:

1. Section 21: Introduction to TCP parallel streams
2. Section 22: Lab topology
3. Section 23: Parallel streams in a high-latency high-bandwidth WAN
4. Section 24: Parallel streams with packet loss

21   Introduction to TCP Parallel Streams

21.1   Parallel Stream Fundamentals

Parallel streams are multiple TCP connections opened by an application to increase performance and maximize the throughput between communicating hosts. With parallel streams, data blocks for a single file transmitted from a sender to a receiver are distributed over the multiple streams. Figure 122 shows the basic model. A control channel is established between the sender and the receiver to coordinate the data transfer. The actual transfer occurs over the parallel streams, collectively referred to as data channels. In this context, the term stream is a synonym of flow and connection.

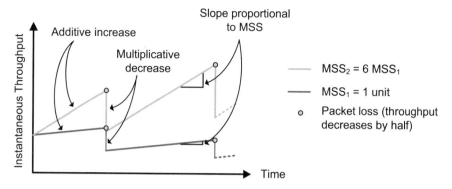

Fig. 122 Data transfer model with parallel streams

21.2   Advantages of Parallel Streams

Transferring large files over high-latency WANs with parallel streams has multiple benefits, as described next.

Combat Random Packet Loss Not due Congestion assume that packet loss occurs randomly rather than due congestion. In steady state, the average throughput of a single TCP stream is given by

$$\textbf{Average throughput} \approx \frac{\textbf{MSS}}{\textbf{RTT}\sqrt{\textbf{L}}} \textbf{ bytes per sound,}$$

where MSS is the maximum segment size and L is the packet loss rate. The above equation indicates that the throughput is directly proportional to the MSS and inversely proportional to RTT and the square root of L. When an application uses K parallel streams and if RTT, packet loss, and MSS are the same in each stream, the aggregate average throughput is the aggregation of the K single stream throughputs:

$$\text{Aggregate average throughput} \approx \sum_{i=1}^{K} \frac{MSS}{RTT\sqrt{L}} = K \cdot \frac{MSS}{RTT\sqrt{L}} \text{ bytes per second.}$$

Thus, an application opening K parallel connections essentially creates a large virtual MSS on the aggregate connection that is K times the MSS of a single connection.

The TCP throughput follows the additive increase multiplicative decrease (AIMD) rule: TCP continuously probes for more bandwidth and increases the throughput of a connection by approximately 1 MSS per RTT as long as no packet loss occurs (additive increase phase). When a packet loss occurs, the throughput is

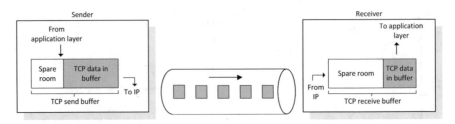

Fig. 123 Additive increase multiplicative decrease (AIMD) behavior. The green curve corresponds to the throughput when the MSS is six times that of the red curve

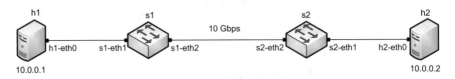

Fig. 124 TCP send and receive buffers

reduced by half (multiplicative decrease event). Figure 123 illustrates the AIMD behavior for two connections with different MSSs. The MSS of the green connection is six times than the MSS of the red connection. Since during the additive increase phase TCP increases the throughput by one MSS every RTT, the speed at which the throughput increases is proportional to the MSS (i.e., the larger the MSS the faster the recovery after a packet loss).

Mitigate TCP Round-Trip Time (RTT) Bias When different flows with different RTTs share a given bottleneck link, TCP's throughput is inversely proportional to the RTT. This is also noted in the equations discussed above. Hence, low-RTT flows get a higher share of the bandwidth than high-RTT flows. Thus, for transfers across high-latency WANs, one approach to combat the higher (unfair) bandwidth allocated to low-latency connections is by using parallel streams. By doing so, even if each high-latency stream receives a less amount of bandwidth than low-latency flows, the aggregate throughput of the parallel streams can be high.

Overcome TCP Buffer Limitation TCP receives data from the application layer and places it in the TCP buffer, as shown in Fig. 124. TCP implements flow control by requiring the receiver indicate how much spare room is available in the TCP receive buffer. For a full utilization of the path, the TCP send and receive buffers must be greater than or equal to the bandwidth-delay product (BDP). This buffer size value is the maximum number of bits that can be outstanding (inflight) if the sender continuously sends segments. If the buffer size is less than the bandwidth-delay product, then throughput will not be maximized. One solution to overcome small TCP buffer size situations is by using parallel streams. Essentially, an application opening K parallel connections creates a large buffer size on the aggregate connection that is K times the buffer size of a single connection.

In this lab, we will explore the use of parallel streams to overcome TCP buffer limitation and to mitigate random packet loss.

22   Lab Topology

Let us get started with creating a simple Mininet topology using MiniEdit. The topology uses 10.0.0.0/8, which is the default network assigned by Mininet (Fig. 125).

The above topology uses 10.0.0.0/8, which is the default network assigned by Mininet.

Fig. 125 Lab topology

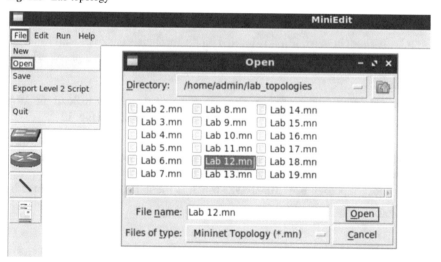

Fig. 126 MiniEdit shortcut

Fig. 127 MiniEdit's *Open* dialog

Fig. 128 Running the emulation

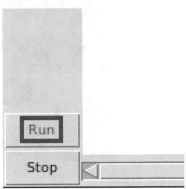

Step 1. A shortcut to MiniEdit is located on the machine's desktop. Start MiniEdit by clicking on MiniEdit's shortcut (Fig. 126). When prompted for a password, type password .

Step 2. On MiniEdit's menu bar, click on *File* and then *Open* to load the lab's topology. Locate the *Lab 12.mn* topology file and click on *Open* (Fig. 127).

Step 3. Before starting the measurements between host h1 and host h2, the network must be started. Click on the *Run* button located at the bottom left of MiniEdit's window to start the emulation (Fig. 128).

The above topology uses 10.0.0.0/8, which is the default network assigned by Mininet.

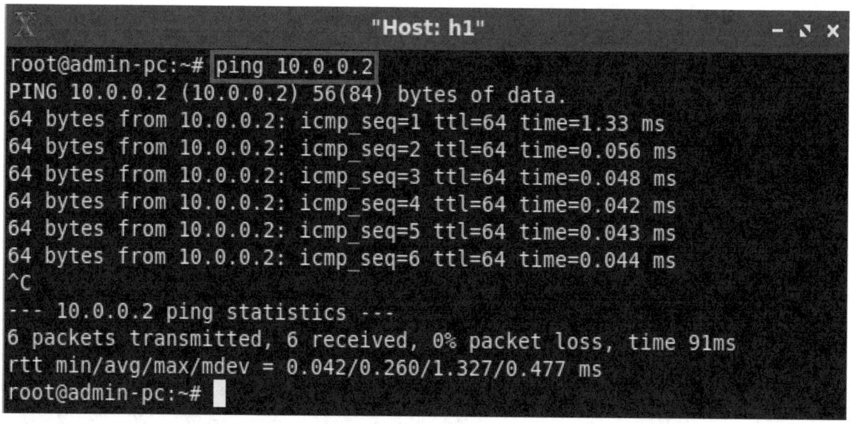

```
"Host: h1"                              -  ↘ ✕
root@admin-pc:~# ping 10.0.0.2
PING 10.0.0.2 (10.0.0.2) 56(84) bytes of data.
64 bytes from 10.0.0.2: icmp_seq=1 ttl=64 time=1.33 ms
64 bytes from 10.0.0.2: icmp_seq=2 ttl=64 time=0.056 ms
64 bytes from 10.0.0.2: icmp_seq=3 ttl=64 time=0.048 ms
64 bytes from 10.0.0.2: icmp_seq=4 ttl=64 time=0.042 ms
64 bytes from 10.0.0.2: icmp_seq=5 ttl=64 time=0.043 ms
64 bytes from 10.0.0.2: icmp_seq=6 ttl=64 time=0.044 ms
^C
--- 10.0.0.2 ping statistics ---
6 packets transmitted, 6 received, 0% packet loss, time 91ms
rtt min/avg/max/mdev = 0.042/0.260/1.327/0.477 ms
root@admin-pc:~# ▊
```

Fig. 129 Opening a terminal on host h1

Fig. 130 Connectivity test using ping command

22.1   Starting Host h1 and Host h2

Step 1. Hold the right-click on host h1 and select *Terminal*. This opens the terminal of host h1 and allows the execution of commands on that host (Fig. 129).

Step 2. Apply the same steps on host h2 and open its *Terminals*.

Step 3. Test connectivity between the end-hosts using the ping command. On host h1, type the command ping 10.0.0.2 . This command tests the connectivity between host h1 and host h2. To stop the test, press Ctrl+c . The figure below shows a successful connectivity test (Fig. 130).

Figure 130 indicates that there is connectivity between host h1 and host h2. Thus, we are ready to start the throughput measurement process.

Fig. 131 Shortcut to open a Linux terminal

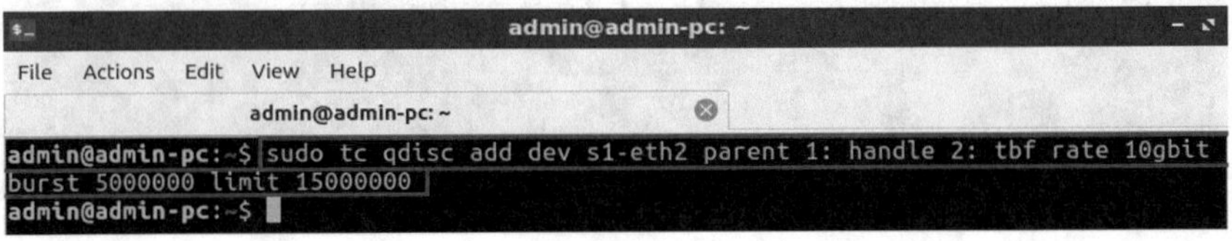

Fig. 132 Adding delay of 20 ms to switch S1's *s1-eth2* interface

22.2   Emulating 10 Gbps High-Latency WAN

This section emulates a high-latency WAN. We will first emulate 20 ms delay between switch S1 and switch S2 to measure the throughput. Then, we will set the bandwidth between host h1 and host h2 to 10 Gbps.

Step 1. Launch a Linux terminal by holding the Ctrl+Alt+T keys or by clicking on the Linux terminal icon (Fig. 131).

The Linux terminal is a program that opens a window and permits you to interact with a command-line interface (CLI). A CLI is a program that takes commands from the keyboard and sends them to the operating system for execution.

Step 2. In the terminal, type the command below. When prompted for a password, type password and hit enter. This command introduces 20 ms delay on switch S1's *s1-eth2* interface (Fig. 132).

```
sudo tc qdisc add dev s1-eth2 root handle 1: netem delay
20 ms
```

Step 3. Modify the bandwidth of the link connecting the switch S1 and switch S2: on the same terminal, type the command below. This command sets the bandwidth to 10 Gbps on switch S1's *s1-eth2* interface (Fig. 133). The tbf parameters are the following:

- rate : 10gbit
- burst : 5,000,000
- limit : 15,000,000

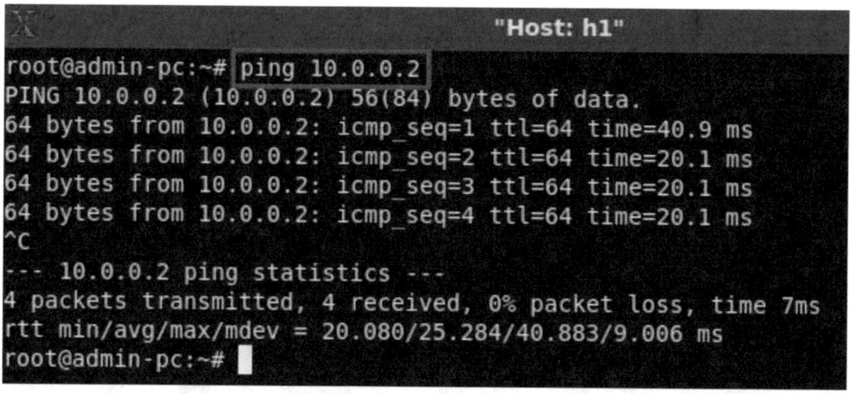

```
"Host: h1"
root@admin-pc:~# ping 10.0.0.2
PING 10.0.0.2 (10.0.0.2) 56(84) bytes of data.
64 bytes from 10.0.0.2: icmp_seq=1 ttl=64 time=40.9 ms
64 bytes from 10.0.0.2: icmp_seq=2 ttl=64 time=20.1 ms
64 bytes from 10.0.0.2: icmp_seq=3 ttl=64 time=20.1 ms
64 bytes from 10.0.0.2: icmp_seq=4 ttl=64 time=20.1 ms
^C
--- 10.0.0.2 ping statistics ---
4 packets transmitted, 4 received, 0% packet loss, time 7ms
rtt min/avg/max/mdev = 20.080/25.284/40.883/9.006 ms
root@admin-pc:~# █
```

Fig. 133 Limiting the bandwidth to 10 Gbps on switch S1's *s1-eth2* interface

```
"Host: h2"                                          — ⌕ ✕
root@admin-pc:~# iperf3 -s
- - - - - - - - - - - - - - - - - - - - - - - - - - - - - - - -
Server listening on 5201
- - - - - - - - - - - - - - - - - - - - - - - - - - - - - - - -
█
```

Fig. 134 Output of ping 10.0.0.2 command

```
sudo tc qdisc add dev s1-eth2 parent 1: handle 2: tbf rate
10gbit burst 5000000 limit 15000000
```

22.3   Testing Connection

To test connectivity, you can use the command ping .

Step 1. On the terminal of host h1, type ping 10.0.0.2 . To stop the test, press
Ctrl+c . The figure below shows a successful connectivity test. Host h1 (10.0.0.1)
sent four packets to host h2 (10.0.0.2), successfully receiving responses back
(Fig. 134).

The result above indicates that all four packets were received successfully (0%
packet loss) and that the minimum, average, maximum, and standard deviation of the
Round-Trip Time (RTT) were 20.080, 25.284, 40.883, and 9.006 ms, respectively.
The output above verifies that delay was injected successfully, as the RTT is
approximately 20 ms.

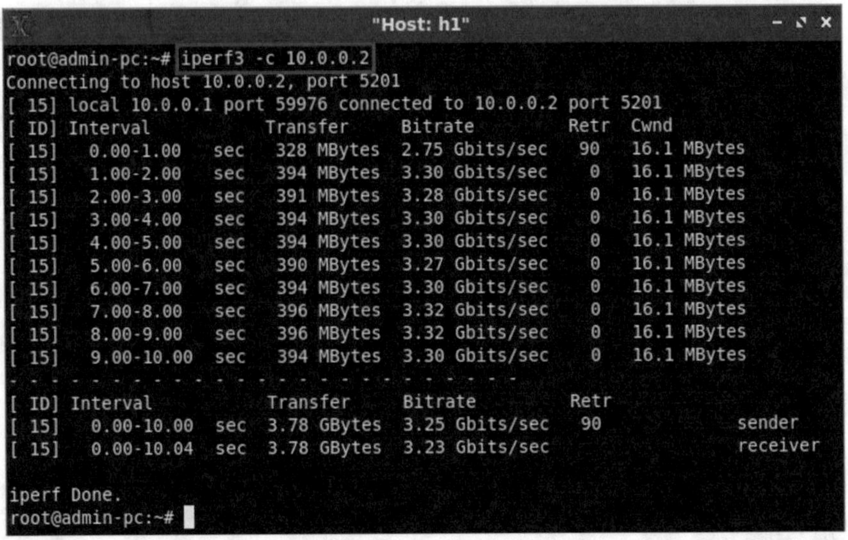

```
X                              "Host: h1"                        -  ⬩ ✕
root@admin-pc:~# iperf3 -c 10.0.0.2
Connecting to host 10.0.0.2, port 5201
[ 15] local 10.0.0.1 port 59976 connected to 10.0.0.2 port 5201
[ ID] Interval           Transfer     Bitrate        Retr  Cwnd
[ 15]   0.00-1.00   sec   328 MBytes  2.75 Gbits/sec   90   16.1 MBytes
[ 15]   1.00-2.00   sec   394 MBytes  3.30 Gbits/sec    0   16.1 MBytes
[ 15]   2.00-3.00   sec   391 MBytes  3.28 Gbits/sec    0   16.1 MBytes
[ 15]   3.00-4.00   sec   394 MBytes  3.30 Gbits/sec    0   16.1 MBytes
[ 15]   4.00-5.00   sec   394 MBytes  3.30 Gbits/sec    0   16.1 MBytes
[ 15]   5.00-6.00   sec   390 MBytes  3.27 Gbits/sec    0   16.1 MBytes
[ 15]   6.00-7.00   sec   394 MBytes  3.30 Gbits/sec    0   16.1 MBytes
[ 15]   7.00-8.00   sec   396 MBytes  3.32 Gbits/sec    0   16.1 MBytes
[ 15]   8.00-9.00   sec   396 MBytes  3.32 Gbits/sec    0   16.1 MBytes
[ 15]   9.00-10.00  sec   394 MBytes  3.30 Gbits/sec    0   16.1 MBytes
- - - - - - - - - - - - - - - - - - - - - - - - -
[ ID] Interval           Transfer     Bitrate        Retr
[ 15]   0.00-10.00  sec  3.78 GBytes  3.25 Gbits/sec   90            sender
[ 15]   0.00-10.04  sec  3.78 GBytes  3.23 Gbits/sec                 receiver

iperf Done.
root@admin-pc:~# ▮
```

Fig. 135 Starting iPerf3 server on host h2

```
X                              "Host: h2"                        -  ⬩ ✕
root@admin-pc:~# iperf3 -s
------------------------------------------------------------
Server listening on 5201
------------------------------------------------------------
▮
```

Fig. 136 Running iPerf3 client on host h1

Step 2. On the terminal of host h2, type ping 10.0.0.1 . The ping output in this test should be relatively close to the results of the test initiated by host h1 in Step 1. To stop the test, press Ctrl+c . ,

Step 3. Launch iPerf3 in server mode on host h2's terminal (Fig. 135).

```
iperf3 -s
```

Step 4. Launch iPerf3 in client mode on host h1's terminal. To stop the test, press Ctrl+c (Fig. 136).

```
iperf3 -c 10.0.0.2
```

Although the link was configured to 10 Gbps, the test results show that the achieved throughput is 3.22 Gbps. This is because the TCP buffer size is less than the bandwidth-delay product. In the upcoming section, we run a throughput test without modifying the TCP buffer size, but with multiple parallel streams.

Step 5. In order to stop the server, press Ctrl+c in host h2's terminal. The user can see the throughput results in the server side too.

23 Parallel Streams to Overcome TCP Buffer Limitation

In this section, parallel streams are specified by the client when executing the throughput test in iPerf3. The iPerf3 server should start as usual, without specifying any additional options or parameters.

Step 1. To launch iPerf3 in server mode, run the command iperf3 -s in host h2's terminal as shown the figure below (Fig. 137):

```
iperf3 -s
```

Step 2. Now the iPerf3 client should be launched with the -P option specified (not to be confused with the -p option that specifies the listening port number). This option specifies the number of parallel streams. Run the following command in host h1's terminal (Fig. 138):

```
                                    "Host: h1"                          _  ☐ ✗
root@admin-pc:~# iperf3 -c 10.0.0.2 -P 8
Connecting to host 10.0.0.2, port 5201
[ 15] local 10.0.0.1 port 60000 connected to 10.0.0.2 port 5201
[ 17] local 10.0.0.1 port 60002 connected to 10.0.0.2 port 5201
[ 19] local 10.0.0.1 port 60004 connected to 10.0.0.2 port 5201
[ 21] local 10.0.0.1 port 60006 connected to 10.0.0.2 port 5201
[ 23] local 10.0.0.1 port 60008 connected to 10.0.0.2 port 5201
[ 25] local 10.0.0.1 port 60010 connected to 10.0.0.2 port 5201
[ 27] local 10.0.0.1 port 60012 connected to 10.0.0.2 port 5201
[ 29] local 10.0.0.1 port 60014 connected to 10.0.0.2 port 5201
[ ID] Interval         Transfer     Bitrate         Retr  Cwnd
[ 15]  0.00-1.00   sec   221 MBytes  1.85 Gbits/sec     5   13.2 MBytes
[ 17]  0.00-1.00   sec   206 MBytes  1.73 Gbits/sec  1477   9.61 MBytes
[ 19]  0.00-1.00   sec   139 MBytes  1.16 Gbits/sec  1935   3.23 MBytes
[ 21]  0.00-1.00   sec   138 MBytes  1.16 Gbits/sec  4151   5.16 MBytes
[ 23]  0.00-1.00   sec  64.3 MBytes   539 Mbits/sec  2630   2.39 MBytes
[ 25]  0.00-1.00   sec   104 MBytes   874 Mbits/sec  2823   3.57 MBytes
[ 27]  0.00-1.00   sec  64.3 MBytes   539 Mbits/sec  1815   1.95 MBytes
[ 29]  0.00-1.00   sec  89.3 MBytes   748 Mbits/sec  2105   2.88 MBytes
[SUM]  0.00-1.00   sec  1.00 GBytes  8.60 Gbits/sec 16941
```

Fig. 137 Host h2 running iPerf3 as server

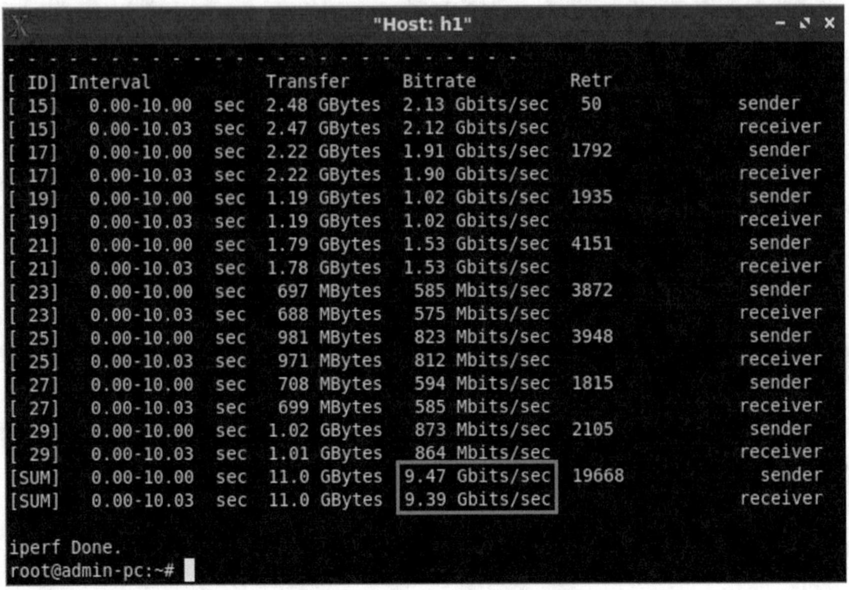

```
X                           "Host: h1"                              -  ⌁  x
- - - - - - - - - - - - - - - - - - - - - - - - -
[ ID] Interval           Transfer     Bitrate         Retr
[ 15]  0.00-10.00  sec  2.48 GBytes  2.13 Gbits/sec   50            sender
[ 15]  0.00-10.03  sec  2.47 GBytes  2.12 Gbits/sec                 receiver
[ 17]  0.00-10.00  sec  2.22 GBytes  1.91 Gbits/sec  1792           sender
[ 17]  0.00-10.03  sec  2.22 GBytes  1.90 Gbits/sec                 receiver
[ 19]  0.00-10.00  sec  1.19 GBytes  1.02 Gbits/sec  1935           sender
[ 19]  0.00-10.03  sec  1.19 GBytes  1.02 Gbits/sec                 receiver
[ 21]  0.00-10.00  sec  1.79 GBytes  1.53 Gbits/sec  4151           sender
[ 21]  0.00-10.03  sec  1.78 GBytes  1.53 Gbits/sec                 receiver
[ 23]  0.00-10.00  sec   697 MBytes   585 Mbits/sec  3872           sender
[ 23]  0.00-10.03  sec   688 MBytes   575 Mbits/sec                 receiver
[ 25]  0.00-10.00  sec   981 MBytes   823 Mbits/sec  3948           sender
[ 25]  0.00-10.03  sec   971 MBytes   812 Mbits/sec                 receiver
[ 27]  0.00-10.00  sec   708 MBytes   594 Mbits/sec  1815           sender
[ 27]  0.00-10.03  sec   699 MBytes   585 Mbits/sec                 receiver
[ 29]  0.00-10.00  sec  1.02 GBytes   873 Mbits/sec  2105           sender
[ 29]  0.00-10.03  sec  1.01 GBytes   864 Mbits/sec                 receiver
[SUM]  0.00-10.00  sec  11.0 GBytes  9.47 Gbits/sec  19668          sender
[SUM]  0.00-10.03  sec  11.0 GBytes  9.39 Gbits/sec                 receiver

iperf Done.
root@admin-pc:~#
```

Fig. 138 iPerf3 throughput test with parallel streams

```
$_
File    Actions   Edit   View   Help
                  admin@admin-pc: ~                          ⊗
admin@admin-pc:~$ sudo tc qdisc del dev s1-eth2 root
admin@admin-pc:~$
```

Fig. 139 iPerf3 throughput test with parallel streams summary output

```
iperf3 -c 10.0.0.2 -P 8
```

The above command uses 8 parallel streams. Note that 8 sockets are now opened on different local ports, and their streams are connected to the server, ready for transmitting data and performing the throughput test (Fig. 139).

Note the measured throughput now is approximately 9.5 Gbps, which is close to the value assigned in the tbf rule (10 Gbps).

Step 3. In order to stop the server, press Ctrl+c in host h2's terminal. The user can see the throughput results in the server side too.

24 Parallel Streams to Combat Packet Loss

Packet loss is inevitable in real-world networks. This section explores the use of parallel streams to mitigate packet loss not due congestion (i.e., random packet loss) and compares the performance of single and parallel streams.

24.1 Limit Rate and Add Packet Loss on Switch S1's s1-eth2 Interface

In this topology, rate limiting is applied on switch S1's interface that connects it to switch S2 (*s1-eth2*) and 1% packet loss is introduced.

Step 1. Before applying any additional configuration, the previous rules assigned on the switch's interface must be deleted. To remove these, type the following command on the Client's terminal. When prompted for a password, type password and hit enter (Fig. 140).

```
sudo tc qdisc del dev s1-eth2 root
```

Step 2. On the same terminal, type the below command to add 1% packet loss (Fig. 141).

```
sudo tc qdisc add dev s1-eth2 root handle 1: netem loss 1%
```

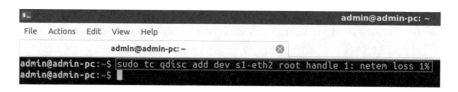

Fig. 140 Deleting previous rules on switch S1's *s1-eth2* interface

Fig. 141 Adding 1% packet loss to switch S1's *s1-eth2* interface

```
                    "Host: h2"                    –  ⤡  ✕
root@admin-pc:~# iperf3 -s
- - - - - - - - - - - - - - - - - - - - - - - - - - - - - - - - - -
Server listening on 5201
- - - - - - - - - - - - - - - - - - - - - - - - - - - - - - - - - -
▊
```

Fig. 142 Limiting the bandwidth to 10 Gbps on switch S1's *s1-eth2* interface

Step 3. Modify the bandwidth of the link connecting the switch S1 and switch S2: on the same terminal, type the command below. This command sets the bandwidth to 10 Gbps on switch S1's *s1-eth2* interface (Fig. 142). The tbf parameters are the following:

- rate : 10gbit
- burst : 5,000,000
- limit : 15,000,000

```
sudo tc qdisc add dev s1-eth2 parent 1: handle 2: tbf rate
10gbit burst 5000000 limit 15000000
```

Step 3. The user can now verify the rate limit configuration by using the iperf3 tool to measure throughput. To launch iPerf3 in server mode, run the command iperf3 -s in host h2's terminal as shown in the figure below (Fig. 143):

```
iperf3 -s
```

Step 4. Launch iPerf3 in client mode on host h1's terminal. To stop the test, press Ctrl+c (Fig. 144).

```
iperf3 -c 10.0.0.2
```

Note the measured throughput now is approximately 7.6 Gbps, which is different than the value assigned in the tbf rule (10 Gbps).

Step 5. In order to stop the server, press Ctrl+c in host h2's terminal. The user can see the throughput results in the server side too.

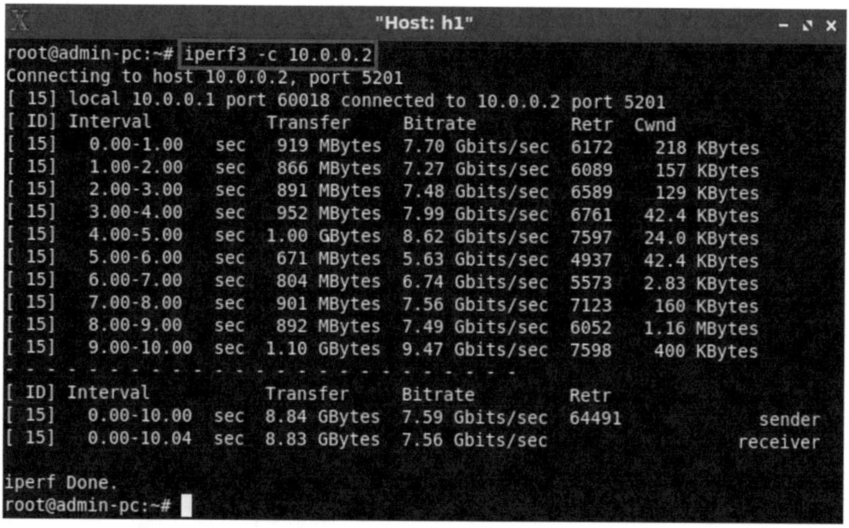

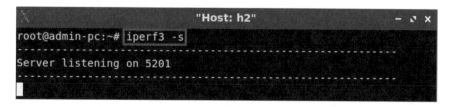

Fig. 143 Starting iPerf3 server on host h2

Fig. 144 Running iPerf3 client on host h1

24.2   Test with Parallel Streams

Step 1. Now the test is repeated while using parallel streams. To launch iPerf3 in server mode, run the command iperf3 -s in host h2's terminal as shown in Fig. 145:

```
iperf3 -s
```

Step 2. Now the iPerf3 client should be launched with the -P option specified (not to be confused with the -p option that specifies the listening port number). This option specifies the number of parallel streams. Run the following command in host h1's terminal (Fig. 146):

```
iperf3 -c 10.0.0.2 -P 8
```

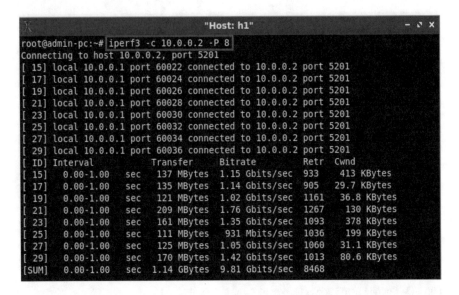

```
                              "Host: h1"                           -  ⤢  ✕
root@admin-pc:~# iperf3 -c 10.0.0.2 -P 8
Connecting to host 10.0.0.2, port 5201
[ 15] local 10.0.0.1 port 60022 connected to 10.0.0.2 port 5201
[ 17] local 10.0.0.1 port 60024 connected to 10.0.0.2 port 5201
[ 19] local 10.0.0.1 port 60026 connected to 10.0.0.2 port 5201
[ 21] local 10.0.0.1 port 60028 connected to 10.0.0.2 port 5201
[ 23] local 10.0.0.1 port 60030 connected to 10.0.0.2 port 5201
[ 25] local 10.0.0.1 port 60032 connected to 10.0.0.2 port 5201
[ 27] local 10.0.0.1 port 60034 connected to 10.0.0.2 port 5201
[ 29] local 10.0.0.1 port 60036 connected to 10.0.0.2 port 5201
[ ID] Interval           Transfer     Bitrate         Retr  Cwnd
[ 15]   0.00-1.00   sec   137 MBytes  1.15 Gbits/sec   933     413 KBytes
[ 17]   0.00-1.00   sec   135 MBytes  1.14 Gbits/sec   905    29.7 KBytes
[ 19]   0.00-1.00   sec   121 MBytes  1.02 Gbits/sec  1161    36.8 KBytes
[ 21]   0.00-1.00   sec   209 MBytes  1.76 Gbits/sec  1267     130 KBytes
[ 23]   0.00-1.00   sec   161 MBytes  1.35 Gbits/sec  1093     378 KBytes
[ 25]   0.00-1.00   sec   111 MBytes   931 Mbits/sec  1036     199 KBytes
[ 27]   0.00-1.00   sec   125 MBytes  1.05 Gbits/sec  1060    31.1 KBytes
[ 29]   0.00-1.00   sec   170 MBytes  1.42 Gbits/sec  1013    80.6 KBytes
[SUM]   0.00-1.00   sec  1.14 GBytes  9.81 Gbits/sec  8468
```

Fig. 145 Host h2 running iPerf3 as server

```
                              "Host: h1"                           -  ⤢  ✕
- - - - - - - - - - - - - - - - - - - - - - - - - -
[ ID] Interval           Transfer     Bitrate         Retr
[ 15]   0.00-10.00  sec  1.48 GBytes  1.27 Gbits/sec  10341           sender
[ 15]   0.00-10.02  sec  1.47 GBytes  1.26 Gbits/sec                receiver
[ 17]   0.00-10.00  sec  1.34 GBytes  1.15 Gbits/sec   9173           sender
[ 17]   0.00-10.02  sec  1.33 GBytes  1.14 Gbits/sec                receiver
[ 19]   0.00-10.00  sec  1.35 GBytes  1.16 Gbits/sec  11049           sender
[ 19]   0.00-10.02  sec  1.34 GBytes  1.15 Gbits/sec                receiver
[ 21]   0.00-10.00  sec  1.41 GBytes  1.21 Gbits/sec  10069           sender
[ 21]   0.00-10.02  sec  1.41 GBytes  1.20 Gbits/sec                receiver
[ 23]   0.00-10.00  sec  1.34 GBytes  1.15 Gbits/sec   9948           sender
[ 23]   0.00-10.02  sec  1.34 GBytes  1.15 Gbits/sec                receiver
[ 25]   0.00-10.00  sec  1.53 GBytes  1.31 Gbits/sec  10783           sender
[ 25]   0.00-10.02  sec  1.52 GBytes  1.31 Gbits/sec                receiver
[ 27]   0.00-10.00  sec  1.33 GBytes  1.14 Gbits/sec  10676           sender
[ 27]   0.00-10.02  sec  1.32 GBytes  1.13 Gbits/sec                receiver
[ 29]   0.00-10.00  sec  1.41 GBytes  1.21 Gbits/sec  10025           sender
[ 29]   0.00-10.02  sec  1.40 GBytes  1.20 Gbits/sec                receiver
[SUM]   0.00-10.00  sec  11.2 GBytes  9.60 Gbits/sec  82064           sender
[SUM]   0.00-10.02  sec  11.1 GBytes  9.55 Gbits/sec                receiver

iperf Done.
root@admin-pc:~# █
```

Fig. 146 Host h1 running iPerf3 as client with 8 parallel streams

The above command uses 8 parallel streams. Note that 8 sockets are now opened on different local ports, and their streams are connected to the server, ready for transmitting data and performing the throughput test.

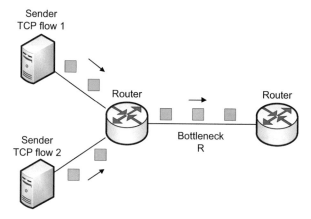

Fig. 147 iPerf3 throughput test with parallel streams summary output

Figure 147 shows that the measured throughput now is approximately 96 Gbps, which is close to the value assigned in our ⟨tbf⟩ rule (10 Gbps). In conclusion, parallel streams are beneficial when the packet loss rate is high. As shown in the previous test, when using parallel streams, the host was able to achieve the maximum theoretical bandwidth.

25   Chapter 4—Lab 13: Measuring TCP Fairness

Overview
To conduct the experiment described in this section, please login into the Academic Cloud at http://highspeednetworks.net/, and reserve a pod for Lab 13.

This lab introduces TCP fairness in Wide Area Networks (WAN) and explains how competing TCP connections converge to fairness. The lab describes how to calculate the TCP fairness index, a metric that quantifies how fair the aggregate connection is divided between active connections. Finally, the lab conducts throughput tests in an emulated high-latency network and derives the fairness index.

Objectives
By the end of this lab, students should be able to:

1. Define TCP fairness.
2. Calculate TCP fairness index.
3. Emulate a WAN and calculate fairness index among parallel streams.
4. Emulate a WAN and calculate fairness index among competing TCP connections.

Lab Settings
The information in Table 6 provides the credentials of the machine containing Mininet.

Table 6 Credentials to
access Client1 machine

Device	Account	Password
Client1	admin	password

Lab Roadmap

This lab is organized as follows:

1. Section 26: Fairness concepts
2. Section 27: Lab topology
3. Section 28: Calculating fairness among parallel flows
4. Section 29: Calculating fairness index with different congestion control algorithms

26 Fairness Concepts

26.1 TCP Bandwidth Allocation

Many networks do not use any bandwidth reservation mechanism for TCP flows passing through a router. Instead, routers simply make forwarding decisions based on the destination field of the IP header. As a result, flows may attempt to use as much bandwidth as possible. In this situation, it is the TCP congestion control algorithm that allocates bandwidth to the competing flows.

Consider the scenario where two TCP flows share a bottleneck link with bandwidth capacity R, as illustrated in Fig. 148. Assume that the two senders are in equal conditions (Round-Trip Time, maximum segment size, configuration parameters) and that they use the same congestion control algorithm. Furthermore, assume that the two flows are in steady state and that the congestion control algorithm operates according to the additive increase multiplicative decrease (AIMD) rule. A fair bandwidth allocation would result in a bandwidth partition of R/2 to each flow.

Figure 149 shows the bandwidth allocation region for the two flows. The bandwidth allocation to flow 1 is on x-axis and to flow 2 is on the y-axis. If TCP is to share the bottleneck bandwidth equally between the two flows, then the bandwidth will fall along the fairness line emanating from the origin. Note that the origin (0, 0) is a fair but undesirable solution. When the allocations sum to 100% of the bottleneck capacity, the allocation is efficient. This is shown by the efficiency line. Note that potential *efficient* solutions include points A (R, 0) and points B (0, R). On point A, flow 1 receives 100% of the capacity, and on point B flow 2 receives 100% of the capacity. Clearly, these solutions are not desirable, as they lead to starvation and unfairness.

Assume that the sending rates of senders 1 and 2 at a given time are indicated by point p_1. As the amount of aggregate bandwidth jointly consumed by the two

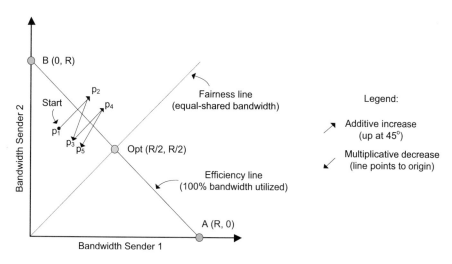

Fig. 148 Two TCP flows that share a bottleneck link of capacity R

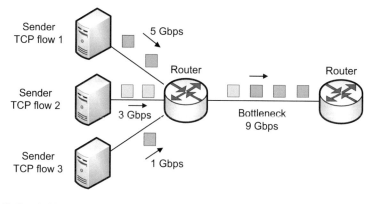

Fig. 149 Bandwidth allocation region realized by two competing TCP flows

flows is less than R, no loss will occur, and TCP will gently increase the bandwidth allocation (this process is called additive increase phase). Eventually, the bandwidth jointly consumed by the two connections will be greater than R, and a packet loss will occur at a point, say p_2. TCP reacts to a packet loss by aggressively decreasing the sending rate by half (this operation is called multiplicate decrease). The resulting bandwidth allocations are realized at point p_3. Since the joint bandwidth use is less than R at point p_3, TCP will again increase the allocation to flows 1 and 2. Eventually, the TCP additive increase phase will lead to the operating point p_4, where a loss will again occur, and the two flows again will see a decrease in the bandwidth allocation, and so on. The bandwidth realized by the two flows eventually will fluctuate along the fairness line, near the optimal operating point Opt (R/2, R/2). Chiu and Jain describe the reasons of why TCP converges to a fair and efficient allocation. This convergence occurs independently of the starting point.

26.2 TCP Fairness Index Calculation

A useful index to quantify fairness is Jain's index. The index has the following properties:

1. *Population size independence:* The index is applicable to any number of flows.
2. *Scale and metric independence:* The index is independent of scale, i.e., the unit of measurement does not matter.
3. *Boundedness:* The index is bounded between 0 and 1. A totally fair system has an index of 1, and a totally unfair system has an index of 0.
4. *Continuity:* The index is continuous. Any change in allocation is reflected in the fairness index.

Jain's fairness index is given by the following equation:

$$I = \frac{\left(\sum_{i=1}^{n} T_i\right)^2}{n \sum_{i=1}^{n} T_i^2},$$

where

- I is the fairness index, with values between 0 and 1.
- n is the total number of flows.
- $T_1 T_2, \ldots, T_n$ are the measured throughput of individual flows.

As an example of fairness index calculation, consider the three flows shown in Fig. 150. Given the bottleneck capacity of 9 Gbps, assume that the bandwidth allocations for flows 1, 2, and 3 are 5 Gbps, 3 Gbps, and 1 Gbps. The fairness index for this allocation is

$$I = \frac{\left(\sum_{i=1}^{3} T_i\right)^2}{3 \sum_{i=1}^{3} T_i^2} = \frac{\left(5 \cdot 10^9 + 3 \cdot 10^9 + 1 \cdot 10^9\right)^2}{3 \cdot \left((5 \cdot 10^9)^2 + (3 \cdot 10^9)^2 + (1 \cdot 10^9)^2\right)} = 0.77$$

Note that by property 2 (scale and metric independence), the fairness index of the above example is the same as that of an allocation of 5 Mbps, 3 Mbps, and 1 Mbps (or more generally, an allocation of 5, 3, and 1 units). Also, note that an optimal fair allocation of 3 Gbps to each flow would produce a fairness index of 1.

27 Lab Topology

Let us get started with creating a simple Mininet topology using MiniEdit. The topology uses 10.0.0.0/8, which is the default network assigned by Mininet (Fig. 151).

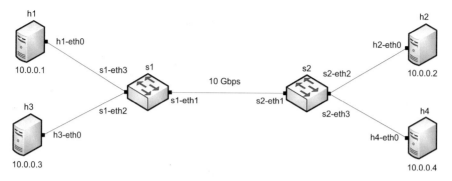

Fig. 150 Three TCP flows that share a bottleneck link of capacity 9 Gbps

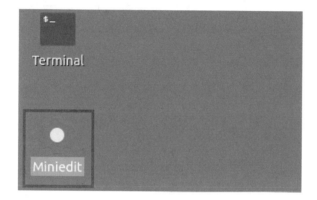

Fig. 151 Lab topology

The above topology uses 10.0.0.0/8, which is the default network assigned by Mininet.

Step 1. A shortcut to MiniEdit is located on the machine's Desktop. Start MiniEdit by clicking on MiniEdit's shortcut (Fig. 152). When prompted for a password, type password .

Step 2. On MiniEdit's menu bar, click on *File* and then *Open* to load the lab's topology. Locate the *Lab 13.mn* topology file and click on *Open* (Fig. 153).

Step 3. Before starting the measurements between host h1 and host h2, the network must be started. Click on the *Run* button located at the bottom left of MiniEdit's window to start the emulation (Fig. 154).

The above topology uses 10.0.0.0/8, which is the default network assigned by Mininet.

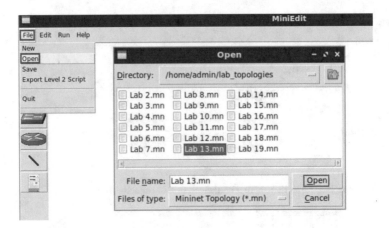

Fig. 152 MiniEdit shortcut

Fig. 153 MiniEdit's *Open* dialog

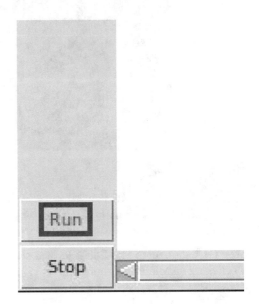

27.1   Starting Host h1 and Host h2

Step 1. Hold the right-click on host h1 and select *Terminal*. This opens the terminal of host h1 and allows the execution of commands on that host (Fig. 155).

Step 2. Apply the same steps on host h2 and open its *Terminal*.

Step 3. Test connectivity between the end-hosts using the ping command. On host h1, type the command ping 10.0.0.2 . This command tests the connectivity

Fig. 154 Running the
emulation

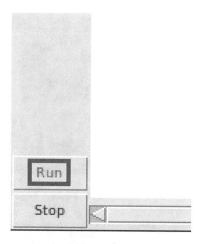

```
 X                            "Host: h1"                           - ⌃ x
root@admin-pc:~# ping 10.0.0.2
PING 10.0.0.2 (10.0.0.2) 56(84) bytes of data.
64 bytes from 10.0.0.2: icmp_seq=1 ttl=64 time=1.33 ms
64 bytes from 10.0.0.2: icmp_seq=2 ttl=64 time=0.056 ms
64 bytes from 10.0.0.2: icmp_seq=3 ttl=64 time=0.048 ms
64 bytes from 10.0.0.2: icmp_seq=4 ttl=64 time=0.042 ms
64 bytes from 10.0.0.2: icmp_seq=5 ttl=64 time=0.043 ms
64 bytes from 10.0.0.2: icmp_seq=6 ttl=64 time=0.044 ms
^C
--- 10.0.0.2 ping statistics ---
6 packets transmitted, 6 received, 0% packet loss, time 91ms
rtt min/avg/max/mdev = 0.042/0.260/1.327/0.477 ms
root@admin-pc:~# █
```

Fig. 155 Opening a terminal on host h1

between host h1 and host h2. To stop the test, press Ctrl+c . The figure below
shows a successful connectivity test (Fig. 156).

Figure 156 indicates that there is connectivity between host h1 and host h2. Thus,
we are ready to start the throughput measurement process.

27.2   *Emulating 10 Gbps High-Latency WAN*

This section emulates a high-latency WAN. We will first emulate 20 ms delay
between switch S1 and switch S2 and measure the throughput. Then, we will set
the bandwidth between host h1 and host h2 to 10 Gbps.

Fig. 156 Connectivity test using ping command

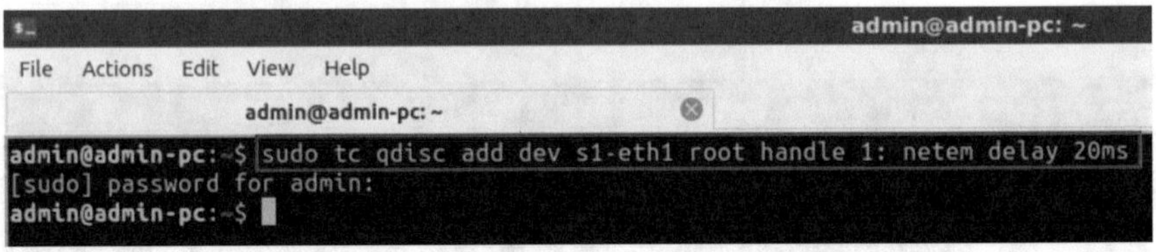

Fig. 157 Shortcut to open a Linux terminal

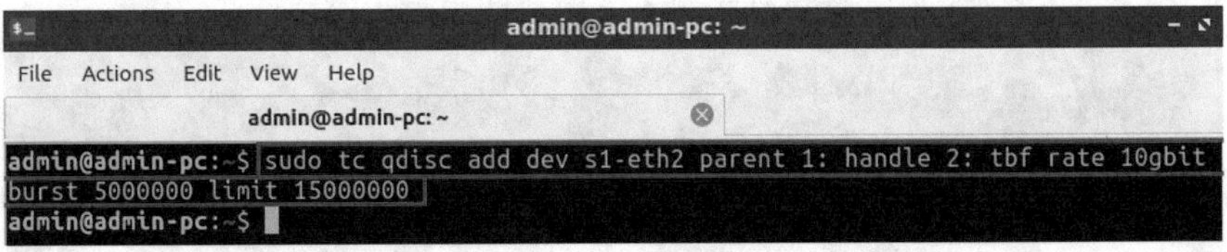

Fig. 158 Adding delay of 20 ms to switch S1's *s1-eth1* interface

Step 1. Launch a Linux terminal by holding the Ctrl+Alt+T keys or by clicking on the Linux terminal icon (Fig. 157).

The Linux terminal is a program that opens a window and permits you to interact with a command-line interface (CLI). A CLI is a program that takes commands from the keyboard and sends them to the operating system for execution.

Step 2. In the terminal, type the command below. When prompted for a password, type password and hit *Enter*. This command introduces 20 ms delay on switch S1's *s1-eth1* interface (Fig. 158).

```
sudo tc qdisc add dev s1-eth1 root handle 1: netem delay
20 ms
```

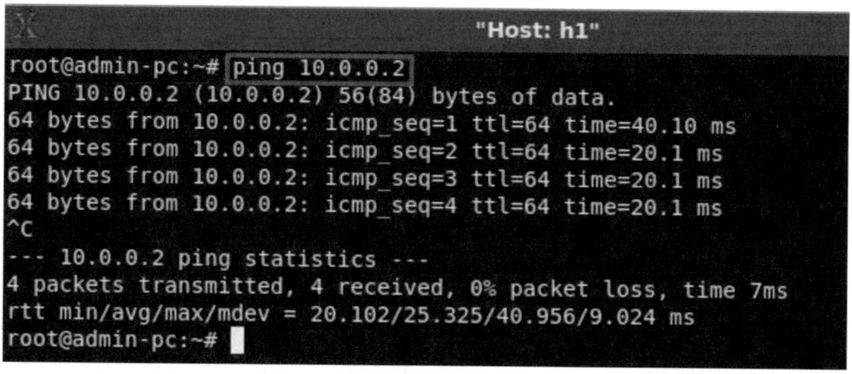

```
X                                          "Host: h1"
root@admin-pc:~# ping 10.0.0.2
PING 10.0.0.2 (10.0.0.2) 56(84) bytes of data.
64 bytes from 10.0.0.2: icmp_seq=1 ttl=64 time=40.10 ms
64 bytes from 10.0.0.2: icmp_seq=2 ttl=64 time=20.1 ms
64 bytes from 10.0.0.2: icmp_seq=3 ttl=64 time=20.1 ms
64 bytes from 10.0.0.2: icmp_seq=4 ttl=64 time=20.1 ms
^C
--- 10.0.0.2 ping statistics ---
4 packets transmitted, 4 received, 0% packet loss, time 7ms
rtt min/avg/max/mdev = 20.102/25.325/40.956/9.024 ms
root@admin-pc:~# ▊
```

Fig. 159 Limiting the bandwidth to 10 Gbps on switch S1's *s1-eth1* interface

Step 3. Modify the bandwidth of the link connecting the switch S1 and switch S2: on the same terminal, type the command below. This command sets the bandwidth to 10 Gbps on switch S1's *s1-eth2* interface (Fig. 159). The tbf parameters are the following:

- rate : 10gbit
- burst : 5,000,000
- limit : 15,000,000

```
sudo tc qdisc add dev s1-eth1 parent 1: handle 2: tbf rate
10gbit burst 5000000 limit 15000000
```

27.3   Testing Connection

To test connectivity, you can use the command ping .

Step 1. On the terminal of host h1, type ping 10.0.0.2 . To stop the test, press Ctrl+c . The figure below shows a successful connectivity test. Host h1 (10.0.0.1) sent four packets to host h2 (10.0.0.2), successfully receiving responses back (Fig. 160).

The result above indicates that all four packets were received successfully (0% packet loss) and that the minimum, average, maximum, and standard deviation of the Round-Trip Time (RTT) were 20.102, 25.325, 40.956, and 9.024 ms, respectively. The output above verifies that delay was injected successfully, as the RTT is approximately 20 ms.

```
                          "Host: h2"                        -  ↙ ✕
root@admin-pc:~# iperf3 -s
- - - - - - - - - - - - - - - - - - - - - - - - - - - - - - - - - - - - - - -
Server listening on 5201
- - - - - - - - - - - - - - - - - - - - - - - - - - - - - - - - - - - - - - -
```

Fig. 160 Output of ping 10.0.0.2 command

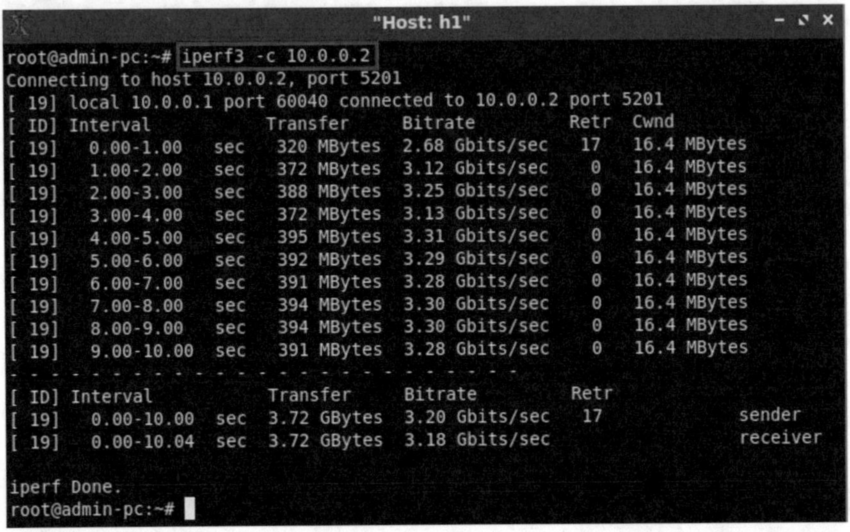

```
                          "Host: h1"                        -  ↙ ✕
root@admin-pc:~# iperf3 -c 10.0.0.2
Connecting to host 10.0.0.2, port 5201
[ 19] local 10.0.0.1 port 60040 connected to 10.0.0.2 port 5201
[ ID] Interval           Transfer     Bitrate          Retr  Cwnd
[ 19]   0.00-1.00   sec   320 MBytes  2.68 Gbits/sec    17   16.4 MBytes
[ 19]   1.00-2.00   sec   372 MBytes  3.12 Gbits/sec     0   16.4 MBytes
[ 19]   2.00-3.00   sec   388 MBytes  3.25 Gbits/sec     0   16.4 MBytes
[ 19]   3.00-4.00   sec   372 MBytes  3.13 Gbits/sec     0   16.4 MBytes
[ 19]   4.00-5.00   sec   395 MBytes  3.31 Gbits/sec     0   16.4 MBytes
[ 19]   5.00-6.00   sec   392 MBytes  3.29 Gbits/sec     0   16.4 MBytes
[ 19]   6.00-7.00   sec   391 MBytes  3.28 Gbits/sec     0   16.4 MBytes
[ 19]   7.00-8.00   sec   394 MBytes  3.30 Gbits/sec     0   16.4 MBytes
[ 19]   8.00-9.00   sec   394 MBytes  3.30 Gbits/sec     0   16.4 MBytes
[ 19]   9.00-10.00  sec   391 MBytes  3.28 Gbits/sec     0   16.4 MBytes
- - - - - - - - - - - - - - - - - - - - - - - - - - - - - -
[ ID] Interval           Transfer     Bitrate          Retr
[ 19]   0.00-10.00  sec  3.72 GBytes  3.20 Gbits/sec    17            sender
[ 19]   0.00-10.04  sec  3.72 GBytes  3.18 Gbits/sec                  receiver

iperf Done.
root@admin-pc:~# ▊
```

Fig. 161 Starting iPerf3 server on host h2

Step 2. On the terminal of host h2, type ping 10.0.0.1 . The ping output in this test should be relatively close to the results of the test initiated by host h1 in Step 1. To stop the test, press Ctrl+c .

Step 3. Launch iPerf3 in server mode on host h2's terminal (Fig. 161).

```
iperf3 -s
```

Step 4. Launch iPerf3 in client mode on host h1's terminal (Fig. 162).

```
iperf3 -c 10.0.0.2
```

Although the link was configured to 10 Gbps, the test results show that the achieved throughput is 3.20 Gbps. This is because the TCP buffer size was not modified at this point.

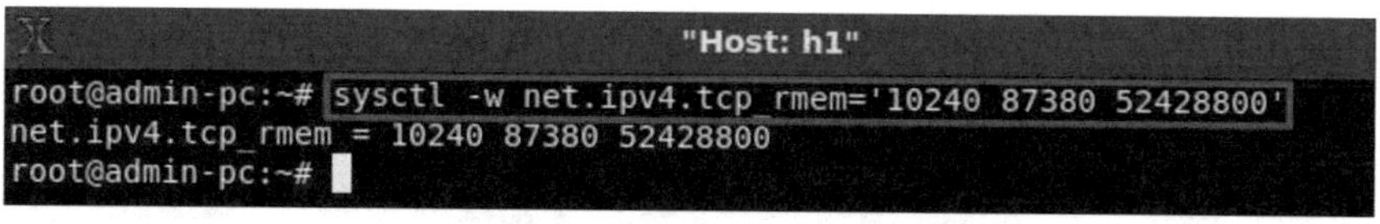

Fig. 162 Running iPerf3 client on host h1

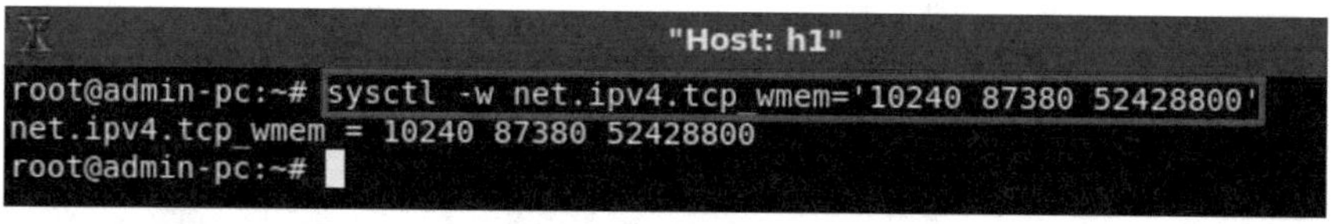

Fig. 163 Receive window change in sysctl

Step 5. In order to stop the server, press Ctrl+c in host h2's terminal. The user can see the throughput results in the server side too.

Step 6. To change the current receive window size value(s), we calculate the bandwidth-delay product by performing the following calculation:

$$\text{BW} = 10,000,000,000 \text{ bits/second}$$

$$\text{RTT} = 0.02 \text{ seconds}$$

$$\text{BDP} = 10,000,000,000 * 0.02 = 200,000,000 \text{ bits}$$

$$= 25,000,000 \text{ bytes} \approx 25 \text{ Mbytes}$$

The send and receive buffer sizes should be set to $2 \cdot$ BDP. We will use the 25 Mbytes value for the BDP instead of 25,000,000 bytes

$$1 \text{ Mbyte} = 1024^2 \text{ bytes}$$

$$\text{BDP} = 25 \text{ Mbytes} = 25 * 1024^2 \text{ bytes} = 26,214,400 \text{ bytes}$$

$$2 \cdot \text{BDP} = 2 \cdot 26,214,400 \text{ bytes} = 52,428,800 \text{ bytes}$$

Now, we have calculated the maximum value of the TCP sending and receiving buffer size. In order to apply the new values on host h1's terminal type the command showed down below. The values set are: 10,240 (minimum), 87,380 (default), and 52,428,800 (maximum, calculated by doubling the BDP) (Fig. 163).

```
sysctl -w net.ipv4.tcp_rmem='10240 87380 52428800'
```

```
                                    "Host: h2"
root@admin-pc:~# sysctl -w net.ipv4.tcp_rmem='10240 87380 52428800'
net.ipv4.tcp_rmem = 10240 87380 52428800
root@admin-pc:~#
```

Fig. 164 Send window change in sysctl

```
                                    "Host: h2"
root@admin-pc:~# sysctl -w net.ipv4.tcp_wmem='10240 87380 52428800'
net.ipv4.tcp_wmem = 10240 87380 52428800
root@admin-pc:~#
```

Fig. 165 Receive window change in sysctl

```
                          "Host: h2"                          —  ↘  ✕
root@admin-pc:~# iperf3 -s
- - - - - - - - - - - - - - - - - - - - - - - - - - - - - - - - - - - -
Server listening on 5201
- - - - - - - - - - - - - - - - - - - - - - - - - - - - - - - - - - - -
```

Fig. 166 Send window change in sysctl

Step 7. To change the current send window size value(s), use the following command on host h1's terminal. The values set are: 10,240 (minimum), 87,380 (default), and 52,428,800 (maximum, calculated by doubling the BDP) (Fig. 164).

```
sysctl -w net.ipv4.tcp_wmem='10240 87380 52428800'
```

Next, the same commands must be configured on host h2.

Step 8. To change the current receive window size value(s), use the following command on host h2's terminal. The values set are: 10,240 (minimum), 87,380 (default), and 52,428,800 (maximum, calculated by doubling the BDP) (Fig. 165).

```
sysctl -w net.ipv4.tcp_rmem='10240 87380 52428800'
```

Step 9. To change the current send window size value(s), use the following command on host h2's terminal. The values set are: 10,240 (minimum), 87,380 (default), and 52,428,800 (maximum, calculated by doubling the BDP) (Fig. 166)

```
sysctl -w net.ipv4.tcp_wmem='10240 87380 52428800'
```

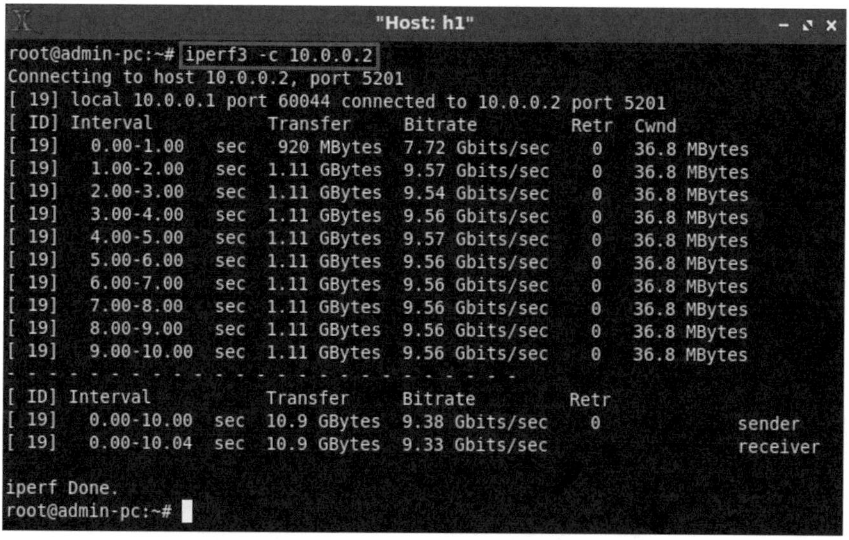

```
                              "Host: h1"                        - ∿ ×
root@admin-pc:~# iperf3 -c 10.0.0.2
Connecting to host 10.0.0.2, port 5201
[ 19] local 10.0.0.1 port 60044 connected to 10.0.0.2 port 5201
[ ID] Interval           Transfer     Bitrate         Retr  Cwnd
[ 19]   0.00-1.00   sec   920 MBytes  7.72 Gbits/sec    0   36.8 MBytes
[ 19]   1.00-2.00   sec  1.11 GBytes  9.57 Gbits/sec    0   36.8 MBytes
[ 19]   2.00-3.00   sec  1.11 GBytes  9.54 Gbits/sec    0   36.8 MBytes
[ 19]   3.00-4.00   sec  1.11 GBytes  9.56 Gbits/sec    0   36.8 MBytes
[ 19]   4.00-5.00   sec  1.11 GBytes  9.57 Gbits/sec    0   36.8 MBytes
[ 19]   5.00-6.00   sec  1.11 GBytes  9.56 Gbits/sec    0   36.8 MBytes
[ 19]   6.00-7.00   sec  1.11 GBytes  9.56 Gbits/sec    0   36.8 MBytes
[ 19]   7.00-8.00   sec  1.11 GBytes  9.56 Gbits/sec    0   36.8 MBytes
[ 19]   8.00-9.00   sec  1.11 GBytes  9.56 Gbits/sec    0   36.8 MBytes
[ 19]   9.00-10.00  sec  1.11 GBytes  9.56 Gbits/sec    0   36.8 MBytes
- - - - - - - - - - - - - - - - - - - - - - - - - - - -
[ ID] Interval           Transfer     Bitrate         Retr
[ 19]   0.00-10.00  sec  10.9 GBytes  9.38 Gbits/sec    0           sender
[ 19]   0.00-10.04  sec  10.9 GBytes  9.33 Gbits/sec                receiver

iperf Done.
root@admin-pc:~# █
```

Fig. 167 Host h2 running iPerf3 as server

```
                              "Host: h2"                        - ∿ ×
root@admin-pc:~# iperf3 -s
- - - - - - - - - - - - - - - - - - - - - - - - - - - - - - - - - - - - - - -
Server listening on 5201
- - - - - - - - - - - - - - - - - - - - - - - - - - - - - - - - - - - - - - -
█
```

Fig. 168 iPerf3 throughput test

Step 10. The user can now verify the rate limit configuration by using the `iperf3` tool to measure throughput. To launch iPerf3 in server mode, run the command `iperf3 -s` in host h2's terminal (Fig. 167):

```
iperf3 -s
```

Step 11. Now to launch iPerf3 in client mode again by running the command `iperf3 -c 10.0.0.2` in host h1's terminal (Fig. 168):

```
iperf3 -c 10.0.0.2
```

Note the measured throughput now is approximately 9.38 Gbps, which is close to the value assigned in our `tbf` rule (10 Gbps).

Step 12. In order to stop the server, press Ctrl+c in host h2's terminal. The user can see the throughput results in the server side too.

28   Calculating Fairness Among Parallel Flows

In this section, an iPerf3 test that includes several parallel streams is conducted, followed by the calculation of the fairness index.

Step 1. Now a test is conducted using parallel streams. To launch iPerf3 in server mode, run the command iperf3 -s in host h2's terminal as shown in Fig. 169:

```
iperf3 -s
```

Step 2. Now the iPerf3 client should be launched with the -P option specified to start parallel streams. The -J option is also specified to indicate that JSON output is desired, and the redirection operator > to store the output in a file. Run the following command in host h1's terminal as shown in Fig. 170:

```
iperf3 -c 10.0.0.2 -P 8 -J >out.json
```

```
                          "Host: h1"
root@admin-pc:~# iperf3 -c 10.0.0.2 -P 8 -J > out.json
root@admin-pc:~# █
```

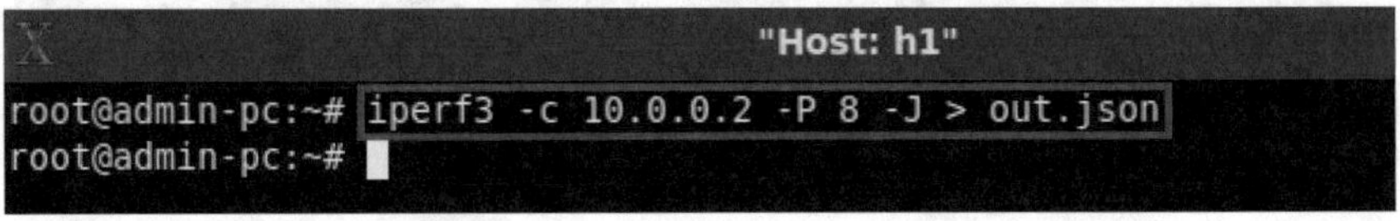

Fig. 169 Host h2 running iPerf3 as server

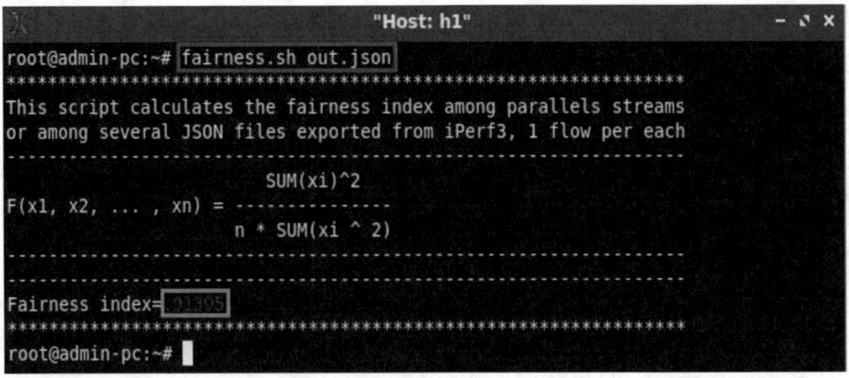

```
                          "Host: h1"                    – ⚙ ✕
root@admin-pc:~# fairness.sh out.json
****************************************************************
This script calculates the fairness index among parallels streams
or among several JSON files exported from iPerf3, 1 flow per each
----------------------------------------------------------------
                          SUM(xi)^2
F(x1, x2, ... , xn) = ----------------
                        n * SUM(xi ^ 2)

----------------------------------------------------------------
Fairness index= 0.1395
****************************************************************
root@admin-pc:~# █
```

Fig. 170 Host h1 running iPerf3 as client with 8 parallel streams and saving output in file

Fig. 171 Calculating the fairness index between the parallel streams

Step 3. The client includes a script called fairness.sh . Basically, this script accepts as input the JSON file exported by iPerf3 and calculates the fairness index. Run the following command to calculate the fairness index (Fig. 171):

```
fairness.sh out.json
```

In the above test, the fairness index is 0.91395 or 91% fair. Note that this result may vary according to the result of your emulation test.

Step 4. In order to stop the server, press Ctrl+c in host h2's terminal. The user can see the throughput results in the server side too.

29   Calculating Fairness Among Several Hosts with the Same Congestion Control Algorithm

In the previous section, we calculated the fairness index among several parallel streams, all initiated by a single host. In this section we calculate the fairness index among two transmitting devices. Specifically, an iPerf3 client is executed on host h1 and connected to host h2 (iPerf3 server); another iPerf3 client is executed on host h3 and connected to host h4 (iPerf3 server).

To calculate the fairness index, the transmitting hosts should initiate their transmissions simultaneously. Since it is difficult to start the clients at the same time, the client's machine provides a script that automates this process.

Step 1. Close the terminals of host h1 and host h2.

Step 2. Go to Mininet's terminal, i.e., the one launched when MiniEdit was started (Figs. 172 and 173).

Step 3. Issue the following command on Mininet's terminal as shown in the figure below (Fig. 174).

```
source concurrent_same_algo
```

The above graph (Fig. 175) shows that the throughput of host h1 is close to that of host h3. Therefore, the fairness index should be close to 1 (100%). Note that this result may vary according to the result of your emulation test.

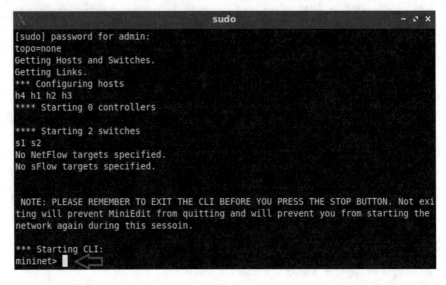

Fig. 172 Opening Mininet's terminal

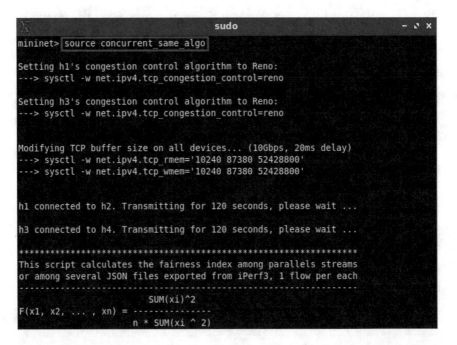

Fig. 173 Mininet's terminal

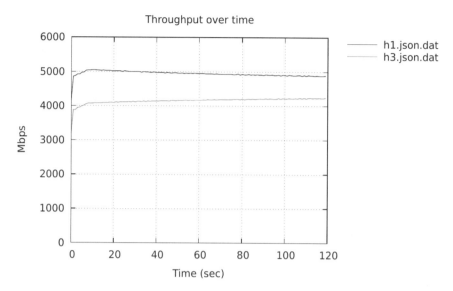

Fig. 174 Running the tests simultaneously for 120 s. Both host h1 and host h3 are using Reno

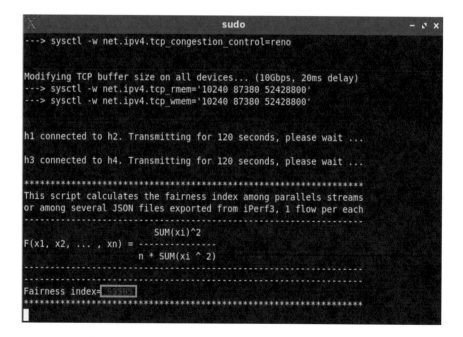

Fig. 175 Throughput of host h1 and host h3

Fig. 176 Calculated fairness index

Step 4. Close the graph window and go back to Mininet's terminal. The fairness index is displayed at the end as shown in the figure below (Fig. 176).

The above figure shows a fairness index of 0.99595. This value indicates that the bottleneck bandwidth was 99% fairly shared among host h1 and host h3. Note that this result may vary according to the result of your emulation test.

30 Calculating Fairness Among Hosts with Different Congestion Control Algorithms

In the previous test, we calculated the fairness index while using the same congestion control algorithm (Reno). In this section we repeat the test, but with host h1 using Reno and host h3 using BBR.

Step 1. Go back to Mininet's terminal, i.e., the one launched when MiniEdit was started (Fig. 177).

Step 2. Issue the following command on Mininet's terminal as shown in the figure below (Fig. 178).

```
source concurrent_diff_algo
```

The above graph (Fig. 179) shows that the device configured with BBR has a larger bandwidth allocation than that configured with Reno. Therefore, the fairness index will not be close to 1 (100%).

Step 3. Close the graph window and go back to Mininet's terminal. The fairness index is displayed at the end as shown in the figure below (Fig. 180).

The above figure shows a fairness index of 0.86036 (~86%), which is very far from 100%. This value indicates that the bottleneck bandwidth was 86% fairly shared among host h1 and host h3. Note that this result may vary according to the result of your emulation test.

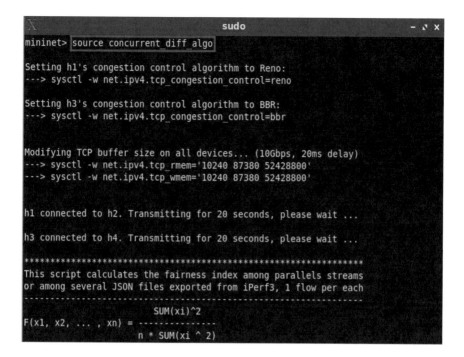

Fig. 177 Opening Mininet's terminal

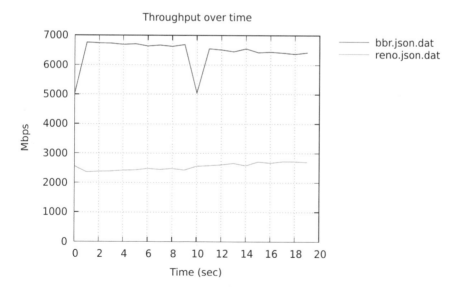

Fig. 178 Running the tests simultaneously for 20 s. Host h1 is using Reno, while host h3 is using BBR

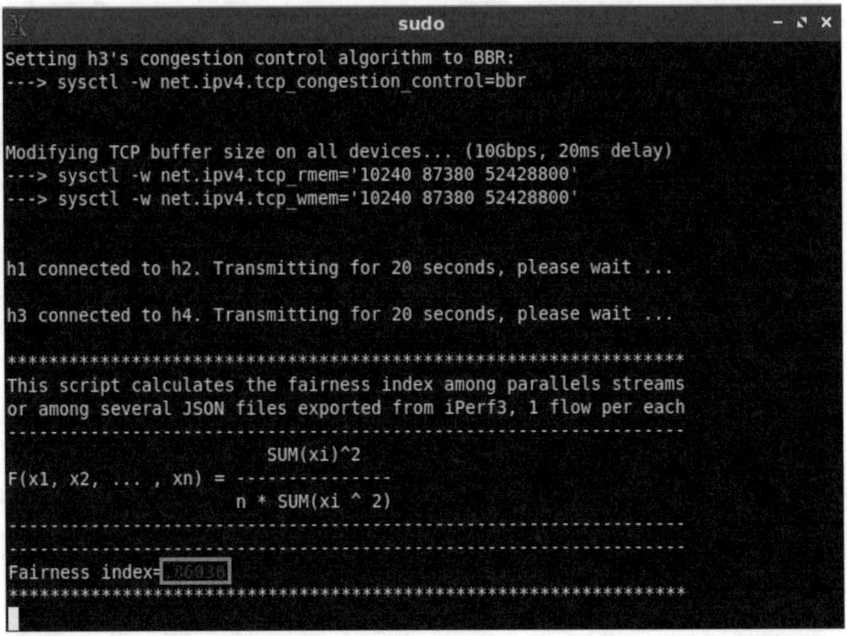

```
X                               sudo                        - ↗ ×
Setting h3's congestion control algorithm to BBR:
---> sysctl -w net.ipv4.tcp_congestion_control=bbr


Modifying TCP buffer size on all devices... (10Gbps, 20ms delay)
---> sysctl -w net.ipv4.tcp_rmem='10240 87380 52428800'
---> sysctl -w net.ipv4.tcp_wmem='10240 87380 52428800'


h1 connected to h2. Transmitting for 20 seconds, please wait ...

h3 connected to h4. Transmitting for 20 seconds, please wait ...

****************************************************************
This script calculates the fairness index among parallels streams
or among several JSON files exported from iPerf3, 1 flow per each
----------------------------------------------------------------
                        SUM(xi)^2
F(x1, x2, ... , xn) = --------------
                        n * SUM(xi ^ 2)
----------------------------------------------------------------
----------------------------------------------------------------
Fairness index=.86038
****************************************************************
```

Fig. 179 Throughput of host h1 and host h3

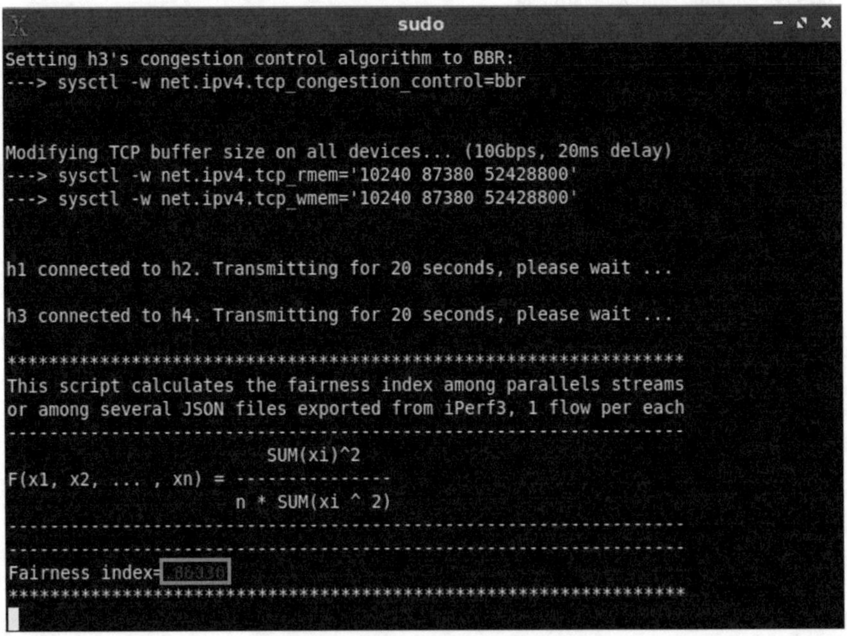

```
X                               sudo                        - ↗ ×
Setting h3's congestion control algorithm to BBR:
---> sysctl -w net.ipv4.tcp_congestion_control=bbr


Modifying TCP buffer size on all devices... (10Gbps, 20ms delay)
---> sysctl -w net.ipv4.tcp_rmem='10240 87380 52428800'
---> sysctl -w net.ipv4.tcp_wmem='10240 87380 52428800'


h1 connected to h2. Transmitting for 20 seconds, please wait ...

h3 connected to h4. Transmitting for 20 seconds, please wait ...

****************************************************************
This script calculates the fairness index among parallels streams
or among several JSON files exported from iPerf3, 1 flow per each
----------------------------------------------------------------
                        SUM(xi)^2
F(x1, x2, ... , xn) = --------------
                        n * SUM(xi ^ 2)
----------------------------------------------------------------
----------------------------------------------------------------
Fairness index=.86336
****************************************************************
```

Fig. 180 Calculated fairness index

References

1. K. Chard, S. Tuecke, I. Foster, Globus: recent enhancements and future plans, in *Proceedings of the XSEDE16 Conference on Diversity, Big Data, and Science at Scale* (2016)
2. W. Allcock, J. Bresnahan, R. Kettimuthu, M. Link, The globus striped GridFTP framework and server, in *Proceedings of the 2005 ACM/IEEE Conference on Supercomputing* (2005)
3. G. Vardoyan, R. Kettimuthu, M. Link, S. Tuecke, Characterizing throughput bottlenecks for secure GridFTP transfers, in *IEEE International Conference on Computing, Networking and Communications (ICNC)* (2013)
4. D. Borman, B. Braden, V. Jacobson, R. Scheffenegger, TCP extensions for high performance, in *Internet Request for Comments, RFC 7323* (2014). https://tools.ietf.org/html/rfc7323#section-4.2
5. Z. Liu, P. Balaprakash, R. Kettimuthu, I. Foster, Explaining wide area data transfer performance, in *IEEE/ACM International Symposium on High-Performance Distributed Computing (HPDC)* (2017)
6. N. Mills, A. Feltus, W. Ligon III, Maximizing the performance of scientific data transfer by optimizing the interface between parallel file systems and advanced research networks. J. Fut. Gener. Comput. Syst. (2017). https://doi.org/10.1016/j.future.2017.04.030
7. F. Feltus, J. Breen, J. Deng, R. Izard, C. Konger, W. Ligon, D. Preuss, K. Ching, The widening gulf between genomics data generation and consumption: a practical guide to big data transfer technology. J. Bioinform. Biol. Insights 9(1), BBI-S28988 (2015)
8. T. Hacker, B. Athey, B. Noble, The end-to-end performance effects of parallel TCP sockets on a lossy wide-area network, in *Proceedings of the Parallel and Distributed Processing Symposium* (2001)
9. A. Aggarwal, S. Savage, T. Anderson, Understanding the performance of TCP pacing, in *Proceedings of the International Conference on Computer Communications (INFOCOM)* (2000)
10. B. Tierney, N. Hanford, D. Ghosal, Optimizing data transfer nodes using packet pacing: a journey of discovery, in *Workshop on Innovating the Network for Data-Intensive Science* (2015)
11. M. Ghobadi, Y. Ganjali, TCP pacing in data center networks, in *IEEE Annual Symposium on High-Performance Interconnects (HOTI)* (2013)
12. The CentOS Project. https://www.centos.org/
13. N. Cardwell, Y. Cheng, C. Gunn, S. Yeganeh, V. Jacobson, BBR: congestion-based congestion control. Commun. ACM 60(2), 58–66 (2017)
14. J. Corbet, TSO sizing and the FQ scheduler. LWN.net Online Mag. (2013). https://lwn.net/Articles/564978
15. B. Tierney, Improving performance of 40G/100G data transfer nodes, in *2016 Technology Exchange Workshop* (2016). https://meetings.internet2.edu/2016-technology-exchange/detail/10004333/
16. K. Fall, S. Floyd, Simulation-based comparisons of Tahoe, Reno, and Sack TCP. Comput. Commun. Rev. 26(3), 5–21 (1996)
17. I. Rhee, L. Xu, CUBIC: a new TCP-friendly high-speed TCP variant. ACM Spec. Interest Group Oper. Syst. Rev. 42(5), 64–74 (2008)
18. D. Leith, R. Shorten, Y. Lee, H-TCP: a framework for congestion control in high-speed and long-distance networks. Hamilton Institute Technical Report, 2005. http://www.hamilton.ie/net/htcp2005.pdf
19. T. Dierks, E. Rescorla, The transport layer security protocol version 1.2. Internet Request for Comments, RFC 5246 (2008). https://tools.ietf.org/html/rfc5246
20. A. Freier, P. Karlton, P. Kocher, The secure sockets layer protocol version 3.0. Internet Request for Comments, RFC 6101 (2011). https://tools.ietf.org/html/rfc6101

Application and Security Aspects for Large Flows

This chapter provides an overview of two common types of application-layer tools used in high-speed networks and Science DMZs: file transfer tools and monitoring application tools. File transfer tools are used by researchers and practitioners to share data. Historically, applications were built around the File Transport Protocol (FTP). While FTP-based applications work well in enterprise networks, their performance in high-throughput, high-latency environments is often poor. In relation to this, monitoring applications tools are essential to identify problems causing poor performance and address them. The chapter describes perfSONAR, the most widely deployed monitoring tool for multi-domain environments.

1 Application-Layer Tools

The essential end devices inside a Science DMZ are the DTNs and the performance monitoring stations. DTNs run a data transfer tool, while monitoring stations run a performance monitoring application, typically perfSONAR. Other useful tools at deployment and evaluation times are WAN emulation and throughput measurement applications. These tools are convenient because they facilitate early performance evaluation without a need of connecting the Science DMZ to a real WAN. Additionally, in contrast to enterprise networks, virtualization technologies have not been adopted in Science DMZs, because of performance limitations.

This section provides an overview of application-layer tools used in Science DMZs. The section also discusses the performance limitations of virtualization technologies preventing their adoption in Science DMZs.

© The Author(s), under exclusive license to Springer Nature Switzerland AG 2022
J. Crichigno et al., *High-Speed Networks*, Practical Networking,
https://doi.org/10.1007/978-3-030-88841-1_5

2 File Transfer Applications

File transfer applications are used by researchers and practitioners to share data. Historically, applications were built around the File Transport Protocol (FTP) [1]. While FTP-based applications work well in enterprise networks, their performance in high-throughput, high-latency environments is often poor. Fig.1 presents a taxonomy for file transfer applications

2.1 Traditional File Transfer Applications

Figure 2a shows the basic FTP model. It uses two TCP connections: a data channel and a control channel. FTP has several limitations when used in Science DMZs, including limited negotiation capability of the TCP buffer size, poor throughput performance in long fat networks, a lack of uniform interfaces between data transfer processes and sources and sinks (local hard disks, parallel file systems, distributed data sources, etc.), and a lack of support for partial file transfer and restartable transfer.

Other file transfer protocols used in enterprise networks include Secure Copy (SCP) and Secure FTP (SFTP). These protocols are implemented above the Secure Shell protocol (SSH) [2], which in turn is implemented above TCP. Figure 2b shows their respective location in the protocol stack. When a file transfer is performed by SCP or SFTP, an SSH channel is open between the end points. This channel uses a window-based flow control. Even though this feature works well for enterprise file transfers, it constitutes another rate limitation for large science flows.

2.2 File Transfer Applications for Science DMZs

The prevalent tool for science data transfers is Globus GridFTP. As of 2017, there are over 40,000 Globus end points deployed [3]. While the following description corresponds to Globus, many of its features apply to other applications recommended for Science DMZs.

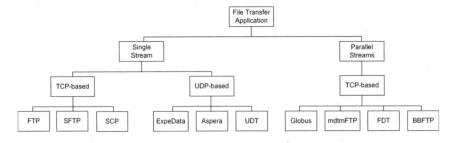

Fig. 1 Taxonomy for data transfer applications

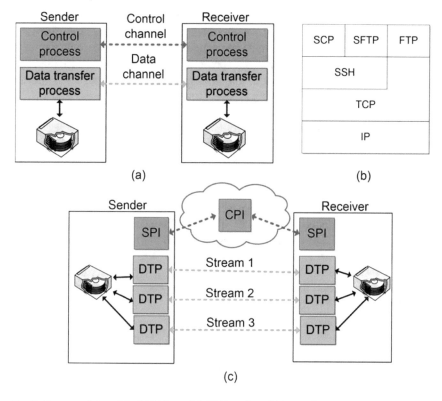

Fig. 2 Data transfer models. (**a**) FTP model. (**b**) Location of file transfer protocols in the protocol stack. (**c**) Globus model. Control information is exchanged between Client Protocol Interpreter (CPI) and Server Protocol Interpreter (SPI). Data transfer processes (DTPs) exchange actual data

GridFTP is an extension of FTP for high-speed networks. Globus is an implementation of GridFTP [3, 4], and its architecture is shown in Fig. 2c. Globus has the following features:

- The control channel is established between the client protocol interpreter (CPI), a third party located in the cloud, and the server protocol interpreters (SPIs).
- Multiple parallel TCP connections are supported. These connections are referred to as streams and constitute the data channels. Typical values are between 2 and 8 streams.
- Globus includes support for partial and restartable file transfers [5]. Science DMZs are used for transferring large data files, which may take hours. If a disruption occurs momentarily, it is beneficial to transfer just the remaining portion of a file.
- The maximum size of the TCP buffer can be explicitly adjusted.

Other file transfer applications for big data are Multicore-Aware Data Transfer Middleware FTP (mdtmFTP) [6] and Fast Data Transport (FDT) [7]. mdtmFTP is designed to efficiently use the multiple computing cores (multicore CPUs) on a single chip that are common in modern computer systems. mdtmFTP also improves the throughput in DTNs that use a non-uniform memory access (NUMA) model. In the traditional Uniform Memory Access (UMA) model, the access to the RAM from any CPU or core takes the same amount of time. However, with NUMA, accessing some parts of memory by a core may take longer than other parts, creating a performance penalty. FDT is an application optimized for the transfer of a large number of files [7]. Hence, thousands of files can be sent continuously, without restarting the network transfer between files. However, FDT and mdtmFTP have not been widely adopted despite encouraging performance results [6].

Table 1 lists additional data transfer applications and implemented features. Besides TCP-based applications, three UDP-based applications are listed. Here, a common feature in UDP-based applications is the use of a single UDP stream, rather than multiple streams. A key limitation observed in Science DMZs when using a single UDP stream is CPU-related. Namely, in multicore processor architectures, a UDP stream is adhered to a single core, which may become saturated. As a result, the UDP transmission rate can be lower than the available bandwidth. At the same time, other cores may be idle and underutilized [13]. UDP-based applications do not use congestion windows for congestion control. Instead, they use rate-based congestion control, similar to BBR [14]. Here, congestion is signaled by an increase in the RTT, which triggers a decrease in the transmission rate. ExpeData [10] and Aspera Fast [8] are two proprietary implementations and some details are not available. Some specialized applications are used in enterprise environments, but their performance for big science data transfers is not available.

3 Virtual Machines and Science DMZs

The idea behind a virtual machine is to abstract the hardware of a computer into several execution environments. As a physical resource, access to a NIC is also shared. In this context, Fig. 3 shows a sample topology with three hosts. Host 1 contains three virtual machines connected by a virtual switch. One virtual machine is a DTN. Hosts 2 and 3 are native (non-virtual) DTNs. A virtual switch is implemented inside the hypervisor. Similar to a physical switch, the virtual switch constructs its own forwarding table and forwards frames at the data-link layer. The virtual switch also connects to the external network through a physical NIC.

A virtual machine is connected to the internal network through a virtual NIC. There are different types of virtual NICs, including the following:

- E1000: An emulated version of the Intel 82545EM Gigabit Ethernet NIC. Older (Linux and Windows) guest operating systems use this virtual NIC.

Table 1 Features of various data transfer applications. U indicates unknown

Application	Transport protocol	Adjustable buffer size	Parallel streams	Partial file transfer	Restartable file transfer	Security	Sharing and publishing	Adoption	SDMZ recommended
FTP, SCP, SFTP [2]	TCP	No	No	No	No	Yes; SCP, SFTP	No	High; enterprise networks	No
Globus [3, 4]	TCP	Yes	Yes	Yes	Yes	Yes, algorithm is determined via openSSL, based on DTNs capability	Yes	High; universities and research centers	Recommended, high adoption and available support
mdtmFTP [6]	TCP	Yes	Yes	No	No	Yes; U	No	Low	Acceptable; limited support
FDT [7]	TCP	Yes	Yes	No	Yes	No[a]	No	Low	Acceptable; limited support
Aspera Fast [8]	UDP	No	Yes	No	No	Yes; Advanced Encryption Standard (AES) 128	No	Medium; enterprise networks	Unknown performance
BBFTP [9]	TCP	Yes	Yes	No	No	No	No	Low	Unknown performance
ExpeData [10]	UDP	U	U	U	U	Yes; AES 128	No	Medium; enterprise networks	Unknown performance
UDT [11, 12]	UDP	No	Yes	No	No	No	No	Low	No; a lack of parallel streams

U: Unknown

[a] Security can be incorporated via third party software package

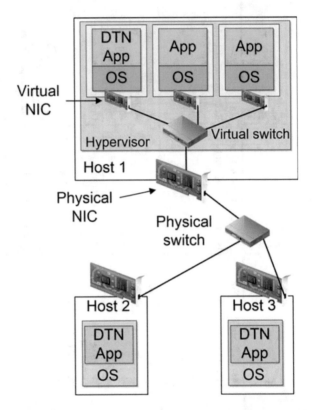

Fig. 3 Network topology including a virtual DTN contained in host 1 and two native (non-virtual) DTNs, host 2 and host 3

- E1000e: This virtual NIC emulates newer models of Intel Gigabit NICs. It is the default virtual NIC for newer (Windows) guest operating systems.
- VMXNET: This virtual NIC has no physical counterpart. There are two enhanced versions, VMXNET2 and VMXNET3. The latter is recommended for high-speed data transfers [15].

While virtual technologies have been widely adopted in enterprise networks, their use in Science DMZs has been discouraged for several reasons. First, the hypervisor represents a software layer that adds processing overhead. Second, the physical NIC is potentially shared among multiple virtual machines. Third, even if the virtual DTN is the only virtual machine running on a physical server, the CPU must be shared with the hypervisor and the virtual switch. Moreover, commercial vendors may not disclose important attributes of the virtual switch, such as buffer size and switching architecture.

Based on the above limitations, virtualization is not recommended for Science DMZs operating at speeds above 10 Gbps. For Science DMZs operating at 10 Gbps, preliminary results in Section VII suggest that virtual DTNs may achieve an acceptable performance, provided the physical server they run on has a high CPU capacity and the workload is controlled.

4 Monitoring and Performance Applications for Science DMZs

One of the essential elements of a Science DMZ is the performance measurement and monitoring point. The monitoring process in Science DMZs focuses on multi-domain end-to-end performance metrics. On the other hand, the monitoring process in enterprise networks focuses on single-domain performance metrics. Accordingly, Fig. 4 presents monitoring applications: perfSONAR [16, 17], Simple Network Management Protocol (SNMP) [18], Syslog [19], and Netflow [20]. The latter is also used for security purposes.

4.1 perfSONAR

perfSONAR [16, 17] is an application that helps locate network failures and maintain optimal end-to-end usage expectations. Each organization deciding to use this tool is required to install a measurement point in its network, as shown in Fig. 5a. The service providers 1, 2, and 3 provide connectivity to campus networks 1 and 2. A measurement point is a Linux machine running the perfSONAR application. perfSONAR offers several services, including automated bandwidth tests and diagnostic tools.

One of the main features of perfSONAR is its cooperative nature by which an institution can measure several metrics (e.g., throughput, latency, packet loss) to different intermediary domains and to a destination network. Using the example of Fig. 5a, campus network 1 can measure metrics from itself to campus network 2. Campus network 1 can also measure metrics to the service providers. Figure 5b

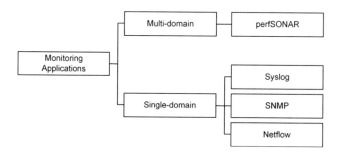

Fig. 4 Monitoring applications

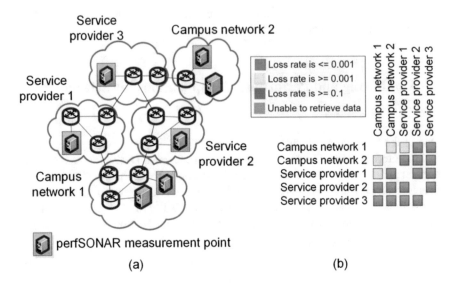

Fig. 5 perfSONAR application. (**a**) Multi-domain topology. Each network has a perfSONAR node. (**b**) Corresponding perfSONAR dashboard

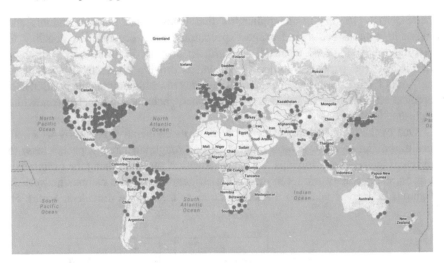

Fig. 6 perfSONAR nodes deployed as of June 2017

shows a sample dashboard view for packet loss rate for the perfSONAR node at campus network 1.

Given the increasing number of Science DMZs, perfSONAR has seen a steady increase in deployments. Currently, there are more than 2000 perfSONAR measurement points deployed around the world. In the U.S., most Science DMZs include at least one perfSONAR node. Figure 6 shows the location of perfSONAR nodes as of June 2017.

4.2 Comparison of Monitoring Applications in Enterprise Networks and Science DMZs

The ubiquitous SNMP protocol is also widely used for monitoring purposes. Accordingly, Table 2 compares SNMP and perfSONAR. Overall, their functionalities are complementary, and a well-monitored Science DMZ may include both. SNMP is used to monitor a single administrative domain; thus it lacks ability to detect failures beyond the local domain. Also SNMP can only infer, to some extent, a performance metric based on polling of individual network elements. Meanwhile, perfSONAR includes a set of active testing tools to measure performance via probing. End-to-end and soft failures can be detected with perfSONAR because of its multi-domain characteristic, sub-path testing, and end-to-end tools. On the other hand, some hard failures are easily detected by SNMP, while they may not be detected quickly by perfSONAR. While reporting applications are available for both, perfSONAR's reports include multi-domain results. Regarding security, SNMPv3 includes confidentiality, integrity, and authentication.

4.3 WAN Emulation and Other Performance Applications

When deploying a Science DMZ, routers, switches, and DTNs should be tested. Problems associated with routers and switches may not be observed in a testing environment unless WAN conditions, such as delay and jitter, are introduced. Thus, inadequate buffer sizes can easily be overlooked. Hence for testing purposes, in the absence of a WAN, a useful alternative is a network emulator. With such a tool, applications and devices can be tested over a virtual network. Now, two applications widely used to emulate a WAN are netem [28] and iPerf [29]. netem is a Linux application that emulates the properties of a WAN and permits to vary parameters such as delay, jitter, packet loss, and duplication and reordering of packets. Meanwhile, iPerf measures memory-to-memory throughput from a client (sender) to a server (receiver). The client generates dummy application-layer data in main memory, which is then moved down through the protocol stack and over the network media. The server receives the data and moves it up through the protocol stack. The two applications, netem and iPerf, can be used together to emulate data transfers between DTNs and test TCP parameters (congestion control algorithms, buffer size, TCP extensions), routers, and switches.

5 Applications in Enterprise Networks and Science DMZs

Table 3 compares data transfer and monitoring applications used in enterprise networks and Science DMZs. The use of virtualization in both environments is

Table 2 Comparison between SNMP and perfSONAR

Feature	SNMP	perfSONAR
Main uses	Enterprise networks: offices, campuses, commercial ISPs	Science DMZ, RENs
Scope	Single domain	Multi-domain
Network monitoring under controlled load	Difficult; SNMP agents can collect statistics or report events	Easy; perfSONAR is composed of several active testing tools
Performance instrumentation	Difficult; SNMP uses polling to track individual network elements rather than end-to-end performance	Easy, perfSONAR's probing tests measure end-to-end performance
Soft failure detection	Difficult; failures could be inferred locally only through polling byte counters	Easier; multi-domain visibility and active monitoring from the local network to any deployed perfSONAR node
End-to-end failure detection	Difficult; limited multi-domain visibility	Easier; a variety of end-to-end tools for performance and troubleshooting; e.g., One-Way Active Measurement Protocol (OWAMP), Bandwidth Test Controller (BWCTL), Network Path and Application Diagnostics (NPAD)
Sub-path testing	No	Yes; perfSONAR's NPAD tool allows the testing of portions of paths.
Hard failure	Easier; SNMP can report on asynchronous events via trap messages	Difficult; perfSONAR does not report asynchronous events
Measurable variables	CPU usage, packet counters, dropped packets, the number of flows	Bandwidth, latency, packet loss, jitter
Schedulable tests	No	Yes; pScheduler
Programmability in configuration and task specification	Commercial products are available, but custom coding to automatically configure/test devices may be required	Easier; it supports jq [21], a command-line JSON [22] processor for parsing and processing commands
Confidentiality, integrity, and authentication	Yes; SNMPv3	No
Reporting	Yes; multiple tools for automatic generation of reports. Usually, reports are for single domain only	Yes; automatic generation of reports and dashboards for end-to-end multi-domain paths; esmond stores and reports time-series measurements

also compared. Owing to the nature and duration of data transfers, Science DMZ applications should incorporate features such as partial and restartable transfers. In addition, features to combat packet losses are important, including the use and orchestration of parallel streams. On the other hand, data synchronization is a mature feature already implemented by applications used in enterprise networks.

Table 3 Comparison of data transfer and monitoring applications, and virtualization use in enterprise networks and Science DMZ

Feature	Enterprise network	Science DMZ
Data transfer application		
Rates	Tens of Mbps to few Gbps	10 Gbps and above
Transport protocol	UDP, TCP, TSL/SSL	TCP
Partial and restartable transfer	Usually not required	Highly desirable
Management of parallel streams	Not required	Required
Parallel file system	Typically not used nor required	Highly desirable; provides parallelism opportunities and higher rates
Sharing and publishing	Mature tools, high adoption, e.g., Google drive, Dropbox	Maturing feature, in developing phase, e.g., Globus
Security	Mature feature, supported with HTTPS and TLS/SSL	Maturing feature, not fully in compliance with rules and regulations yet
Data synchronization between repositories	Supported (e.g., Google drive, Dropbox)	Minimal supported; manual procedure required
Monitoring application		
Monitoring scope	Single domain	Multi-domain
Soft failure detection	Desirable but not essential	Highly desirable
Sub-path testing	Not required; paths in typical switched LAN environments often are single hop.	Highly desirable; paths are typically composed of many hops in multiple domains
Hard failure detection	Easy, highly granular, e.g., more than 6000 Syslog events and 90 SNMP trap notifications in enterprise devices [23]	Few available features
Monitored network type	Focus on LANs and/or interconnected LANs	Focus on inter-networks composed of LANs and WANs
Virtualization technology		
Virtual host	High adoption: server consolidation, multiple execution environments, mobility	Low adoption, limited need for consolidation (often data transfer and perfSONAR applications only); performance penalty
Virtual switch	High adoption; virtual switch used with VLANs to isolate VMs	Low adoption, unavailability of buffer capability, and configuration; performance penalty
Virtual router	Medium adoption; new technology (e.g., NSX [24]), suitable for east–west traffic routing in datacenters	Low adoption, not required; performance penalty
virtual NIC	High adoption; E1000e, VMXNET, VMXNET2, VMXNET3 [15]	Low adoption, e.g., VMXNET3 [15] supports 10 Gbps rates
Protocols for network virtualization	Used for LAN management, e.g., 802.1Q, overlay VXLAN [25]	Used for resource reservation in WANs, e.g., OSCARS [26], MPLS [27]

Virtualization has not been adopted for Science DMZ deployments. The main concern is the performance penalty associated with virtual devices. Additionally, although products such as NSX [24] perform well in enterprise networks, the capacity and architecture of virtual routers and switches (switching rate, buffer size, fabric) are often not available.

In enterprise networks, protocols for network virtualization are mostly used for LAN management. Examples include 802.1Q and Virtual Extensible LAN (VXLAN) [25]. In Science DMZs, protocols and platforms such as On-demand Secure Circuits and Reservation System (OSCARS) [26] and Multi-Protocol Label Switching (MPLS) [27] are used for resource reservation and creation of virtual circuits across WANs.

References

1. J. Postel, J. Reynolds, File transfer protocol. Internet Request for Comments, RFC Editor, RFC 959, Oct. 1985. [Online]. Available: https://tools.ietf.org/html/rfc959
2. T. Ylonen, C. Lonvick, The secure shell connection protocol. Internet Request for Comments, RFC 4254, Jan. 2006. [Online]. Available: https://tools.ietf.org/html/rfc4254
3. K. Chard, S. Tuecke, I. Foster, Globus: recent enhancements and future plans, in *Proceedings of the XSEDE16 Conference on Diversity, Big Data, and Science at Scale* (2016)
4. W. Allcock, J. Bresnahan, R. Kettimuthu, M. Link, The globus striped GridFTP framework and server, in *Proceedings of the 2005 ACM/IEEE Conference on Supercomputing* (2005)
5. Z. Liu, P. Balaprakash, R. Kettimuthu, I. Foster, Explaining wide area data transfer performance, in *IEEE/ACM International Symposium on High-Performance Distributed Computing (HPDC)* (2017)
6. L. Zhang, W. Wu, P. DeMar, E. Pouyoul, mdtmFTP and its evaluation on ESNET SDN testbed. Futur. Gener. Comput. Syst. **79**, 199–204 (2018 Feb 1)
7. Fast data transfer (FDT). [Online]. Available: http://monalisa.cern.ch/FDT
8. Ultra high-speed transport technology. Aspera White Paper. [Online]. Available: http://asperasoft.com/resources/white-papers/ultra-high-speed-transport-technology/
9. High-end computing capability using BBFTP for remote file transfers. [Online]. Available: https://www.nas.nasa.gov/hecc/support/kb/using-bbftp-for-remote-file-transfers_147.html
10. ExpeData, a multipurpose transaction protocol. ExpeDat White Paper, Jan. 2017. [Online]. Available: http://www.dataexpedition.com/expedat/Docs/
11. Y. Gu, R. Grossman, UDT: UDP-based data transfer for high-speed wide area networks. Comput. Netw. **51**(7), 1777–1799 (2007)
12. D. Bernardo, D. Hoang, Empirical survey: experimentation and implementations of high speed protocol data transfer for grid, in *IEEE International Conference on Advance Information Networking and Application Workshops* (2011)
13. UDP tuning in Science DMZs. [Online]. Available: https://fasterdata.es.net/network-tuning/udp-tuning/#toc-anchor-1
14. N. Cardwell, Y. Cheng, C. Gunn, S. Yeganeh, V. Jacobson, BBR: congestion-based congestion control. Commun. ACM **60**(2), 58–66 (2017)
15. Performance evaluation of vmxnet3 virtual network device. VMware Technical Report. [Online]. Available: https://www.vmware.com/pdf/vsp_4_vmxnet3_perf.pdf
16. J. Zurawski, S. Balasubramanian, A. Brown, E. Kissel, A. Lake, M. Swany, B. Tierney, M. Zekauskas, perfSONAR: on-board diagnostics for big data, in *Workshop on Big Data and Science: Infrastructure and Services* (2013)

17. A. Hanemann, J. Boote, E. Boyd, J. Durand, L. Kudarimoti, R. Lapacz, D. Swany, J. Zurawski, S. Trocha, perfSONAR: a service oriented architecture for multi-domain network monitoring, in *Proceedings of the Third International Conference on Service-Oriented Computing* (2005), pp. 241–254

18. D. Levi, P. Meyer, B. Stewart, Simple network management protocol (SNMP) applications. Internet Request for Comments, RFC Edit, RFC 3413, Dec. 2002 [Online]. Available: https:// tools.ietf.org/html/rfc3413

19. C. Lonvick, The BSD syslog protocol. Internet Request for Comments, RFC 3164, Aug. 2001. [Online]. Available: https://www.ietf.org/rfc/rfc3164.txt

20. B. Claise, Cisco Systems NetFlow Services Export Version 9. Internet Request for Comments, RFC Editor, RFC 3954, Oct. 2004. [Online]. Available: https://www.ietf.org/rfc/rfc3954.txt

21. jq command-line JSON processor. [Online]. Available https://stedolan.github.io/jq/

22. T. Bray, The JavaScript Object Notation Data Interchange Format. Internet Request for Comments, RFC 7159, Mar. 2014. [Online]. Available: https://tools.ietf.org/html/rfc7159

23. Building scalable syslog management solutions, Cisco White Paper. [Online]. Available: https://www.cisco.com/c/en/us/products/collateral/services/high-availability/white_paper_ c11-557812.html#wp9000392

24. R. Mijumbi, J. Serrat, J. Gorricho, N. Bouten, F. De Turck, R. Boutaba, Network function virtualization: state-of-the-art and research challenges. IEEE Commun. Surv. Tutorials **18**(1), 236–262 (2016)

25. M. Mahalingam, D. Dutt, K. Duda, P. Agarwal, L. Kreeger, T. Sridhar, M. Bursell, C. Wright, Virtual eXtensible Local Area Network (VXLAN): a framework for overlaying virtualized layer 2 networks over layer 3 networks. Internet Request for Comments, RFC 7348, Aug. (2014)

26. T. Orawiwattanakul, H. Otsuki, E. Kawai, S. Shimojo, Multiple classes of service provisioning with bandwidth and delay guarantees in dynamic circuit network, in *IEEE International Symposium on Integrated Network Management* (2015)

27. E. Rosen, A. Viswanathan, R. Callon, Multiprotocol label switching architecture. Internet Request for Comments, RFC Edit, RFC 3031, Jan. 2001 [Online]. Available: https://tools.ietf. org/html/rfc3031.txt

28. S. Hemminger, Network emulation with netem, in *Australia's National Linux Conference* (2005)

29. iperf3. [Online]. Available: http://software.es.net/iperf/

Security Aspects

This section discusses security aspects in high-speed networks. The section pays particular attention to operational security, which addresses potential attackers attempting unauthorized access, introducing malware into devices, and conducting denial of service (DoS) attacks. The chapter describes router's access-control, firewalls, intrusion prevention systems, and intrusion detection systems.

Security is a growing concern in Science DMZs. Hence, associated security problems can be divided into operational security, confidentiality, integrity, and authentication:

- Operational security: Attackers can attempt unauthorized access, introduce malware into devices, and conduct denial of service (DoS) attacks. ACLs, firewalls, IPSs, and IDSs are commonly used to counter attacks.
- Confidentiality: Only the sender and the intended receiver should understand the contents of the transmitted message. This requires that the message be encrypted.
- Integrity: The content of the communication between the sender and the intended receiver must not be altered, maliciously or by accident. Hash functions are used for integrity control.
- Authentication: The sender and the receiver should confirm the identity of the other party. Authentication methods typically rely on pre-shared key and digital signatures.

Of the above four areas, operational security is the most relevant at the time of designing and deploying a Science DMZ. The remaining three areas (confidentiality, integrity, and authentication) can be implemented at different layers, including relying on the application layer for these services.

Table 1 lists security-related differences between enterprise networks and Science DMZs. The volume distribution differs substantially, as typically there are few simultaneous flows in a Science DMZ, whereas there can be thousands or millions of small flows in an enterprise network. There are a variety of applications in enterprise networks while there are only a couple in Science DMZs, namely, data transfer

J. Crichigno et al., *High-Speed Networks*, Practical Networking,
https://doi.org/10.1007/978-3-030-88841-1_6

Table 1 Security-related differences between enterprise networks and Science DMZs

Feature	Enterprise network	Science DMZ
Volume	Thousands of concurrent small flows	Typically few concurrent large flows
Application type	Web, emails, HTML, XML, mobile applications, media content, SQL, etc.	Data transfers, performance monitoring
Most used ports	80 (HTTP), 443 (HTTPS)	2811 (Globus control channel), 50,000 to 51,000 (Globus data channels)
Operations over data on used ports	Multimedia, image processing, games, mobile code execution, HTML, XML, SQL operations	File operations: open, read, write, close
Number of devices	Typically hundreds to thousands	Few, it could be even a single DTN
Bring-your-own-device policy	Yes	No
Operating systems and platforms	A variety of OSs and platforms, including Windows, Linux, MAC, RIM Blackberry, Android, Windows Mobile, Oracle, Kindle	Typically only Linux (e.g., CentOS) for all DTNs and perfSONAR nodes
Application changes and updates	Continuous changes in applications and operating systems updates	Changes are not frequent

and performance monitoring tools. As a result, delivery attack options are abundant in enterprise networks (e.g., cross-site scripting, SQL injection, XML injection, etc.). By contrast, Science DMZs only see specific data transfer and performance monitoring tools, such as Globus and perfSONAR. Hence, the number of open ports is not an indicator of risk, as a large number of exploits in enterprise networks are delivered via ports 80 and 443. On the other hand, Globus often requires hundreds of open ports for data channels. Also, the number of operations executed on data in a Science DMZ is small. In addition to having hundreds or thousands of servers and desktops, most enterprise networks adopt a bring-your-on-device (BYOD) policy [1], which allows users to use their personal mobile device. BYOD represents additional risks, because mobile devices include a large variety of applications and operating systems with their respective vulnerabilities.

Inline devices are discouraged in Science DMZs, since they check each packet in real-time. On the other hand, offline devices operate with copies of packets and do not interfere with traffic flows. Figure 1 illustrates the typical placement of security appliances. A taxonomy of security appliances discussed next is shown in Fig. 2.Security Aspects

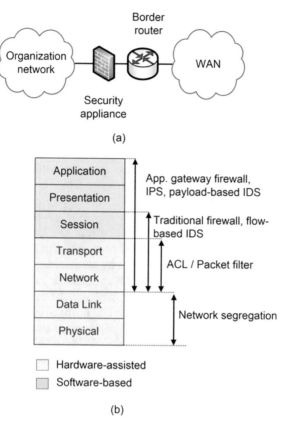

(a)

Hardware-assisted

Software-based

(b)

Fig. 1 Physical and logical locations of security appliances. (**a**) Security appliance (ACL, IPS, IDS, and/or firewall) co-located with the border router. (**b**) Security appliances in the OSI model

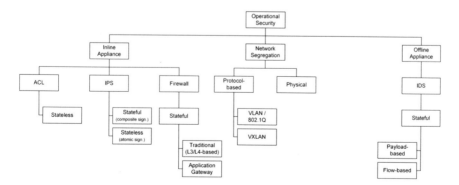

Fig. 2 Operational security appliances and techniques

1 Operational Security for Science DMZs

1.1 Network Segregation

Since traffic characteristics and security policies differ between a Science DMZ and an enterprise network, their segregation is natural. Hence, when implemented contiguously, the two networks must either be physically or logically separated. Figure 3 shows an example of physical separation, where the two networks are attached to different interfaces at the border router. Note that traffic flowing into the campus enterprise network is subject to inspection by a firewall and other security appliances, while traffic flowing into the Science DMZ is subject to a minimal inspection by the border router only.

An alternative to physical segregation is logical segregation by using VLAN technology. A VLAN is a logical subgroup within the LAN that is created at the switch via software rather than physical hardware. The Science DMZ can be isolated from the campus enterprise network by establishing one or more VLANs assigned to the Science DMZ. Since a VLAN can have its own IP address scheme, different access and security policies can be implemented based on IP. However, a disadvantage of a VLAN-based segregation is that the bandwidth of the interface to which both the Science DMZ and enterprise network are attached must be shared. Hence, if there is no mechanism to control the bandwidth allocation to each network, the enterprise network may starve when DTNs are actively transferring data. Additionally, switches must be dimensioned based on Science DMZ requirements.

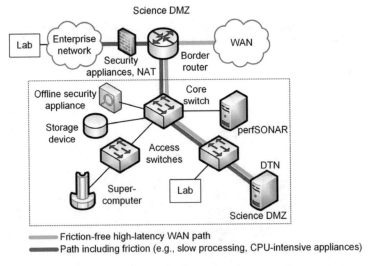

Fig. 3 A Science DMZ co-located with an enterprise network. Notice the absence of firewall or any stateful inline security appliance in the friction-free path

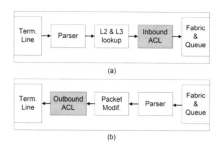

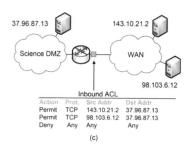

Fig. 4 Implementation and use of ACLs. (**a**) Diagram of the input and (**b**) output ports of a router, and the placement of inbound and outbound ACLs. (**c**) A Science DMZ protected by an inbound ACL. Notice the targeted security by which only specific collaborators' DTNs at 143.10.21.2 and 98.103.6.12 are permitted to connect to the DTN 37.96.87.13

1.2 Access-Control List

ACLs are used to control the access to a Science DMZ. Since ACLs are implemented in the forwarding plane of routers and switches, they do not compromise performance. Additionally, as collaborators' IP addresses may be known in advance, a targeted security policy can be used. Figure 4 shows an example of an ACL implementation [3]. Figure 4a shows the input pipeline, where a packet arrives at the input port and the termination line performs physical-layer functions. The parser engine parses the incoming packets and extracts the fields required for lookup. The lookup process results in an output port the packet will be forwarded to. At this moment, the inbound ACL is applied to the packet. The packet is then switched through the fabric and buffered. The output pipeline, shown in Fig. 4b, follows a similar scheme.

Figure 4c shows an ACL used to protect a Science DMZ. The DTN with IP address 37.96.87.13 is located in the protected Science DMZ. The IP addresses of other DTNs from collaborators' networks are 143.10.21.2 and 98.103.6.12. The ACL is applied in the inbound direction at the interface facing the WAN. The ACL has three rules: the first two rules permit any TCP segment coming from the collaborators' addresses and going to the local DTN. The last rule denies any other packets from entering the Science DMZ. Note that stricter rules can also be applied, even incorporating port information (e.g., an ACL may only permit TCP segments from collaborators at the ports used by Globus).

1.3 Firewalls

These devices are capable of processing a large number of small flows characterized by short durations and low transfer rates. Additionally, firewalls typically have small

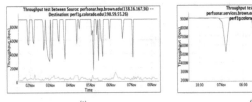 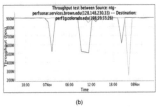 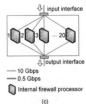

| (a) | (b) | (c) |

Fig. 5 Impact of a firewall on a data transfer. Throughput performance between a DTN at the University of Colorado in Boulder, Colorado, and a DTN at Brown University in Providence, Rhode Island. The blue curve is the throughput from the DTN at Brown University to the DTN at University of Colorado. The green curve is the throughput from the DTN at the University of Colorado to the DTN at Brown University. (a) Data transfer when a firewall is located in the path of the two hosts. (b) Data transfer when the firewall is removed from the path. The results of (a) and (b) are reproduced from [2]. (c) Conceptual 10 Gbps firewall architecture

buffers [4]. Clearly, Science DMZ flows do not match this traffic profile. As a result, when a large flow crosses a firewall, the throughput of the flow deteriorates rapidly.

Consider Fig. 5, which shows the throughput for data transfers between a DTN located at Brown University in Providence, Road Island, and a DTN located at the University of Colorado in Boulder, Colorado [2]. These two DTNs are connected by a 1 Gbps path. Figure 5a shows the throughput achieved when there is one firewall located at Brown University. The blue curve is the throughput from the DTN at Brown University to the DTN at the University of Colorado. This traffic is referred to as outbound, and the firewall is not intended to inspect this flow. The green curve is the throughput from the DTN at the University of Colorado to the DTN at Brown University. This traffic is referred to as inbound, and the firewall inspects each packet of this flow.

While both curves show that throughput is affected, the inspection impact on the inbound traffic is critical. For example, the inbound throughput does not even reach 50 Mbps, or 5% of the 1 Gbps path capacity. Figure 5b shows the performance between the same two DTNs, but for the case where the firewall is removed from the path. In this instance, the throughput is approximately 900 Mbps, or 90% of the capacity. The reason of this performance difference is related to TCP retransmissions. Namely, every time that TCP receives a triple duplicate ACK for a packet that is lost, a fast retransmission is triggered and the congestion window is reduced by half, thus reducing throughput.

Figure 5c illustrates a generic architecture of a 10 Gbps enterprise firewall. Internally, load balancing among 20 firewall processors is achieved on a per-flow basis (each firewall processor has a capacity of 0.5 Gbps). Note that the maximum throughput is not determined by a single large flow; instead, the maximum throughput is the aggregate throughput of thousands of small flows. When a flow with a rate above 0.5 Gbps arrives at the input interface, all packets of the flow are processed by the same firewall processor. Eventually, incoming packets are dropped as a consequence of the low capacity of the individual firewall processor.

Note that data transfers in LANs may still achieve reasonable performance in the presence of firewalls. Namely, since the latency is small, the TCP throughput can increase quickly after a packet loss. Specifically, after reducing the congestion window by half, TCP increases the congestion window again in a time that is proportional to the RTT.

1.4 Intrusion Prevention System

One of the main features of an IPS is the database containing attack signatures. However, the process of matching a signature with the content of the packet in real-time is time consuming. Even new IPS devices such as the Next Generation IPS (NGIPS) [5], which advertise throughputs of tens to hundreds of Gbps, are not suitable for processing large flows. Akin to firewalls, they are designed for processing thousands of small flows simultaneously. For example, the underlying technology of the NGIPS is Snort [6], an open-source IPS engine. For an NGIPS appliance rated at 10 Gbps with 20 internal processors, its maximum throughput is only achieved by aggregating the individual throughput of these 20 internal Snort instances. Since each instance has a capacity of 0.5 Gbps and packets belonging to the same flow are processed by the same Snort instance, inspecting a 10 Gbps science flow is not feasible here [7].

1.5 Intrusion Detection System

For Science DMZs, IDSs represent a better option than IPSs. These systems can be classified based on the information used to detect attacks: payload-based IDS and flow-based IDS [8, 9]. Payload-based IDSs inspect the content of every packet. For high-speed networks, the main challenge of this approach is the processing capability. Meanwhile, flow-based IDSs analyze the communication patterns within the network rather than the contents of individual packets. These devices are attractive for Science DMZs, because of the substantial processing reduction.

Figure 6a illustrates the deployment of a payload-based IDS. The border router forwards traffic to a switch. Packets addressed to the protected network are copied and sent to the IDS. The copy is typically done by a switch with a feature called Switched Port Analyzer (SPAN). SPAN copies network traffic from a selected source port in the switch to a selected destination port. The latter is connected to the IDS.

A popular payload-based IDS is Bro [10]. Bro is well-suited for use in a Science DMZ for several reasons: (1) flexibility in defining security policies, which can be granularly customized by using a domain-specific scripting language interpreted by a policy script interpreter layer; (2) incorporation of hundreds of protocol analyzers in the event engine, allowing the IDS to detect anomalies carried in practically all

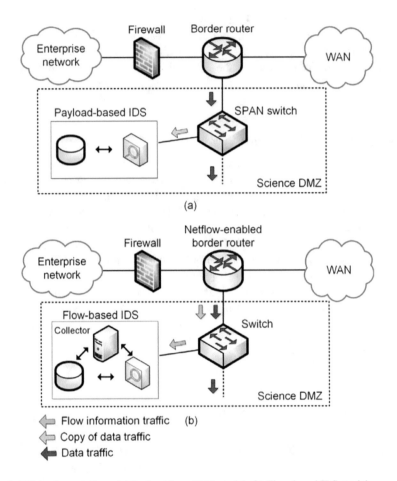

Fig. 6 IDS implementations. (**a**) Payload-based IDS model. (**b**) Flow-based IDS model

existent protocols; and (3) scalability. As science flows are characterized at rates of tens of Gbps or more, the potential traffic volumes surpass the capacity of a single-instance IDS. Hence, Bro nodes can be organized in clusters, where clusters of nodes cooperate seamlessly [11]. However, at very high rates, the amount of processing may become excessive, even for a clustered IDS [12].

Flow-based IDSs track the lifetime of a flow and characterize its behavior [8], [9]. This characterization may incorporate several attributes such as the time the exchange of data started, the time it ended, the number of transferred bytes, etc. Figure 6b shows a Netflow-based IDS protecting a Science DMZ. The Netflow-enabled router collects statistical information of all incoming flows passing through the interface facing the WAN. These statistics are collected in hardware by an interface' network processor. The router then extracts the packet header from each packet seen on the monitored interface and marks the header with a timestamp. It then proceeds to update a flow entry in the flow cache of the router. Once a flow record expires (typically seconds or few minutes), it is sent to a flow collector. Note that the volume of information sent to and stored by the flow collector is several orders of magnitude lower than the actual traffic.

For campuses operating at 100 Gbps, the sampling flow (sFlow) technique [13] is a more scalable solution than Netflow. Here, for a given flow, instead of processing each packet, sFlow can process 1 out of a packets, where a is a configurable parameter. It should be noted, however, that sampling not only lowers the demands put on the flow exporter, but also could make the detection of intrusions harder [8].

1.6 Response Plan

In general, at least two actions that can be taken once an anomaly is detected are black hole routing and ACL blocking. The black hole routing approach drops packets coming from a suspicious source IP address (e.g., an attacker identified by an IDS) by installing a particular entry in the routing table. The mechanism used is called Unicast Reverse Path Forwarding (uRPF) [14]. This information can be disseminated to other routers via BGP [15]. The ACL blocking technique creates and installs an ACL in the border router when an offender is identified.

Black hole routing is more effective if the information is disseminated to other routers, thus the attack is prevented before packets reach the Science DMZ. On the other hand, ACL blocking is simpler and effective, but the router must still process each offender packet.

2 Confidentiality, Integrity, and Authentication

Confidentiality, integrity, and authentication services are typically provided by the application layer, specifically, by the data transfer tool. These security aspects are required for certain applications. For example, medical Science DMZs [16, 17] transport medical information that must adhere to security and privacy laws and regulations.

Globus [18, 19] provides authentication on the control channel by default. Confidentiality and integrity are both supported on the data channel but are not enabled by default. Vardoyan et al. [20] showed that by using Globus with multiple threads, the encryption of the data channel has a minimal performance impact. Globus also includes a feature called striped configuration, which is illustrated in Fig. 7a. In this configuration, multiple cooperating DTNs can exchange data with remote DTNs [19]. The DTNs are coordinated by a server protocol interpreter (SPI), which implements the control channel. Transfers are then divided over all available DTNs, thus allowing the combined bandwidth of all DTNs to be used. An advantage

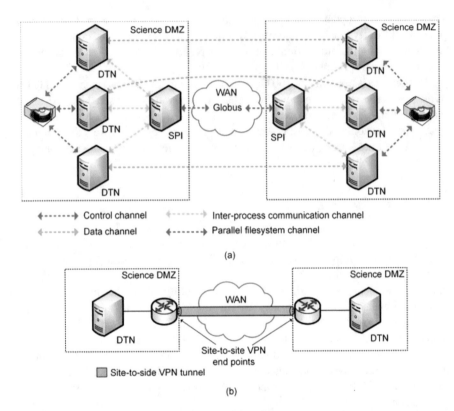

Fig. 7 Configuration options for confidentiality, integrity, and authentication. (**a**) Cluster-to-cluster configuration of the Globus striped configuration. The server protocol interpreter (SPI) implements the control channel and coordinates the load distribution among DTNs. The communication between SPI and DTNs is via an inter-process communication channel. Data is read/write from/to a storage device using a parallel file system channel. (**b**) Site-to-site virtual private network (VPN) tunnel. Confidentiality, integrity, and authentication are implemented by the routers located at the points of the tunnel

of this configuration when implementing full encryption is the distribution of processing load.

Modern symmetric-key algorithms can also efficiently encrypt and decrypt data. Some manufacturers, such as Intel and AMD, now offer hardware-based instructions to improve the encryption and decryption throughput of some algorithms, such as the Advanced Encryption Standard (AES). Also, block ciphers provide abundant parallelism's opportunities. For example, when operating in counter mode, AES can encrypt and decrypt blocks in parallel, and the throughput can be increased according to the amount of parallelism provided by multiple cores. Current encryption technology is suitable for 10 Gbps rates.

On the other side of encryption is the file integrity check (FIC). To verify the integrity of a file, the entire file must be received first. Only then can the destination DTN run a cryptographic hash function on the received file. Thus, FIC may represent a larger performance penalty than encryption and decryption.

For authentication purposes, an industry standard that is increasingly being adopted is OAuth 2.0 [21]. Consider a client DTN attempting to download a large file from a server DTN. Here the client DTN is provided with a delegated access to the file resting at the server DTN without sharing credentials.

Confidentiality, integrity, and authentication can also be implemented via a site-to-site virtual private network (VPN). Here, the sender router encrypts the traffic before it enters the WAN. The receiver router then decrypts the traffic upon arrival to the destination Science DMZ. Figure 7b illustrates this alternative. Now, most routers implement VPNs based on the IP security (IPsec) architecture [22]. While IPsec is a well-proven technology, its main disadvantage is the additional processing overhead at the router.

3 Security Summary

Table 2 summarizes the various security techniques. Clearly, securing a Science DMZ cannot be done with a single device or technique. ACLs are strongly recommended for protecting Science DMZ; however, other offline devices, such as IDSs, should also be implemented to supplement an ACL's lack of context. Inline devices must be avoided.

Academic Cloud and Virtual Laboratories

The book is accompanied by hands-on virtual laboratory experiments conducted in a cloud system, referred to as the Academic Cloud. Access to the Academic Cloud is available for a fee (six-month access) and includes all material needed to conduct the experiments. The URL is

http://highspeednetworks.net/

Chapter 6—Lab 14: Introduction to the Capabilities of Zeek

Overview
To conduct the experiment described in this section, please login into the Academic Cloud at http://highspeednetworks.net/ and reserve a pod for Lab 14.

This lab introduces Zeek, an open-source network analysis framework primarily used in security monitoring and traffic analysis. The primary focus of this lab is to explain Zeek's layered architecture while demonstrating Zeek's capabilities toward performing network traffic analysis.

Table 2 Security considerations in Science DMZs

Device/Technique	Advantage	Disadvantage	SDMZ recommended
Physical segregation	Easy bandwidth allocation for each network; equipment specifically dimensioned for each network: enterprise network and Science DMZ; easy to apply different security policies for each network	More expensive; having two physical infrastructures may require higher maintenance and operation costs	Yes
VLAN segregation	Only one physical infrastructure is required; cheaper	Shared infrastructure between enterprise network and Science DMZ; more complex allocation of resources (bandwidth, buffer, etc.); potential bandwidth starvation in enterprise network, if resources are not adequately allocated	Acceptable, if shared resources are appropriately allocated
ACL	Very scalable; it is implemented in router's forwarding plane; minimal performance impact; easy to implement	Decisions are not based on context; addresses of collaborators must be identified in advance; fragmented packets are unreliably filtered; susceptible to IP spoofing	Yes
Firewalls	Session tracking adds context to decisions; they are robust against IP spoofing; rich data log	Inspection capacity is below required rates; for science flows, throughput is severely impacted; it represents a bottleneck for Science DMZs and leads to packet losses and/or out-of-order delivery	No
Payload-based IDS	Payload inspection provides full application-layer information; state information is collected without interference with flows; no performance impact on switches or routers	Additional resources for scalability may be needed (e.g., cluster of servers); peak of traffic inspection may be very large at times; attacks might only be stopped from reaching target after they occur; without large clusters, monitoring 100 Gbps links is very difficult	Yes
Flow-based IDS	The most scalable IDS solution; ability to inspect hundreds of Gbps with a single CPU (sFlow); state information is collected without interference; minimal performance impact on routers and switches	Application-layer payload is not inspected, but only flow information; flows represent aggregate information only; when sampling is used (sFlow), flow information may be lost; attacks might only be stopped from reaching target after they occur	Yes

(continued)

Table 2 (continued)

Device/Technique	Advantage	Disadvantage	SDMZ rec-ommended
Confidentiality, integrity, authentication at application layer	Encryption at modern DTNs can now be achieved at high rates; scalable alternatives are available, if needed (e.g., Globus' striped configuration)	There is a throughput degradation when file integrity check is performed; additional resources (e.g., CPU) may be needed for scalability	Yes
IPS	Inspection of application-layer payload provides full information; attacks can be detected and stopped immediately	Inspection capacity under single large flow is well below required rates; for science flows, throughput is severely impacted	No
Confidentiality, integrity, authentication with IPsec	Integrity is checked on a per-packet basis at the router, avoiding a resource-expensive file integrity check at the end of the transfer; well-known, proven technology (IPsec)	If router is overloaded, the additional processing overhead may lead to packet losses	Not recom-mended, but acceptable if the router has sufficient CPU capability

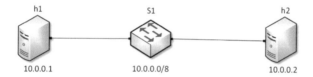

Fig. 8 Lab topology

Objectives

By the end of this lab, students should be able to:

1. Understand Zeek's layered architecture.
2. Start and manage a Zeek instance using the *ZeekControl* utility.
3. Use Zeek to process packet captures files.
4. Generate and analyze live network traffic in Zeek.

Lab Topology

Figure 8 shows the lab topology. The topology uses 10.0.0.0/8, which is the default network assigned by Mininet. The *h1* and *h2* virtual machines will be used to generate and collect network traffic.

Lab Settings

The information (case-sensitive) in the table below provides the credentials necessary to access the machines used in this lab (Table 3).

Table 3 Credentials to access the Client machine	Device	Account	Password
	Client	admin	password

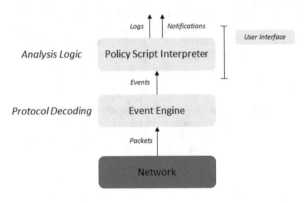

Fig. 9 Zeek's architecture

Lab Roadmap

This lab is organized as follows:

1. Section 4: Introduction to Zeek.
2. Section 5: Using *ZeekControl* to update the status of Zeek.
3. Section 6: Introduction to Zeek's traffic analysis capabilities.

4 Introduction to Zeek

Zeek is a passive, open-source network traffic analyzer. It is primarily used as a security monitor that inspects all traffic on a network link for signs of suspicious activity. It can run on commodity hardware with standard UNIX-based systems and can be used as a passive network monitoring tool.

Setting Zeek as a node with an assigned IP address on the monitored network is not mandatory. Figure 9 shows Zeek's layered architecture. Once Zeek receives packets, its *eventengine* converts them into *events*. The *events* are then forwarded to the policy script interpreter, which generates logs, notifications, and/or actions.

Zeek uses the standard ⎡libpcap⎤ library for capturing packets to be used in network monitoring and analysis.

4.1 The Zeek Event Engine

The event engine layer performs low-level network packets analysis. It receives raw packets from the network layer (packet capture), sorts them by connection, reassembles data streams, and decodes application-layer protocols. Whenever it encounters something potentially relevant to the policy layer, it generates an event.

The event engine consists of several analyzers responsible for well-defined tasks. Typical tasks include decoding a specific protocol, performing signature-matching, identifying backdoors, etc. Usually, an analyzer is accompanied by a default script, which implements some general policy adjustable to the local environment. The event engine can be divided into four major parts.

4.1.1 State Management

Zeek's main data structure is a connection, which follows typical flow identification mechanisms, such as 5-tuple approaches. The 5-tuple structure consists of the source IP address/port number, destination IP address/port number, and the protocol in use. For a connection-oriented protocol like TCP, the definition of a connection is more clear-cut; however, for others such as UDP and ICMP, Zeek implements a flow-like abstraction to aggregate packets. Each packet belongs to exactly one connection.

4.1.2 Transport-Layer Analyzers

On the transport layer, Zeek analyzes TCP, UDP packets. In TCP, Zeek's associated analyzer closely follows the various state changes, keeps track of acknowledgments, handles retransmissions and much more.

4.1.3 Application-Layer Analyzers

The analysis of the application-layer data of a connection depends on the service. There are analyzers for a wide variety of different protocols, e.g., HTTP, SMTP, or DNS, that generally conduct detailed analysis of the data stream.

4.1.4 Infrastructure

The general infrastructure of Zeek includes the event and timer management components, the script interpreter, and data structures.

4.2 The Zeek Policy Script Interpreter

While the event engine itself is policy-neutral, the top layer of Zeek defines the environment-specific network security policy. By writing handlers for events that may be raised by the event engine, the user can precisely define the constraints within the given network. If a security breach is detected, the policy layer generates an alert.

New event handlers can be created in Zeek's own scripting language. While providing all expected convenience of a powerful scripting language, it has been designed with network intrusion detection in mind. While it is expected that additional policy scripts are written by the user, there are nevertheless several default scripts included with the initial installation of Zeek. These default scripts already perform a wide range of analyses and are easily customizable.

4.3 Zeek Analyzers

The majority of Zeek's analyzers are in its event engine with accompanying policy scripts that can be customized by the user. Sometimes, however, the analyzer is just a policy script implementing multiple event handlers. The analyzers perform application-layer decoding, anomaly detection, signature matching, and connection analysis. Zeek has been designed so that it is easy to add additional analyzers.

4.4 Signatures

Most network intrusion detection systems (NIDS) match a large set of signatures against the network traffic. Here, a signature is a pattern of bytes that the NIDS tries to locate in the payload of network packets. As soon as a match is found, the system generates an alert.

A well-known IDS system is *Snort*; conversely, Zeek's general approach to intrusion detection has a much broader scope than traditional signature-matching, yet still contains a signature engine providing a functionality that is similar to that of other systems. Furthermore, while Zeek implements its own flexible signature language, there exists a converter that directly translates Snort's signatures into Zeek's syntax, as shown below (Fig. 10):

```
alert tcp any any -> [a.b.0.0/16,c.d.e.0/24] 80
   ( msg:"WEB-ATTACKS conf/httpd.conf attempt";
     nocase; sid:1373; flow:to_server,established;
     content:"conf/httpd.conf"; [...] )
```

(a) Snort

```
signature sid-1373 {
   ip-proto == tcp
   dst-ip == a.b.0.0/16,c.d.e.0/24
   dst-port == 80
   # The payload below is actually generated in a
   # case-insensitive format, which we omit here
   # for clarity.
   payload /.*conf\/httpd\.conf/
   tcp-state established,originator
   event "WEB-ATTACKS conf/httpd.conf attempt"
}%
```

(b) Zeek

Fig. 10 Example of signature conversion. (**a**) Snort's signature. (**b**) Zeek's signature

4.5 ZeekControl

ZeekControl, formerly known as *BroControl*, is an interactive shell for easily operating and managing Zeek installations on a single system or across multiple systems in a traffic-monitoring cluster (Fig. 11).

5 Using ZeekControl to Update the Status of Zeek

Step 1. From the top of the screen, click on the *Client* button as shown below to enter the *Client* machine (Fig. 12).

Step 2. The *Client* machine will now open, and the desktop will be displayed. On the left side of the screen, double click on the Terminal icon as shown below (Fig. 13).

Step 3. Using the Terminal, input the following command to enter the *ZeekControl* directory (Fig. 14). To type capital letters, it is recommended to hold the *Shift* key while typing rather than using the *Caps* key.

```
cd $ZEEK_INSTALL/bin/
```

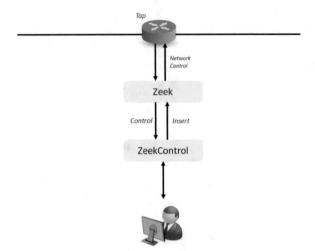

Fig. 11 *ZeekControl* scheme

Fig. 12 Opening the *Client* machine

Fig. 13 Opening the
Terminal

The active directory will change, as seen on the second line of the Terminal. Note that $ZEEK_INSTALL variable was substituted by its value (*/usr/local/zeek*) listed in Table 4.

Step 4. Use the following command to view the contents of the active directory (Fig. 15):

```
ls
```

The directory contents will be displayed. The green file name portrays an executable file.

```
cd $ZEEK_INSTALL/bin/
```

Fig. 14   Navigating into the *ZeekControl* directory

Table 4   Shell variables and their corresponding absolute paths

Variable name	Absolute path
$ZEEK_INSTALL	/usr/local/zeek
$ZEEK_TESTING_TRACES	/home/admin/zeek/testing/btest/Traces
$ZEEK_PROTOCOLS_SCRIPT	/home/admin/zeek/scripts/policy/protocols

```
ls
```

Fig. 15   Listing the directory's files

```
sudo ./zeekctl
```

Fig. 16   Launching the *ZeekControl* tool

Step 5. Use the following command to launch the *ZeekControl* tool. When prompted for a password, type ⟨password⟩ and hit *Enter* (Fig. 16).

```
sudo ./zeekctl
```

Once active, the *ZeekControl* prompt will be displayed within the Terminal. The ⟨help⟩ command will display additional information regarding *ZeekControl*.

5.1   Starting a New Instance of Zeek

Step 1. To initialize Zeek, enter the following command into the *ZeekControl* prompt (Fig. 17):

```
start
```

If you see error messages during the new Zeek instance initializing process, please ignore it.

Step 2. Use the following command to view the status of the currently active Zeek instance to ensure that it is active (Fig. 18):

```
status
```

The *running* status indicates that Zeek is currently active and functioning properly. The output of the status command includes other useful parameters:

- Name : the name of the Zeek instance.
- Type : the type of the instance (standalone in our case).
- Host : the host name (local host).
- Pid : the process ID. This ID can be used with other tools like kill to send a signal to the process.
- Started : the starting date and time of the instance.

Fig. 17   Initializing and starting Zeek

Fig. 18   Displaying the status of Zeek

5.2 Stopping the Active Instance of Zeek

Step 1. To stop Zeek, enter the following command into the *ZeekControl* prompt (Fig. 19):

```
stop
```

Step 2. Use the following command to verify the exit status of Zeek (Fig. 20):

```
status
```

The *stopped* status indicates that Zeek is currently stopped.

Step 3. To restart Zeek, enter the following command into the *ZeekControl* prompt (Fig. 21):

```
start
```

Step 4. Type the following command to check if the Zeek restarted. You will verify that Zeek is running (Fig. 22).

```
status
```

Fig. 19 Stopping Zeek

Fig. 20 Displaying the status of Zeek

Fig. 21 Restarting Zeek

Fig. 22 Displaying the status of Zeek

Fig. 23 Leaving the *ZeekControl* tool

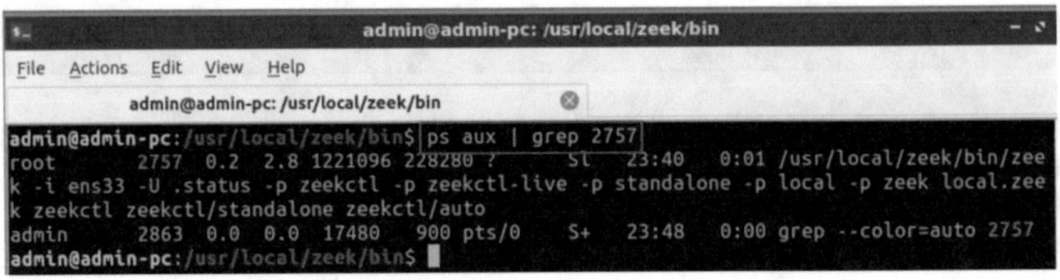

Fig. 24 Displaying Zeek's PID

Step 5. To exit from *ZeekControl* type following command (Fig. 23):

```
exit
```

Note that exiting the *ZeekControl* tool does not stop Zeek. Zeek is only stopped by explicitly using the stop command in the ZeekControl prompt.

Step 6. To verify that Zeek control is not stopped type the following command (Fig. 24):

```
ps aux |grep <PID_number>
```

where *<PID_number>* is the number inside the gray box depicted in step 4.

Notice that the <PID_number> may differ than the figure above.

6   Introduction to Zeek's Traffic Analysis Capabilities

Zeek's broad range of traffic analysis capabilities makes it an exceptional intrusion detection system (IDS) and network analysis framework. Zeek is proficient in processing packet capture (pcap) files and logging traffic on a given network interface.

6.1 Processing Offline Packet Capture Files

Linux-based systems process packet capture (pcap) files using the | libpcap | library. In Zeek, it is possible to capture live traffic and analyze trace files. In the following example, we analyze a pcap file using a premade script that detects brute force attacks.

6.1.1 Command Format for Processing Packet Capture Files

The general format for initializing offline packet capture analysis is as follows:

```
zeek -r <pcap_file_location><script_location>
```

- | zeek | : command to invoke Zeek.
- | -r | : option signifies to Zeek that it will be reading from an offline file.
- | <pcap_file_location> | : indicates the pcap file location.
- | <script_location> | : indicates the script location.

6.1.2 Leveraging a Script to Detect Brute Force Attacks Present in a pcap File

Zeek installs a number of default scripts and trace files that can be used for testing purposes. In this section, we use the *bruteforce.pcap* as the input packet capture file and *ZeekBruteforceDetection.zeek* as the detection script. The packet capture file contains network traffic of a brute force password attack, while the script defines the brute forcing event for the Zeek event engine.

Step 1. Enter the lab workspace directory (Fig. 25). To type capital letters, it is recommended to hold the | Shift | key while typing rather than using the | Caps | key.

```
cd ~/Zeek-Labs/
```

Step 2. Initialize Zeek offline packet parsing on the packet capture file (Fig. 26). It is possible to use the | tab | key to autocomplete the longer paths.

Fig. 25 Navigating into ~/Zeek-Labs/ directory

Fig. 26 Initializing Zeek's offline packet parsing on a packet capture file

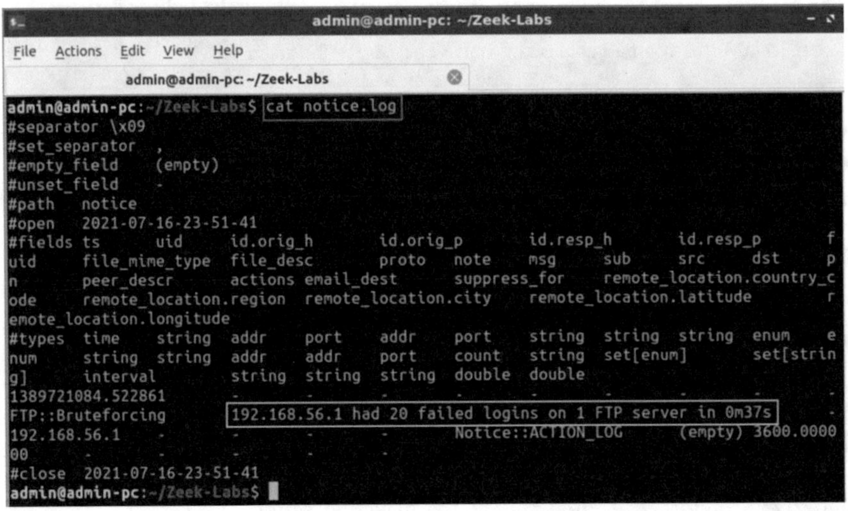

Fig. 27 Showing the contents of *notice.log* file

```
zeek -C -r Sample-PCAP/bruteforce.pcap Lab-
Scripts/ZeekDetectBruteForce.zeek
```

The -C option is included to prevent Zeek from displaying specific warnings.

Step 3. After running the command, if a brute forcing attack was found, it will be logged in the *notice.log* output log file. We will use the cat command to view the file (Fig. 27).

```
cat notice.log
```

Examining the proceeding image, a possible brute force attack was detected. The log file shows that the IP address *192.168.56.1* had 20 or more failed login attempts on the hosted FTP server.

6.2 Launching Mininet

Mininet is a network emulator that creates a network topology consisting of virtual hosts, switches, controllers, and links. Within the Zeek lab series, we will be leveraging Mininet to generate and capture network traffic.

Step 1. From the *Client* machine's desktop, on the left side of the screen, double click on the MiniEdit icon as shown below (Fig. 28). When prompted for a password, type ⟨password⟩ and hit ⟨Enter⟩. The MiniEdit editor will now launch.

Step 2. The MiniEdit editor will now launch and allow for the creation of new, virtualized lab topologies. The image below shows the default MiniEdit display (Fig. 29).

Step 3. A premade topology has already been created for this lab series. To load the correct topology, begin by clicking the ⟨Open⟩ button within the ⟨File⟩ tab on the top left of the MiniEdit editor (Fig. 30).

Step 4. Select the *Lab 14.mn* file by double clicking the *Lab 14.mn* icon, or by clicking the *Open* button (Fig. 31).

Step 5. The lab topology will contain two virtual machines—*h1* and *h2*, which are able to connect and communicate with one another through the *s1* switch, as seen in the image below (Fig. 32).

Fig. 28 Starting MiniEdit

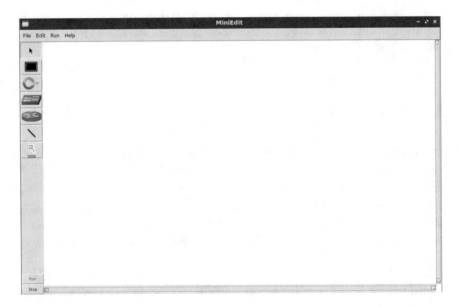

Fig. 29 MiniEdit's interface

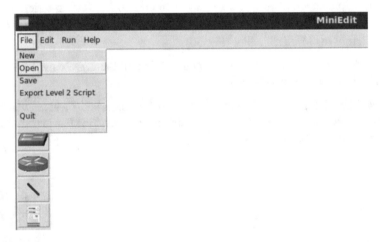

Fig. 30 MiniEdit's menu

Step 6. To begin running the virtual machines, navigate to the *Run* button, found on the bottom left of the Miniedit editor, and select the *Run* button, as seen in the image below (Fig. 33).

Step 7. To access either the *h1* or *h2* terminals for subsequent steps, hold the right mouse button on the desired machine, which will then display a *Terminal* button

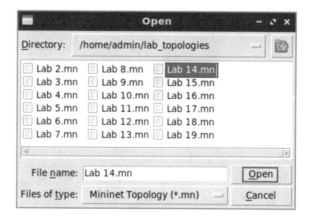

Fig. 31 MiniEdit's *Open* dialog

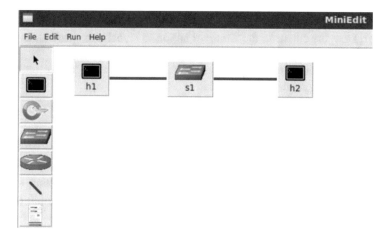

Fig. 32 Lab topology on MiniEdit

Fig. 33 Running the emulation

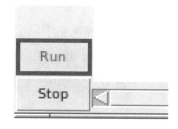

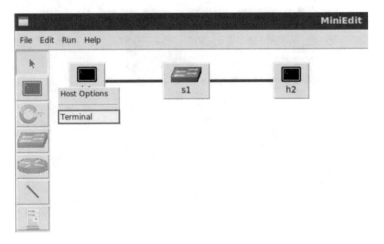

Fig. 34 Opening the host's terminal

(Fig. 34). Drag the cursor to the *Terminal* button to launch the terminal, as seen in the image below.

With the Mininet lab topology loaded, we can now begin to generate and analyze live network traffic capture.

6.3   Generating and Analyzing Live Network Traffic Capture

The tcpdump command utility is a famous network packet analyzing tool that is used to display TCP/IP and other network packets being transmitted over the network.

6.4   Leveraging the Tcpdump Command Utility

The general format for tcpdump is the following command:

```
sudo tcpdump -i <interface_name>-s <num>-w
<pcap_file_location>
```

- sudo : option to enable higher level privileges.
- tcpdump : program for capturing live network traffic.
- -i : option used to specify a network interface.
- <interface_name> : denotes the interface name.

- [-s]: option used to specify number of packets to capture.
- [<num>]: denotes the number of packets to capture. 0 equals infinite.
- [-w]: option used to specify that we will be writing to a new file.
- [<pcap_file_location>]: indicates the file location.

6.5 Capturing Live Network Traffic

The *h2* machine will be used to capture live network traffic, while the *h1* machine will be used to generate live network traffic.

Step 1. Open the *h2* Terminal by holding the right mouse button on the desired machine, which will then display a *Terminal* button (Fig. 35). Drag the cursor to the *Terminal* button to launch the terminal, as seen in the image below.

Step 2. Navigate to the TCP-Traffic directory (Fig. 36). To type capital letters, it is recommended to hold the *Shift* key while typing rather than using the *Caps* key.

```
cd Zeek-Labs/TCP-Traffic/
```

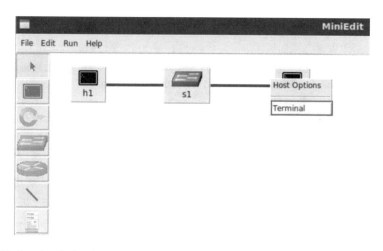

Fig. 35 Opening the host's terminal

Fig. 36 Navigating into *Zeek-Labs/TCP-Traffic/* directory

Fig. 37 Navigating into *Running live packet capture*

Fig. 38 Opening the host's terminal

Step 3. Use the following command to begin live packet capture (Fig. 37). If the Terminal session has not been terminated or closed, you may not be prompted to enter the password. If prompted for a password, type password and hit Enter. Live packet capture will start on interface *h2-eth0*.

```
tcpdump -i h2-eth0 -s -w ntraffic.pcap
```

Step 4. Minimize the *h2 Terminal* and open the *h1 Terminal* (Fig. 38). If necessary, right click within the Miniedit editor to activate your cursor.

Step 5. Generate traffic by using the ping utility. Ping operates by sending Internet Control Message Protocol(ICMP) echo request packets to the target host and waiting for an ICMP echo reply. Issue the following command on the newly opened *h1* Terminal (Fig. 39):

```
ping -c 3 10.0.0.2
```

The -c option is used to indicate the number of packets to send—in this example, 3 packets.

Step 6. Minimize the *h1 Terminal* and open the *h2 Terminal* using the navigation bar at the bottom of the screen (Fig. 40). If necessary, right-click within the Miniedit editor to activate your cursor.

Step 6. Use the Ctrl+c key combination to stop live traffic capture. Statistics of the capture session will be displayed (Fig. 41). 10 packets were recorded by the interface, which were then captured and stored in the new *ntraffic.pcap* file.

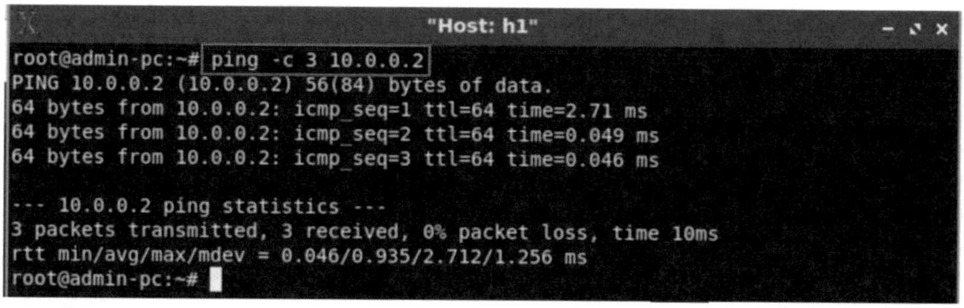

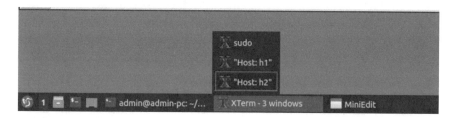

Fig. 39 Generating traffic by using the ping utility

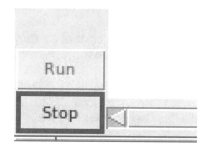

Fig. 40 Opening the host's terminal

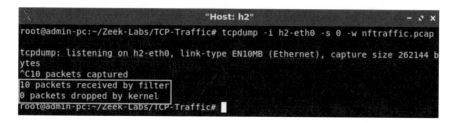

Fig. 41 Statistics of the capture session

Fig. 42 Stopping the emulation

Step 7. Stop the current Mininet session by clicking the *Stop* button on the bottom left of the MiniEdit editor and close the MiniEdit editor by clicking the x on the top right of the editor (Fig. 42).

Fig. 43 Launching the
Terminal

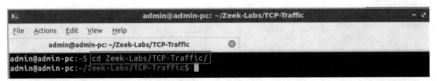

Fig. 44 Navigating into *Zeek-Labs/TCP-Traffic/* directory

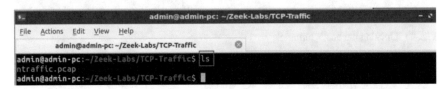

Fig. 45 Viewing the file contents of the TCP-Traffic directory

We will now return to the *Client* machine to process and analyze the newly generated network traffic.

6.5.1 Analyzing the Newly Captured Network Traffic

Step 1. On the left side of the *Client* desktop, double click on the Terminal icon as shown below (Fig. 43).

Step 2. Navigate to the *TCP-Traffic* directory to find the *ntraffic.pcap* file. To type capital letters, it is recommended to hold the *Shift* key while typing rather than using the *Caps* key (Fig. 44).

```
cd Zeek-Labs/TCP-Traffic/
```

Step 3. View the file contents of the *TCP-Traffic* directory (Fig. 45).

```
ls
```

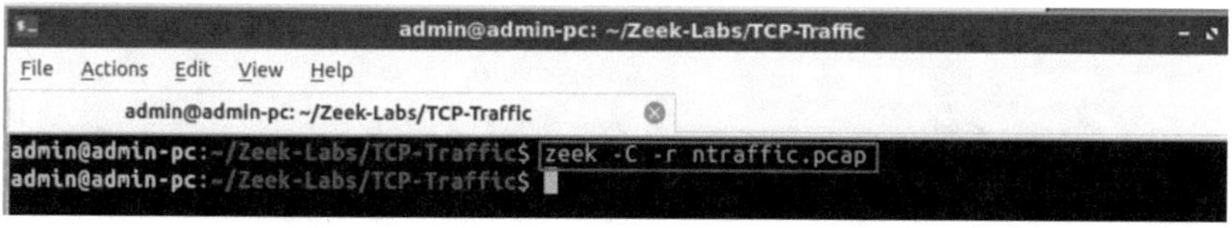
admin@admin-pc: ~/Zeek-Labs/TCP-Traffic
File Actions Edit View Help
admin@admin-pc: ~/Zeek-Labs/TCP-Traffic
admin@admin-pc:~/Zeek-Labs/TCP-Traffic$ zeek -C -r ntraffic.pcap
admin@admin-pc:~/Zeek-Labs/TCP-Traffic$

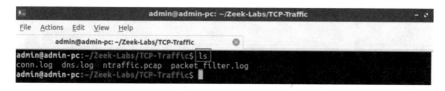

Fig. 46 Initializing Zeek offline packet parsing on the packet capture file

admin@admin-pc: ~/Zeek-Labs/TCP-Traffic
File Actions Edit View Help
admin@admin-pc: ~/Zeek-Labs/TCP-Traffic
admin@admin-pc:~/Zeek-Labs/TCP-Traffic$ ls
conn.log dns.log ntraffic.pcap packet_filter.log
admin@admin-pc:~/Zeek-Labs/TCP-Traffic$

Fig. 47 Listing the newly generated Zeek log files

We can see the *ntraffic.pcap* file that was generated by the host *h2* is now accessible.

Step 4. Initialize Zeek offline packet parsing on the packet capture file. The -r option is used to read from a given pcap file, and the -C option is used to disable checksums validation (Fig. 46).

```
zeek -C -r ntraffic.pcap
```

Step 5. View the newly generated Zeek log files (Fig. 47).

```
ls
```

The generated log files will contain important information regarding the network traffic. For instance, the *conn.log* file will contain connection-based information, specifically the hosts communicating, their IP addresses, protocols and ports. The following labs will offer in-depth insight and examples toward understanding these Zeek log files.

Step 6. Viewing the *conn.log* connection-based log file with the cat command, we can see that the IP address *10.0.0.1* was detected to generate the captured traffic, corresponding to the host *h1* (Fig. 48).

```
cat conn.log
```

Step 7. Stop Zeek by entering the following command on the terminal (Fig. 49). If required, type password as the password. If the Terminal session has not been terminated or closed, you may not be prompted to enter a password. To type capital

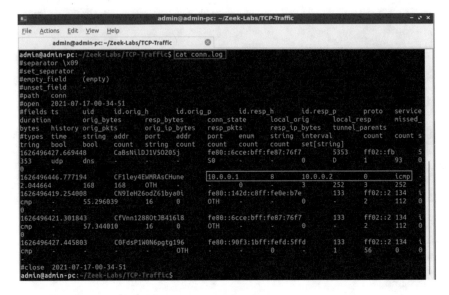

Fig. 48 Viewing the conn.log connection-based log file

Fig. 49 Stopping Zeek

letters, it is recommended to hold the *Shift* key while typing rather than using the *Caps* key.

```
cd $ZEEK_INSTALL/bin && sudo ./zeekctl stop
```

The above command navigates to Zeek's installation directory and executes the stop command in *zeekctl*.

If you see error messages during the new Zeek instance initializing process, please ignore it.

Chapter 6—Lab 15: An Overview of Zeek Logs

Overview

To conduct the experiment described in this section, please login into the Academic Cloud at http://highspeednetworks.net/ and reserve a pod for Lab 15.

This lab covers Zeek's logging files. Zeek's event-based engine will generate log files based on signatures or events found during network traffic analysis. The focus in this lab is on explaining each logging file and introducing some basic analytic functions and tools.

Objectives

By the end of this lab, students should be able to:

1. Generate Zeek log files.
2. Use Linux Terminal tools combined with Zeek's *zeek-cut* utility to customize the output of logs for analysis.

Lab Topology

Figure 50 shows the lab topology. The topology uses 10.0.0.0/8, which is the default network assigned by Mininet. The *h1* and *h2* virtual machines will be used to generate and collect network traffic.

Lab Settings

The information (case-sensitive) in the table below provides the credentials necessary to access the machines used in this lab (Tables 5 and 6).

Lab Roadmap

This lab is organized as follows:

1. Section 7: Introduction to Zeek logs.
2. Section 8: Starting a new instance of Zeek.
3. Section 9: Parsing packet capture files into Zeek log files.
4. Section 10: Analyzing Zeek log files.

Fig. 50 Lab topology

Table 5 Machines credentials

Device	Account	Password
Client	admin	password

Table 6 Shell variables and their corresponding absolute paths

Variable name	Absolute path
$ZEEK_INSTALL	/usr/local/zeek
$ZEEK_TESTING_TRACES	/home/zeek/admin/testing/btest/Traces
$ZEEK_PROTOCOLS_SCRIPT	/home/zeek/admin/scripts/policy/protocols

7   Introduction to Zeek Logs

Zeek's generated log files include a comprehensive record of every connection seen on the wire; this includes application-layer protocols and fields (e.g., Hyper-Text Transfer Protocol (HTTP) sessions, Uniform Resource Locator (URL), key headers, Multi-Purpose Internet Mail Extensions (MIME) types, server responses, etc.), Domain Name Server (DNS) requests and responses, Secure Socket Layer (SSL) certificates, key content of Simple Mail Transfer Protocol (SMTP) sessions, and others.

7.1   Zeek Logs Generated by Packet Analysis

A Zeek log is a stream of high-level entries that correspond to network activities, such as a login to SSH or an email sent using SMTP. In Zeek, each event stream has a dedicated file with its own set of features, fields, or columns.

During capture or analysis, Zeek generates a log determined by the protocol type. Due to this architecture, a Session Initiation Protocol (SIP) log, for instance, does not contain any other protocols' packets information like HTTP. Furthermore, each log file contains case-relative fields (e.g., *from* and *subject* fields in an SMTP log). Some of these log files are large and contain entries that can be either benign or malicious, whereas others are smaller and contain more actionable information.

7.2   Zeek Logs Generated by Recurrent Network Analysis

With every session of packet analysis, either through live packet analysis or the parsing of an offline packet capture file, Zeek generates session-specific log files. In addition to these session-based log files, Zeek creates network-reliant log files as well. These network-reliant files are continually generated and updated when a new session is initialized and started.

The following Zeek log files are updated daily:

- *known_hosts.log*: Log file containing information for hosts that completed TCP handshakes.

- *known_services.log*: Log file containing a list of services running on hosts.
- *known_certs*.log: Log file containing a list of Secure Socket Layer (SSL) certificates.
- *software.log*: Log file containing information about Software being used on the network.

Additionally, a list of detection-based log files is created during each session. The log files relevant to this lab are:

- *notice.log* (Zeek notices): When Zeek detects an anomaly, a corresponding notice will be raised in this file.
- *intel.log* (Intelligence data matches): When Zeek detects traffic flagged with known malicious indicators, a corresponding reference will be logged in this file.
- *signatures.log* (Signature matches): When Zeek detects traffic flagged with known malicious or faulty packet signatures, a corresponding reference will be logged in this file.

7.3   Typical Uses of Zeek Logs

By default, Zeek logs all information into well-structured, tab-separated text files suitable for postprocessing. Users can also choose from a set of alternative output formats and backends such as external databases.

The Zeek-native `zeek-cut` utility can be leveraged to further specify and parse the information within the generated log files.

8   **Starting a New Instance of Zeek**

Step 1. From the top of the screen, click on the *Client* button as shown below to enter the *Client* machine (Fig. 51).

Step 2. The *Client* machine will now open, and the desktop will be displayed. On the left side of the screen, double click on the Terminal icon as shown below (Fig. 52).

Fig. 51 Opening the *Client* machine

Fig. 52 Opening the
Terminal

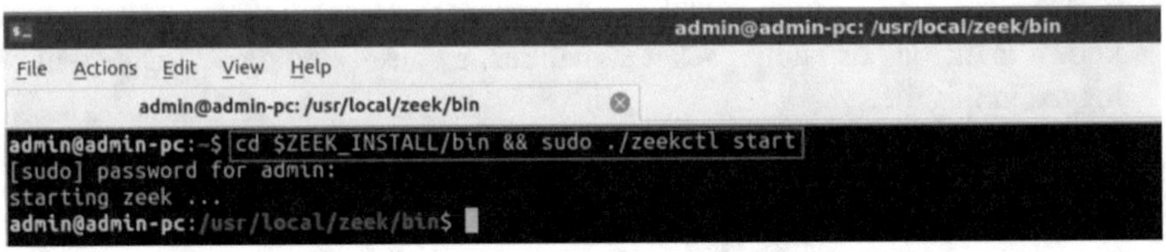

Fig. 53 Initializing and starting Zeek

Step 3. Start Zeek by entering the following command on the terminal (Fig. 53). This command enters Zeek's default installation directory and invokes $\boxed{\text{Zeekctl}}$ tool to start a new instance. When prompted for a password, type $\boxed{\text{password}}$ and hit $\boxed{\text{Enter}}$. To type capital letters, it is recommended to hold the $\boxed{\text{Shift}}$ key while typing rather than using the $\boxed{\text{Caps}}$ key.

```
cd $ZEEK_INSTALL/bin && sudo ./zeekctl start
```

A new instance of Zeek is now active, and we are ready to proceed to the next section of the lab.

If you see error messages during the new Zeek instance initializing process, please ignore it.

9 Parsing Packet Capture Files into Zeek Log Files

In this section we introduce Zeek's capability of generating and viewing log files. Packet capture files used in this lab are preloaded onto the *Client* machine and can be found with the following path:

```
~/Zeek-Labs/Sample-PCAP/
```

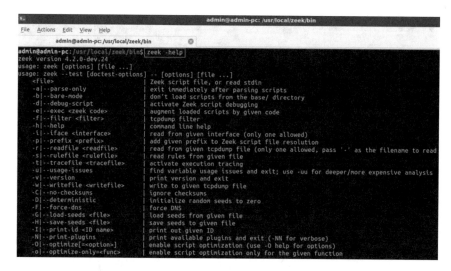

Fig. 54 Zeek's options

These packet capture files were downloaded from Tcpreplay's sample capture collection. To access the following link, users must have access to an external computer connected to the Internet, because the Zeek Lab topology does not have an active Internet connection.

http://tcpreplay.appneta.com/wiki/captures.html

Tcpreplay is a suite of free Open-Source utilities for editing and replaying previously captured network traffic and can be used to test transmissions and network health.

9.1   Overview of Zeek Command Options

When using Zeek, the user specifies a running state option. In this lab, three primarily options are used:

- -C : specifies to ignore checksum warnings, specifically to avoid redundancy since we are focusing on TCP traffic only.
- -r : specifies offline packet capture file analysis.
- -w : specifies live network capture.

Additional Zeek options can be found by passing the -help option to the zeek command (Fig. 54):

```
zeek -help
```

Fig. 55 Navigating into *~/Zeek-Labs/ TCP-Traffic/* directory

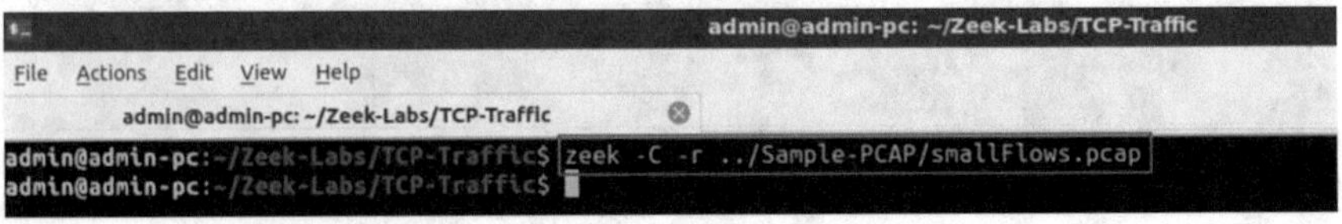

Fig. 56 Processing the *smallFlows.pcap*

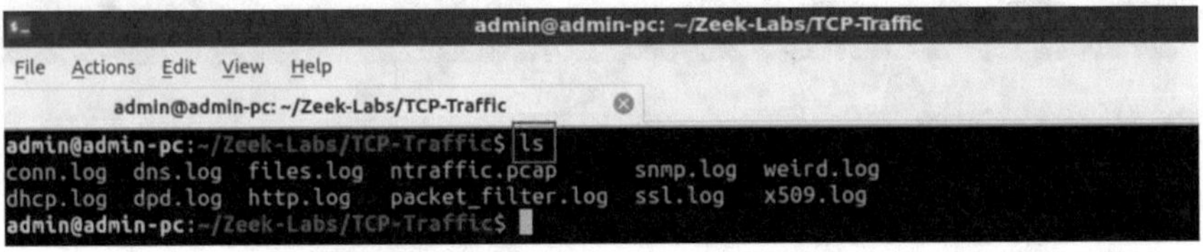

Fig. 57 Listing the generated log files

9.2 Using Zeek to Process Offline Packet Capture Files

In this subsection we will use Zeek to process the existing offline packet capture file *smallFlows.pcap*. By specifying the `-r` option and the directory path to the pcap file, Zeek can generate the corresponding log files.

Step 1. Navigate to the lab workspace directory (Fig. 55). To type capital letters, it is recommended to hold the `Shift` key while typing rather than using the `Caps` key.

```
cd ~/Zeek-Labs/TCP-Traffic/
```

Step 2. Use the following command to process the *smallFlows.pcap* file (Fig. 56). It is possible to use the `tab` key to autocomplete the longer paths.

```
zeek -C -r ../Sample-PCAP/smallFlows.pcap
```

After Zeek finishes processing the packet capture file, it will generate a number of log files.

Step 3. Use the following command to list the generated log files (Fig. 57):

```
ls
```

9.3 Understanding Zeek Log Files

Zeek's generated log files can be summarized as follows:

- *conn.log*: A file containing information pertaining to all TCP/UDP/ICMP connections; this file contains most of the information gathered from the packet capture.
- *files.log*: A file consisting of analytic results of packets' counts and sessions' durations.
- *packet_filter.log*: A file listing the active filters applied to Zeek upon reading the packet capture file.
- *x509.log*: A file containing public key certificates used by protocols.
- *weird.log*: A file containing packet data non-conformant with standard protocols. It also contains packets with possibly corrupted or damaged packet header fields.
- *(protocol).log* (*dns.log, dhcp.log, http.log, snmp.log*): These are files containing information for packets found in each respective protocol. For instance, *dns.log* will only contain information generated by Domain Name Service (DNS) packets.

More information regarding log files is available in the Zeek official documentation, which can be viewed online using an external Internet-connected machine through this link:

```
https://docs.zeek.org/en/stable/script-reference/log-files.html
```

9.4 Basic Viewing of Zeek Logs

In this subsection we examine the generated log files and their contents.

Step 1. Use the following command to display the contents of the *conn.log* file using the `head` command (Fig. 58):

```
head conn.log
```

The topmost rows within the *conn.log* file will be displayed in the Terminal; however, the current formatting wraps around multiple lines, making it unclear and hard to understand. In the following section we introduce the `zeek-cut` utility for enhancing the output of these log files.

Fig. 58 Displaying the contents of the *conn.log*

10 Analyzing Zeek Log Files

In this section, we review the utilities that help in displaying log files with well-formatted outputs, as well as saving output to text files.

10.1 Leveraging Zeek-Cut for a More Refined View of Log Files

Although the produced log file is tab delimited, it is difficult to visualize and parse information from the terminal. The zeek-cut utility can be used to parse the log files by specifying which column data to be displayed in a more organized output.

10.1.1 Using Zeek-Cut in Conjunction with Cat and Head Command Utilities

Generally, the zeek-cut utility is typically coupled with cat using the pipe | command. In Linux, the pipe command sends the output of one command as input to another. Essentially, the output of the left command is passed as input to that on its right, and multiple commands can be chained together.

Step 1. Use the following command to pipe the contents of cat into zeek-cut (Fig. 59):

```
cat conn.log |zeek-cut id.orig_h id.orig_p id.resp_h
id.resp_p
```

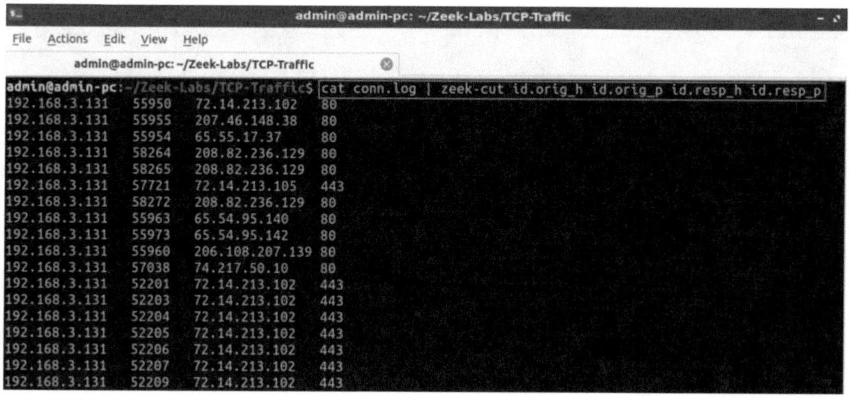

Fig. 59 Using `zeek-cut` utility

Fig. 60 Using `zeek-cut` utility

The options passed into the `zeek-cut` utility represent the column headers to be extracted from the log file:

- `id.orig_h` : Column containing the source IP address.
- `id.orig_p` : Column containing the source port.
- `id.resp_h` : Column containing the destination IP address.
- `id.resp_p` : Column containing the destination port.

Alternatively, instead of using the `cat` command, the `head` command can be used to display the topmost rows of the log file, which can be very useful to view a large file's contents.

Step 2. Use the following command to pipe the contents of `head` into `zeek-cut` (Fig. 60):

```
head conn.log |zeek-cut id.orig_h id.orig_p id.resp_h
id.resp_p
```

Notice that only two records are shown. This is caused by the `head` command taking the 10 topmost rows of *conn.log*, regardless of what that entails, and passing it as input to `zeek-cut`.

Fig. 61 Using zeek-cut utility

Since the log file contains 8 lines of header padding used for displaying the file's format, we will have to specify the first 18 rows of file in order to successfully display the first 10 packets of the log file.

Step 3. Use the following command to pipe the contents of head into zeek-cut (Fig. 61):

```
head -n 18 conn.log |zeek-cut id.orig_h id.orig_p
id.resp_h id.resp_p
```

The -n option can be passed to the head utility to specify the desired number of rows.

10.1.2 Printing the Output of Zeek-Cut to a Text File

While the results displayed in the Terminal after using the zeek-cut utility can be easily viewed for smaller data sets, it is often necessary to save the output into a separate file. Using the > character, we can send the output to a new file for further processing by other applications.

Step 1. Use the following command to change the output location of zeek-cut (Fig. 62):

```
cat conn.log |zeek-cut id.orig_h id.orig_p id.resp_h
id.resp_p >output.txt
```

By including the file extension in *output.txt*, we are choosing to print the output into a plain text file.

Step 2. We can display the topmost contents of the new *output.txt* file by using the head command (Fig. 63).

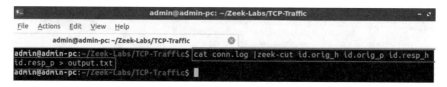

Fig. 62 Redirecting zeek-cut utility's output to a text file

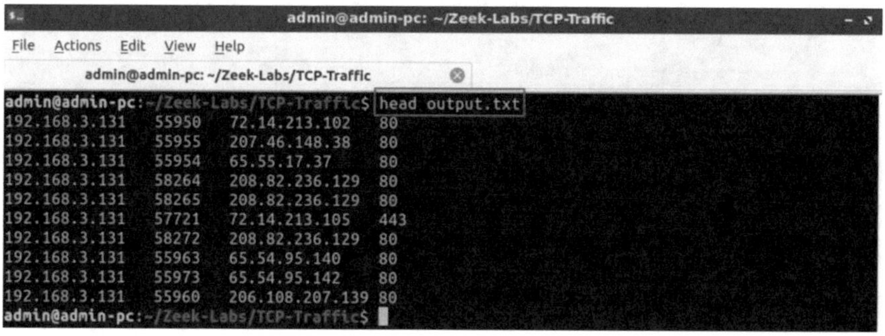

```
admin@admin-pc:~/Zeek-Labs/TCP-Traffic$ head output.txt
192.168.3.131    55950    72.14.213.102      80
192.168.3.131    55955    207.46.148.38      80
192.168.3.131    55954    65.55.17.37        80
192.168.3.131    58264    208.82.236.129     80
192.168.3.131    58265    208.82.236.129     80
192.168.3.131    57721    72.14.213.105      443
192.168.3.131    58272    208.82.236.129     80
192.168.3.131    55963    65.54.95.140       80
192.168.3.131    55973    65.54.95.142       80
192.168.3.131    55960    206.108.207.139    80
admin@admin-pc:~/Zeek-Labs/TCP-Traffic$
```

Fig. 63 Displaying the contents of the new *output.txt* file

```
head output.txt
```

The *output.txt* file contains the same tab-delimited format as shown in previous zeek-cut examples.

10.1.3   Printing the Output of Zeek-Cut to a csv File

In some situations, it is helpful to save the output of zeek-cut in a csv file. In a csv file, data may be imported into other applications, such as databases or machine learning classifiers.

Step 1. The exported output file by zeek-cut is tab delimited due to the default zeek-cut settings (Fig. 64). To export a file with another delimiter, the -F option is used.

```
cat conn.log |zeek-cut -F ',' id.orig_h id.orig_p
id.resp_h id.resp_p >output.csv
```

Step 2. We can now display the topmost contents of the *output.csv* file (Fig. 65).

```
head output.csv
```

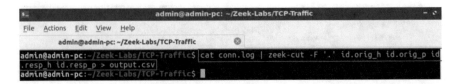

Fig. 64 Redirecting zeek-cut utility's output to a csv file

```
admin@admin-pc: ~/Zeek-Labs/TCP-Traffic                               - ˅

File   Actions   Edit   View   Help

        admin@admin-pc: ~/Zeek-Labs/TCP-Traffic        ⊗

admin@admin-pc:~/Zeek-Labs/TCP-Traffic$ head output.csv
192.168.3.131.55950.72.14.213.102.80
192.168.3.131.55955.207.46.148.38.80
192.168.3.131.55954.65.55.17.37.80
192.168.3.131.58264.208.82.236.129.80
192.168.3.131.58265.208.82.236.129.80
192.168.3.131.57721.72.14.213.105.443
192.168.3.131.58272.208.82.236.129.80
192.168.3.131.55963.65.54.95.140.80
192.168.3.131.55973.65.54.95.142.80
192.168.3.131.55960.206.108.207.139.80
admin@admin-pc:~/Zeek-Labs/TCP-Traffic$ ▮
```

Fig. 65 Displaying the contents of the output.csv file

As shown in the image, the *output.csv* file is in a comma-delimited format, rather than the previous tab-delimited format.

In conclusion, zeek-cut is a flexible tool that can be called to format Zeek log files depending on the user's needs. The zeek-cut utility can be utilized with more advanced commands to further increase customization.

10.2 Closing the Current Instance of Zeek

After you have finished the lab, it is necessary to terminate the currently active instance of Zeek. Shutting down a computer while an active instance persists will cause Zeek to shut down improperly and may cause errors in future instances.

Step 1. Stop Zeek by entering the following command on the terminal (Fig. 66). If required, type password as the password. If the Terminal session has not been terminated or closed, you may not be prompted to enter a password. To type capital letters, it is recommended to hold the *Shift* key while typing rather than using the *Caps* key.

```
cd $ZEEK_INSTALL/bin && sudo ./zeekctl stop
```

Fig. 66 Stopping Zeek

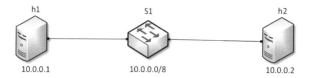

Fig. 67 Lab topology

Table 7 Credentials to access the Client machine

Device	Account	Password
Client	admin	password

Chapter 6—Lab 16: Parsing, Reading, and Organizing Zeek Log Files

Overview

To conduct the experiment described in this section, please login into the Academic Cloud at http://highspeednetworks.net/ and reserve a pod for Lab 16.

This lab explains how to format and organize Zeek's log files by combining zeek-cut utility with basic Linux shell commands. Utilities and tools introduced in this lab provide practical examples for logs customization in a real network environment.

Objectives

By the end of this lab, students should be able to:

1. Use Linux tools and commands for text files processing.
2. Practice Linux shell scripts and the AWK scripting language.
3. Incorporate AWK with zeek-cut to provide formatted logs.

Lab Topology

Figure 67 shows the lab topology. The topology uses 10.0.0.0/8, which is the default network assigned by Mininet.

Lab Settings

The information (case-sensitive) in the table below provides the credentials to access the machines used in this lab (Tables 7 and 8).

Table 8 Shell variables and their corresponding absolute paths

Variable name	Absolute path
$ZEEK_INSTALL	/usr/local/zeek
$ZEEK_TESTING_TRACES	/home/zeek/admin/testing/btest/Traces
$ZEEK_PROTOCOLS_SCRIPT	/home/zeek/admin/scripts/policy/protocols

Lab Roadmap

This lab is organized as follows:

1. Section 11: Introduction to shell scripts.
2. Section 12: Advanced zeek-cut log file analysis.
3. Section 13: Incorporating the AWK scripting language for log file analysis.

11 Introduction to Shell Scripts

A shell script is a text file containing commands to be executed by the Unix command-line interpreter. Shell scripts provide a convenient way to manipulate files and automate programs' executions. Selection and repetition are incorporated into scripts to branch control based on conditioning and looping statements. Running a shell script can immensely save time and prevent manually entering repetitive commands in recurrent tasks.

11.1 Ubuntu Linux Text Editors

Linux-based distributions include pre-installed text editors like nano , vi , vim , gedit , etc. nano is a keyboard-oriented lightweight text editor with a simple Command-Line Interface (CLI). Other editors such as vi and vim are highly customizable and extensible, making them attractive for users that demand a large amount of control and flexibility over their text editing environment. Alternatively, the Graphical User Interface (GUI) text editor gedit can be used to visually work outside of the terminal. More information on these text editors can be found on the Ubuntu help pages. To access the following links, users must have access to an external computer connected to the Internet, because the Zeek Lab topology does not have an active Internet connection.

- Nano – https://help.ubuntu.com/community/Nano
- Vim – https://help.ubuntu.com/community/VimHowto
- Gedit – https://help.ubuntu.com/community/gedit

Fig. 68 Opening the *Client* machine

Fig. 69 Opening the
Terminal

For simplicity, in this lab we use [nano] text editor to view, create, and edit text files.

11.2 Creating a Shell Script

Shell scripts are effective in executing repetitive terminal commands. Unlike executing commands manually in the terminal, scripts can be saved and executed whenever needed simple by invoking their names. We will begin this lab by writing some basic shell scripts.

Step 1. From the top of the screen, click on the *Client* button as shown below to enter the *Client* machine (Fig. 68).

Step 2. The *Client* machine will now open, and the desktop will be displayed. On the left side of the screen, click on the Terminal icon as shown below (Fig. 69).

A new instance of Zeek is now active, and we are ready to proceed to the next section of the lab.

Step 3. In the Linux terminal, navigate to the lab workspace directory by typing the following command (Fig. 70):

```
cd Zeek-Labs/
```

Step 4. Use the [nano] text editor to create the *lab16script.sh* file (Fig. 71).

```
sudo nano lab16script.sh
```

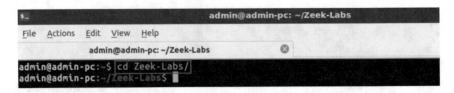

Fig. 70 Navigating into *Zeek-Labs/* directory

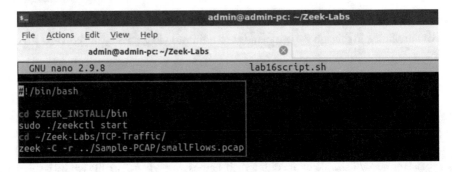

Fig. 71 Opening the *lab16script.sh* file

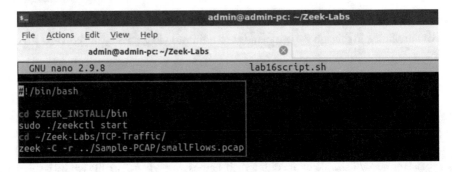

Fig. 72 Editing the *lab16script.sh* file

Step 5. Edit the *lab16script.sh* file contents (Fig. 72).

Once the text editor has opened, we will be able to enter the following commands. Each new line will denote a new Terminal command being passed. To type capital letters, it is recommended to hold the Shift key while typing rather than using the Caps key.

```
cd $ZEEK_INSTALL/bin
sudo ./zeekctl start
cd ~/Zeek-Labs/TCP-Traffic/
zeek -C -r ../Sample-PCAP/smallFlows.pcap
```

The file's content is explained as follows:

- Line 1: changes the current directory to the Zeek's installation directory.
- Line 2: starts a new instance of Zeek through zeekctl .
- Line 3: changes the current directory to the lab workspace.

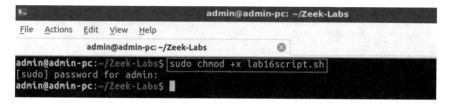

Fig. 73 Making the script executable

Fig. 74 Executing the script

- Line 4: invokes the ⌷zeek⌷ command with the ⌷-r⌷ option to begin processing the *smallFlows.pcap* capture file located in the *Sample-PCAP* directory.

Step 6. When using ⌷nano⌷, the following keyboard shortcuts are used to save a file and then exit the workspace:

- ⌷CTRL + o⌷ ⊢save the file
- ⌷CTRL + x⌷ ⊢save and exit the file, return to terminal

After completing Step 6 and adding the correct commands with proper formatting, we will save and exit the text editor. Press ⌷CTRL + o⌷ and hit ⌷Enter⌷ to save the file's contents, then ⌷CTRL + x⌷ to exit ⌷nano⌷ and return to the terminal.

Step 7. Use the following command to modify the permissions of the script file to make it executable. When prompted for a password, type ⌷password⌷ and hit ⌷Enter⌷ (Fig. 73).

```
sudo chmod +x lab16script.sh
```

Step 8. Execute the *lab16script.sh* shell script by typing the following command (Fig. 74):

```
./lab16script.sh
```

In case there is an error message, please ignore it. Note that this error does not affect the lab functionality.

admin@admin-pc: ~/Zeek-Labs/TCP-Traffic
File Actions Edit View Help
admin@admin-pc: ~/Zeek-Labs/TCP-Traffic
admin@admin-pc:~/Zeek-Labs$ cd ~/Zeek-Labs/TCP-Traffic/
admin@admin-pc:~/Zeek-Labs/TCP-Traffic$

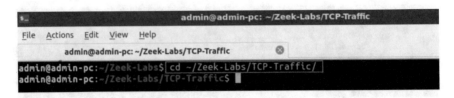

Fig. 75 Navigating into the *~/Zeek-Labs/TCP-Traffic/* directory

admin@admin-pc: ~/Zeek-Labs/TCP-Traffic
File Actions Edit View Help
admin@admin-pc: ~/Zeek-Labs/TCP-Traffic
admin@admin-pc:~/Zeek-Labs/TCP-Traffic$ ls
conn.log dns.log files.log ntraffic.pcap output.txt snmp.log weird.log
dhcp.log dpd.log http.log output.csv packet_filter.log ssl.log x509.log
admin@admin-pc:~/Zeek-Labs/TCP-Traffic$

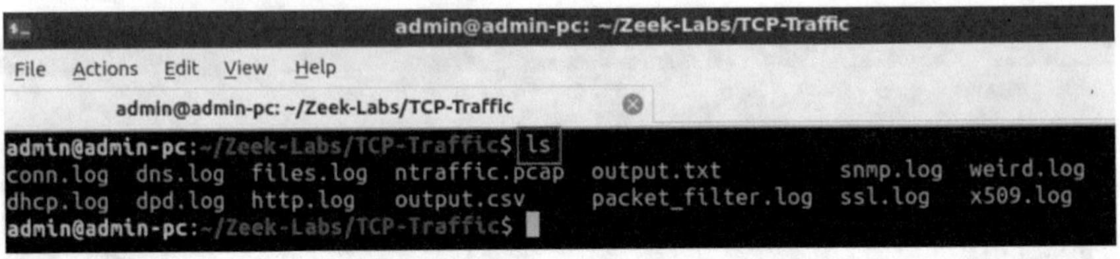

Fig. 76 Listing the contents of the *~/Zeek-Labs/TCP-Traffic/* directory

Step 9. Navigate to the lab workspace directory (Fig. 75).

```
cd ~/Zeek-Labs/TCP-Traffic/
```

Step 10. Verify that the *smallFlows.pcap* file was processed successfully (Fig. 76).

```
ls
```

The above output shows the list of log files generated by Zeek's processing, verifying that the script executed without errors.

12 Advanced Zeek-Cut Log File Analysis

This section introduces more advanced `zeek-cut` functionality to analyze packet capture statistics. These statistics can be used for planning and anomaly analysis. For instance, if a single port has been targeted and received a large number of network traffic, it may highlight a possible vulnerability. We can use the `zeek-cut` utility to determine if a host sends an abnormal number of packets to a specific destination and further analyze this event.

Fig. 77 Opening the *lab16script.sh* script

Fig. 78 Editing the *lab16script.sh* script

12.1   Example 1

Example 1 Show the 10 source IP addresses that generated the most network traffic, organized in descending order.

To solve this example, we will be looking at the id.orig_h column because it contains the source IP addresses from the packet capture file.

Step 1. Open the *lab16script.sh* file with nano text editor (Fig. 77).

```
sudo nano ~/Zeek-Labs/lab16script.sh
```

Step 2. Modify the script file's contents. Delete all the previous content and type the following command (Fig. 78):

```
cd TCP-Traffic/
zeek-cut id.orig_h <conn.log |sort |uniq -c |sort -rn |
head -n 10
```

Press CTRL + o and hit Enter to save the file's contents, then CTRL + x to exit nano and return to the terminal. The above command is explained as follows:

- zeek-cut id.orig_h <conn.log : selects the id.orig_h column from the *conn.log* file.
- |sort : uses the sort command to organize the rows in alphabetical order.
- |uniq -c : uses the uniq command with the -c option to remove duplicates while returning unique instances and their counts.

admin@admin-pc: ~/Zeek-Labs
File Actions Edit View Help
admin@admin-pc: ~/Zeek-Labs
admin@admin-pc:~/Zeek-Labs/TCP-Traffic$ cd ~/Zeek-Labs/
admin@admin-pc:~/Zeek-Labs$

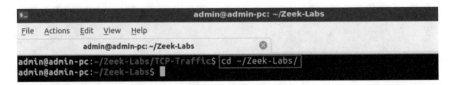

Fig. 79 Navigating into *~/Zeek-Labs/* directory

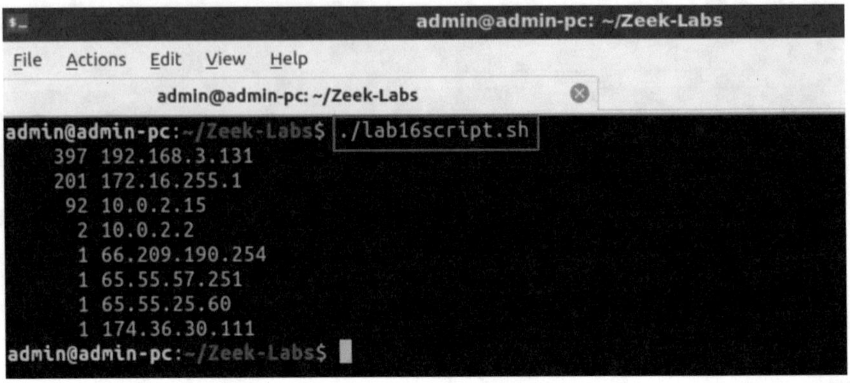

admin@admin-pc: ~/Zeek-Labs
File Actions Edit View Help
admin@admin-pc: ~/Zeek-Labs
admin@admin-pc:~/Zeek-Labs$ ./lab16script.sh
397 192.168.3.131
201 172.16.255.1
92 10.0.2.15
2 10.0.2.2
1 66.209.190.254
1 65.55.57.251
1 65.55.25.60
1 174.36.30.111
admin@admin-pc:~/Zeek-Labs$

Fig. 80 Executing the script

- `|sort -rn`: uses the `sort` command with the `-rn` option to organize the rows in reverse numerical order.
- `|head –n 10`: uses the `head` command with the `-n` option to display the 10 topmost values.

Step 3. Navigate into *Zeek-Labs* folder by issuing the following command (Fig. 79):

```
cd ~/Zeek-Labs
```

Step 4. Execute the modified shell script (Fig. 80).

```
./lab16script.sh
```

The number of duplicates is seen in the left column, while the matching source IP address is seen in the right column. Only 8 unique source addresses were found, and each was returned. From this output, we can conclude that the majority of network traffic was generated by the top 3 source IP addresses.

Fig. 81 Opening the *lab16script.sh* script

Fig. 82 Editing the *lab16script.sh* script

12.2   Example 2

Example 2 Show the 10 destination ports that received the most network traffic, organized in descending order.

To solve this example, we will be looking at the $\boxed{\text{id.resp\_p}}$ column because it contains the destination ports from the packet capture file.

Step 1. Open the *lab16script.sh* file with $\boxed{\text{nano}}$ text editor (Fig. 81).

```
sudo nano lab16script.sh
```

Step 2. Modify the script file's contents. Delete all the previous content and type the following command (Fig. 82):

```
cd TCP-Traffic/
zeek-cut id.resp_p <conn.log |sort |uniq -c |sort -rn |
head -n 10
```

Press $\boxed{\text{CTRL} + \text{o}}$ and hit $\boxed{\text{Enter}}$ to save the file's contents, then $\boxed{\text{CTRL} + \text{x}}$ to exit $\boxed{\text{nano}}$ and return to the terminal. The above command is explained as follows:

- $\boxed{\text{zeek-cut id.resp\_p <conn.log}}$: selects the $\boxed{\text{id.resp\_p}}$ column from the *conn.log* file.
- $\boxed{\text{|sort}}$: uses the $\boxed{\text{sort}}$ command to organize the rows in alphabetical order.
- $\boxed{\text{|uniq -c}}$: uses the $\boxed{\text{uniq}}$ command with the $\boxed{\text{-c}}$ option to remove duplicates while returning unique instances and their counts.

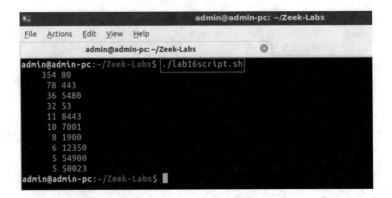

Fig. 83 Executing the *lab16script.sh* script

- |sort -rn| : uses the |sort| command with the |-rn| option to organize the rows in reverse numerical order.
- |head –n 10| : uses the |head| command with the |-n| option to display the 10 topmost values.

Step 3. Execute the modified shell script (Fig. 83).

```
./lab16script.sh
```

The number of duplicates is seen in the left column, while the matching destination port is seen in the right column. More than 10 unique destination ports were found, so only the top 10 were returned. From this output we can conclude that port 80 received the most traffic.

12.3   Example 3

Example 3 Show the number of connections per protocol service.

To solve this example, we will be looking at the |service| column because it contains the destination ports from the packet capture file.

Step 1. Open the *lab16script.sh* file with |nano| text editor (Fig. 84).

```
sudo nano lab16script.sh
```

Step 2. Modify the script file's contents. Delete all the previous content and type the following command (Fig. 85):

Fig. 84 Opening the *lab16script.sh* script

Fig. 85 Editing the *lab16script.sh* script

```
cd TCP-Traffic/
zeek-cut service <conn.log |sort |uniq -c |sort -n
```

Press CTRL + o and hit Enter to save the file's contents, then CTRL + x to exit nano and return to the terminal. The above command is explained as follows:

- zeek-cut service <conn.log : selects the service column from the *conn.log* file.
- |sort : uses the sort command to organize the rows in alphabetical order.
- |uniq -c : uses the uniq command with the -c option to remove duplicates while returning unique instances and their counts.
- |sort -n : uses the sort command with the -n option to organize the rows in numerical order.

Step 3. Execute the modified shell script (Fig. 86).

```
./lab16script.sh
```

The number of duplicates is seen in the left column, while the matching destination port is seen in the right column. From this output we can see that 331 packets did not have a marked protocol. This can be caused by a number of anomalies and is an example of how you can use the zeek-cut utility to return anomalies that require further identification.

Fig. 86 Executing the *lab16script.sh* script

Fig. 87 Opening the *lab16script.sh* script

12.4   Example 4

Example 4 Print the distinct browsers used by the hosts in this packet capture file to a separate file.

To solve this example, we will be looking at the user_agent column because it contains the browser and connection-related information from the packet capture file.

Step 1. Open the *lab16script.sh* file with nano text editor (Fig. 87).

```
sudo nano lab16script.sh
```

Step 2. Modify the script file's contents (Fig. 88).

```
cd TCP-Traffic/
zeek-cut user_agent <http.log |sort -u >browser.txt
```

Press CTRL + o and hit Enter to save the file's contents, then CTRL + x to exit nano and return to the terminal. The above command is explained as follows:

- zeek-cut user_agent <http.log selects the user_agent column from the *http.log* file.

Fig. 88 Editing the *lab16script.sh* script

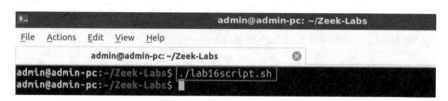

Fig. 89 Executing the *lab16script.sh* script

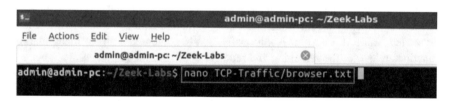

Fig. 90 Opening the *TCP-Traffic/browser.txt*

- |sort -u >browser.txt | uses the | sort | command to sort the lines in the file and the | -u | option checks for strict ordering. The output is then saved into the *browser.txt* file.

Step 3. Execute the modified shell script (Fig. 89).

```
./lab16script.sh
```

Step 4. Use a text editor to view the contents of the *browser.txt* file (Fig. 90).

```
nano TCP-Traffic/browser.txt
```

Step 5. View the distinct browser information (Fig. 91).

Each browser found within the packet capture file is printed with related information extracted from the traffic by Zeek.

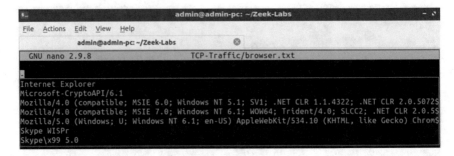

Fig. 91 Displaying the contents of *TCP-Traffic/browser.txt*

13 Incorporating the AWK Scripting Language for Log File Analysis

AWK is a terminal scripting language used to parse, filter, and modify text files. AWK is specifically useful when processing rows and columns found in a Comma Separated Value (CSV) file. Additionally, AWK's integrated string manipulation functions allow for the searching and modifying of specific output.

Like `cat` and `head` commands, AWK output can be piped into the `zeek-cut` utility, allowing more advanced parsing and formatting options. AWK reads each column in a file through its position. The first input column is accessed using $1 while the second column is accessed using $2 and so on. AWK also allows creating simple variables to store and read script values. AWK reads the input data as a loop, starting from the top of the file and finishing at the end of the file. Each row is considered an instance within the script.

13.1 Example 5

Example 5 Find the source and destination IP address of all UDP and TCP connections that lasted more than one minute.

Step 1. Open the *lab16script.sh* file with `nano` text editor (Fig. 92).

```
sudo nano lab16script.sh
```

Step 2. Modify the script file's contents. Delete all the previous content and type the following command (Fig. 93):

```
cd TCP-Traffic/
awk '$9 >60' conn.log |zeek-cut id.orig_h id.resp_h
```

Fig. 92 Opening the *lab16script.sh* script

Fig. 93 Editing the *lab16script.sh* script

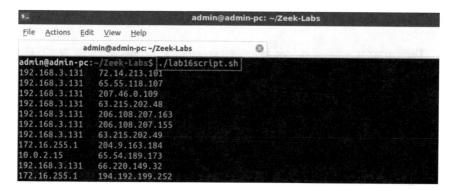

Fig. 94 Executing the *lab16script.sh* script

Press $\boxed{\text{CTRL} + \text{o}}$ and hit $\boxed{\text{Enter}}$ to save the file's contents, then $\boxed{\text{CTRL} + \text{x}}$ to exit $\boxed{\text{nano}}$ and return to the terminal. The above command is explained as follows:

- $\boxed{\text{awk '\$9 >60' conn.log}}$ selects the rows that have their ninth column value greater than 60 from the *conn.log* file. The ninth field represents the connection duration, and we are checking if the value is greater than 60 s (or 1 min).
- $\boxed{\text{|zeek-cut id.orig\_h id.resp\_h}}$ returns the source and destination IP addresses.

Step 3. Execute the modified shell script (Fig. 94).

```
./lab16script.sh
```

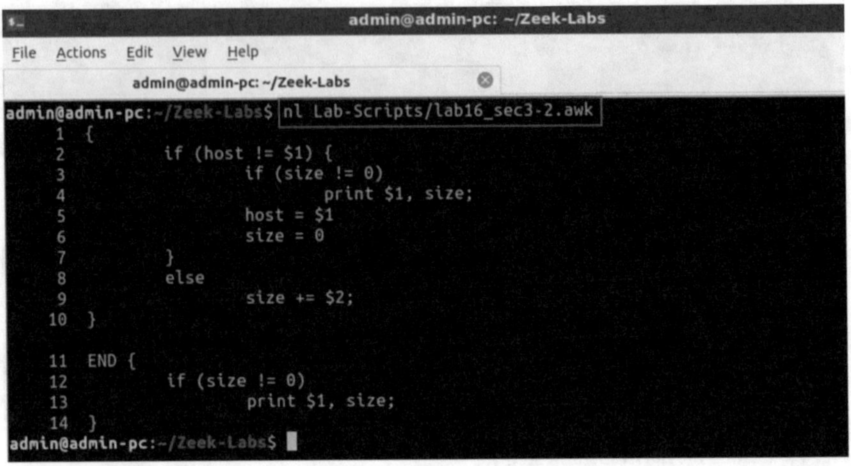

```
admin@admin-pc: ~/Zeek-Labs
File   Actions   Edit   View   Help
        admin@admin-pc: ~/Zeek-Labs                        ⊗
admin@admin-pc:~/Zeek-Labs$ nl Lab-Scripts/lab16_sec3-2.awk
     1  {
     2            if (host != $1) {
     3                    if (size != 0)
     4                            print $1, size;
     5                    host = $1
     6                    size = 0
     7            }
     8         else
     9                    size += $2;
    10  }

    11  END {
    12            if (size != 0)
    13                    print $1, size;
    14  }
admin@admin-pc:~/Zeek-Labs$ █
```

Fig. 95 Displaying the contents of the *Lab-Scripts/lab16_sec3-2.awk*

The source IP address is seen in the left column, while the matching destination IP address is seen in the right column. The pairs will only be displayed if the connection lasted at least one minute.

13.2   Example 6

Example 6 Show the top source host addresses in terms of total traffic (in bytes) sent in descending order.

The *Lab-Scripts* directory contains an AWK script named *lab16_sec3-2.awk* that can be viewed with the following command (Fig. 95):

```
nl Lab-Scripts/lab16_sec3-2.awk
```

The script is explained as follows. Each number represents the respective line number:

1. The `{` character is used to begin nested statements. This instance is the main functionality of the script.
2. The host variable, which will be used to store the source IP addresses found in the first column ($1), is checked against the current data entry in the column. If it is not equal, we will enter the next statement. Because we only want one instance of each source IP address, but the summed value of bytes sent, we will use this check to prevent duplicate entries.
3. This line contains a check to make sure the current packet is not empty and does contain a payload. If the current packet contains a payload of more than 0 bytes, we will proceed to line 4.

Fig. 96 Using the | zeek-cut | utility

4. The current source IP address and its byte payload will be printed or returned to the next statements.
5. Now that we know the current source IP address is not yet stored in the host variable, we will create a new entry into the variable.
6. The size variable is reset back to zero
7. The | } | character is used to end nested statements. Therefore, the first case of a source IP address not being contained in host is complete.
8. If the host variable contains the current data entry, we will proceed to line 9.
9. Here we will sum the unique source IP address' total bytes by adding the payload from the second column ($2).
10. The | } | character is used to end nested statements. This is the ending of the main functionality of the script.
11. The | END | statement denotes what the script will do once it has reached the end of the file, and there are no more input data rows to be read.
12. If a source IP address contains a total payload of more than 0 bytes, we will proceed to line 13.
13. AWK will return the source IP address found in the first column, as well as the size variable, containing the total payload in relation to that source IP address.

Step 1. Input the following command (Fig. 96):

```
zeek-cut id.orig_h orig_bytes <TCP-Traffic/conn.log |sort
|awk -f Lab-Scripts/lab16_sec3-2.awk |sort -k 2 |head -n
10
```

• | zeek-cut id.orig_h orig_bytes <conn.log |: selects the | id.orig_h | and | orig_bytes | columns from the *conn.log* file.
• | |sort |: uses the | sort | command to organize the rows in alphabetical order.
• | |awk -f lab16_sec3-2.awk |: will execute awk with the | -f | option to denote using the script found within the *lab16_sec3-2.awk* file.
• | |sort -k 2 |: uses the | sort | command with the | -k | option to organize the rows based on the values found in the second column—the total number of bytes.

Fig. 97 Opening the *lab16script.sh* script

- |head –n 10 : uses the head command with the -n option to display the 10 topmost values.

The left column contains the source IP address, while the right column contains the number of bytes produced by the paired source IP address.

13.3   Example 7

Example 7 Are there any web servers operating on non-standardized ports?

To solve this example, we will be looking at the service column to view the packets using the Hyper Text Transport Protocol (HTTP) protocol. The standard ports for the HTTP protocol are 80 and 8080, so we will be searching for the network traffic that does not reach those ports.

Step 1. Open the *lab16script.sh* file with nano text editor (Fig. 97).

```
sudo nano lab16script.sh
```

Step 2. Modify the script file's contents. Delete all the previous content and type the following command (Fig. 98):

```
cd TCP-Traffic/
zeek-cut service id.resp_p id.resp_h <conn.log
            \
        |awk '$1 == "http" && ! ($2 == 80 ||$2 == 8080)
{print $3}' \
        |sort -u
```

Press CTRL + o and hit Enter to save the file's contents, then CTRL + x to exit nano and return to the terminal. The above command is explained as follows:

- zeek-cut service id.resp_p id.resp_h <conn.log : selects the service , id.resp_p and id.resp_h columns from the *conn.log* file.

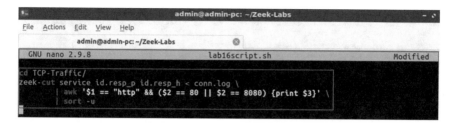

Fig. 98 Editing the *lab16script.sh* script

Fig. 99 Executing the *lab16script.sh* script

- |awk|: passes the input into the following AWK command:

  ○ $1 == "http" : performs a check on the first column to make sure the active data entry is running on the http service.

  ○ && ! ($2 == 80 ||$2 == 8080) : performs a second check if the first check is successfully passed. The ports will be checked and if they are not equal to either of the standard http ports (80 and 8080), they will be passed to the print statement.

  ○ {print $3} : prints the destination IP address of any host that passes both of the previous checks.

- |sort -u|: uses the sort command to sort the lines in the file and the -u option checks for strict ordering.

Step 3. Execute the modified shell script (Fig. 99).

```
./lab16script.sh
```

The destination IP addresses that received traffic on non-standardized ports are displayed.

Fig. 100 Stopping Zeek

13.4 Closing the Current Instance of Zeek

After you have finished the lab, it is necessary to terminate the currently active instance of Zeek. Shutting down a computer while an active instance persists will cause Zeek to shut down improperly and may cause errors in future instances.

Step 1. Stop Zeek by entering the following command on the terminal (Fig. 100). If required, type password as the password. If the Terminal session has not been terminated or closed, you may not be prompted to enter a password. To type capital letters, it is recommended to hold the Shift key while typing rather than using the Caps key.

```
cd $ZEEK_INSTALL/bin && sudo ./zeekctl stop
```

Chapter 6—Lab 17: Generating, Capturing, and Analyzing DoS and DDoS-Centric Network Traffic

Overview
To conduct the experiment described in this section, please login into the Academic Cloud at http://highspeednetworks.net/ and reserve a pod for Lab 17.

This lab covers Denial of Service (DoS)-based network traffic. The lab introduces the generation of DoS-based traffic for testing purposes and uses Zeek to process the collected traffic.

Objective
By the end of this lab, students should be able to:

1. Generate real-time DoS and DDoS traffic.
2. Experiment with the Low Orbit Ion Canon (LOIC) software.
3. Analyze collected DDoS traffic.

Lab Topology
Figure 101 shows the lab workspace topology. This lab primarily uses the host *h1* to generate DoS-based traffic, and the host *h2* to perform live network capture.

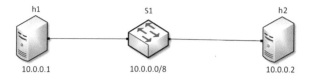

Fig. 101 Lab topology

Table 9 Credentials to access the Client machine

Device	Account	Password
Client	admin	password

Table 10 Shell variables and their corresponding absolute paths

Variable name	Absolute path
$ZEEK_INSTALL	/usr/local/zeek
$ZEEK_TESTING_TRACES	/home/admin/zeek/testing/btest/Traces
$ZEEK_PROTOCOLS_SCRIPT	/home/admin/zeek/scripts/policy/protocols

Lab Settings

The information (case-sensitive) in the table below provides the credentials necessary to access the machines used in this lab (Tables 9 and 10).

Lab Roadmap

This lab is organized as follows:

1. Section 14: Introduction to DoS and DDoS activity.
2. Section 15: Generating real-time DoS traffic.
3. Section 16: Analyzing collected network traffic.

14 Introduction to DoS and DDoS Activity

Denial-of-Service (DoS) is an attack launched by a malicious user to render a target machine or network resource unavailable to its intended users. Distributed Denial-of-Service (DDoS) is an attack originated from different sources to flood the victim's resources. A DDoS attack is more effective than a normal DoS and is harder to mitigate since unlike DoS, it is impossible to stop the attack simply by blocking a single source.

The different types of DoS attacks can be grouped by the traffic they generate, the bandwidth they consume, the services they disrupt, etc. Traffic-based DoS attacks aim at flooding the target with a large volume unsolicited traffic. Bandwidth-based DoS attacks involve transmitting a massive amount of junk data to overload the victim and render its network equipment congested.

14.1   DoS Attack Characteristics

DoS attacks generally involve flooding a targeted victim with network traffic to cause a crash and make it unavailable to benign users. In this lab we explore two common DoS attacks:

- SYN flood : an attacker attempts to overwhelm the server machine by sending a constant stream of TCP connection requests, forcing the server to allocate resources for each new connection until all resources are exhausted.
- ICMP flood : the attacker abuses ICMP Ping and floods the victim computer with Echo Request messages. When a computer receives an ICMP Echo Request message it responds with an ICMP Echo Reply message.

14.2   DDoS Attack Characteristics

DDoS attacks involve using a large number of devices to flood a victim. With an increased number of exploited machines, the amount of resources available to the attacker is far higher. Some relevant DDoS attacks are:

- HTTP flood : simple attack but requires a large number of resources. An attacker who controls several devices (botnet) can continually flood a server with HTTP requests until the server becomes unavailable and unable to respond to additional incoming requests.
- SYN flood : similar to the DoS SYN flood, a botnet initiates several sessions without completing a TCP handshake, causing the victim to consume its available resources.
- Amplification attack : attackers abuse UDP-based network protocols to launch DDoS attacks that exceed hundreds of Gbps in traffic volume. This is achieved via reflective DDoS attacks where an attacker does not directly send traffic to the victim but sends spoofed network packets to a large number of systems that reflect the traffic to the victim. Domain Name System (DNS) and Network Time Protocol (NTP) are examples of application-layer protocols that act as potential amplification attack vectors.

DoS and DDoS attacks can cause catastrophic fallout and monetary losses to a victim.

15   Generating Real-Time DoS Traffic

This lab uses the Low Orbit Ion Canon (LOIC), open-source network stress testing and DoS attack generator. LOIC can be found in the following Github repository. To access the following link, users must have access to an external computer connected to the Internet, because the Zeek Lab topology does not have an active Internet connection.

https://github.com/NewEraCracker/LOIC

Similar to the nmap utility, LOIC can be used to replicate DoS or DDoS activity for testing purposes. LOIC has a Graphical User Interface (GUI), which facilitates the attack's customization.

In this lab, Zeek's default packet capture processing will generate log files containing organized network traffic statistics. In this section, *zeek2* virtual machine is used for live capture and *zeek1* virtual machine is used to generate DoS-related traffic.

15.1   Starting a New Instance of Zeek

Step 1. From the top of the screen, click on the *Client* button as shown below to enter the *Client* machine (Fig. 102).

Step 2. The *Client* machine will now open, and the desktop will be displayed. On the left side of the screen, click on the Terminal icon as shown below (Fig. 103).

Step 3. Start Zeek by entering the following command on the terminal (Fig. 104). This command enters Zeek's default installation directory and invokes Zeekctl tool to start a new instance. To type capital letters, it is recommended to hold the Shift

Fig. 102 Opening the *Client* machine

Fig. 103 Opening the
Terminal

Fig. 104 Starting Zeek

Fig. 105 Launching
MiniEdit

key while typing rather than using the [Caps] key. When prompted for a password,
type [password] and hit [Enter].

```
cd $ZEEK_INSTALL/bin && sudo ./zeekctl start
```

A new instance of Zeek is now active, and we are ready to proceed to the next
section of the lab.

> If you see error messages during the new Zeek instance initializing process, please
> ignore it.

15.2 Launching Mininet

Step 1. From the *Client* machine's desktop, on the left side of the screen, click on
the MiniEdit icon as shown below (Fig. 105). When prompted for a password, type
[password] and hit [Enter]. The MiniEdit editor will now launch.

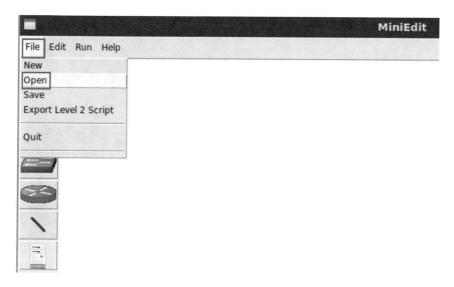

Fig. 106 MiniEdit's menu

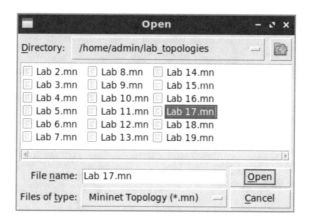

Fig. 107 MiniEdit's Open dialog

Step 2. The MiniEdit editor will now launch and allow for the creation of new, virtualized lab topologies. Load the correct topology by clicking the ⎥Open⎥ button within the ⎥File⎥ tab on the top left of the MiniEdit editor (Fig. 106).

Step 3. Select the *Lab17.mn* file by double clicking the *Lab17.mn* icon, or by clicking the ⎥Open⎥ button (Fig. 107).

Fig. 108 Running the
emulation

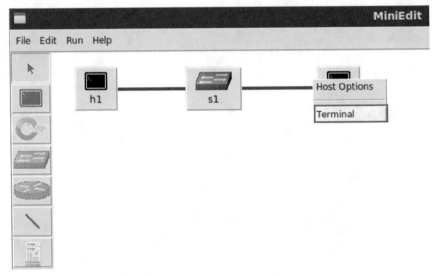

Fig. 109 Opening the host's terminal

Step 4. To begin running the virtual machines, navigate to the Run button, found
on the bottom left of the Miniedit editor, and select the Run button, as seen in the
image below (Fig. 108).

15.3 Setting Up the Zeek2 Machine for Live Network Capture

Step 1. Launch the host *h2* terminal by holding the right mouse button on the
desired machine and clicking the *Terminal* button (Fig. 109).

Step 2. From the *h2* terminal, navigate to the TCP-Traffic directory (Fig. 110).

Fig. 110 Navigating into Zeek-Labs/TCP-Traffic/ directory

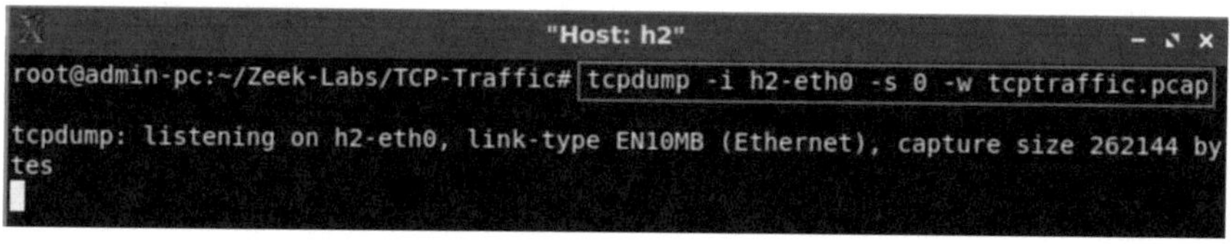

Fig. 111 Running live packet capture

```
cd Zeek-Labs/TCP-Traffic/
```

Step 3. Start live packet capture on interface *h2-eth0* and save the output to a file named *tcptraffic.pcap* (Fig. 111).

```
tcpdump -i h2-eth0 -s -w tcptraffic.pcap
```

The *h2* virtual machine is now ready to begin collecting live network traffic. Next, we will use the *h1* machine to generate scan-based network traffic.

15.4 Launching LOIC

Step 1. Minimize the host *h2 Terminal* and open the host *h1 Terminal* by following the previous steps (Fig. 112). If necessary, right-click within the Miniedit editor to activate your cursor.

Step 2. Execute the *loic.sh* shell script by entering the following command in the terminal (Fig. 113):

```
./loic.sh run
```

Step 3. View the LOIC GUI. If necessary, scale the GUI to a smaller size to fit on the display (Fig. 114).

The figure above shows the LOIC interface. Important features highlighted with colored boxes are explained as follows:

1. Red Box : target IP address. After entering an IP address, clicking the *Lock on* button will select the IP as the target destination address.

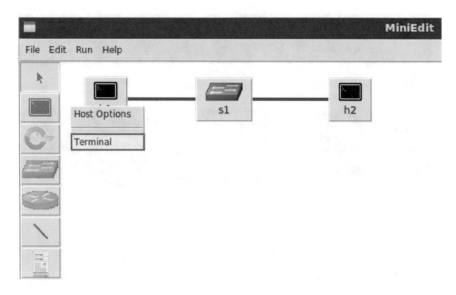

Fig. 112 Opening the host's terminal

Fig. 113 Running LOIC tool

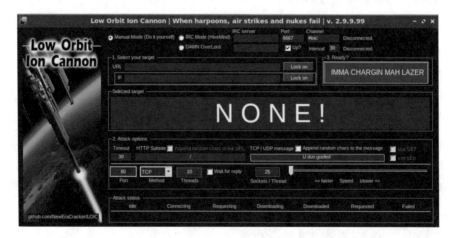

Fig. 114 LOIC's interface

2. Green Box : target port. Can be changed depending on which method is used to launch the DoS attack.
3. Yellow Box : target method. Can be changed to define which protocol is used to launch the DoS attack.
4. Blue Box : number of threads. Indicates the amount of resources *LOIC* will allocate on the host machine.
5. Purple Box : number of sockets per thread. Increasing the number of sockets per thread will exponentially increase the speed of the DoS attack; however, it also requires more resources on the host machine.
6. Brown Box : packet payload. Used to define what each packet will contain as payload.
7. Orange Box : start button. After customizing a desired attack, this button is used to launch the attack.

15.5   Using the Zeek1 Virtual Machine to Launch a TCP-Based DoS Attack

Step 1. Customize the DoS attack by entering the following values in their respective input boxes (Fig. 115):

```
IP:         10.0.0.2
Port:       80
Method:     TCP
Threads:    20
Sockets:    25
Payload:    TCP TEST
```

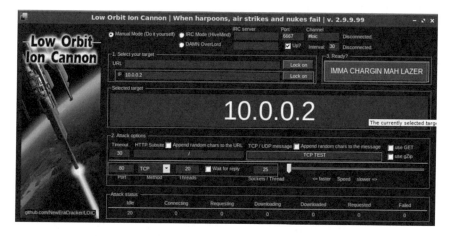

Fig. 115 Customizing the DoS attack on LOIC

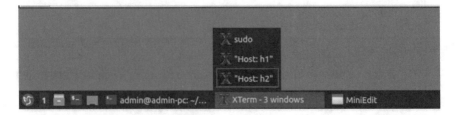

Fig. 116 Maximizing host h2's terminal

```
                              "Host: h2"                              — ⌄ ×
root@admin-pc:~/Zeek-Labs/TCP-Traffic# tcpdump -i h2-eth0 -s 0 -w tcptraffic.pcap

tcpdump: listening on h2-eth0, link-type EN10MB (Ethernet), capture size 262144 by
tes
^C1395162 packets captured
1395162 packets received by filter
0 packets dropped by kernel
root@admin-pc:~/Zeek-Labs/TCP-Traffic#
```

Fig. 117 Stopping the live traffic capture

Step 2. Click the *Lock on* button to save the current configurations. Click the Start (*IMMA CHARGIN MAH LAZER*) button to begin the DoS attack. Wait roughly 10 seconds and click the Stop (*Stop flooding*) button to stop the DoS attack.

Step 3. Minimize the host *h1 Terminal* and open the host *h2 Terminal* using the navigation bar at the bottom of the screen (Fig. 116). If necessary, right-click within the Miniedit editor to activate your cursor.

Step 4. Use the Ctrl+c key combination to stop live traffic capture. Statistics of the capture session will be displayed with network packets being stored in the new *tcptraffic.pcap* file (Fig. 117).

Within the 10 s timeframe, 1,395,162 packets were generated and collected. This number of packets verifies that DoS attacks generate an immense amount of network traffic and can be compared against the much smaller number of packets generated during the previous scan events.

15.6 Using the Zeek1 Virtual Machine to Launch a UDP-Based DoS Attack

Step 1. Using the *zeek2* virtual machine, navigate to the lab workspace directory and enter the *UDP-Traffic* directory (Fig. 118).

```
cd ~/Zeek-Labs/UDP-Traffic/
```

Fig. 118 Navigating into *~/Zeek-Labs/UDP-Traffic/* directory

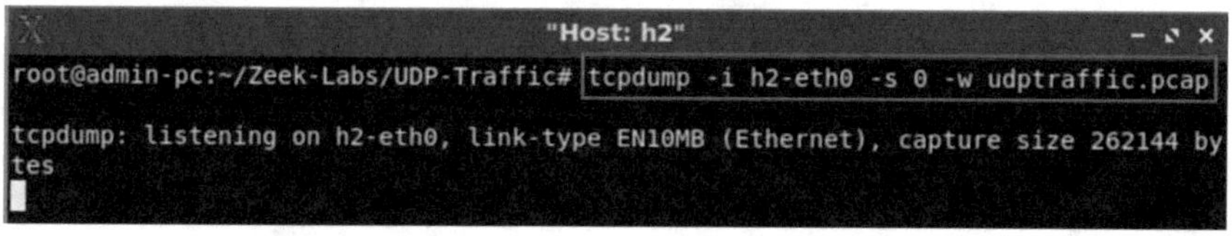

Fig. 119 Running live packet capture

Fig. 120 Maximizing LOIC's interface

Step 2. Start live packet capture on interface *h2-eth0* and save the output to a file named *udptraffic.pcap* (Fig. 119).

```
tcpdump -i h2-eth0 -s -w udptraffic.pcap
```

Step 3. Minimize the *h2 Terminal* and open the *LOIC* GUI using the navigation bar at the bottom of the screen. If necessary, right-click within the Miniedit editor to activate your cursor (Fig. 120).

Step 4. Customize the DoS attack by entering the following values in their respective input boxes (Fig. 121).

```
IP: 10.0.0.2
Port: 20
Method: UDP
Threads: 20
Sockets: 25
Payload: UDP TEST (Must be changed before updating Method
feature)
```

Step 5. Click the *Lock on* button to save the current configurations. Click the *Start (IMMA CHARGIN MAH LAZER)* button to begin the DoS attack. Wait for 10 seconds and click the *Stop (Stop flooding)* button to stop the DoS attack.

Step 6. Minimize the host *h1 Terminal* and open the host *h2 Terminal* using the navigation bar at the bottom of the screen (Fig. 122). If necessary, right-click within the Miniedit editor to activate your cursor.

Fig. 121 Customizing the DoS attack on LOIC

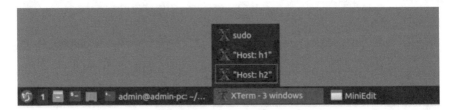

Fig. 122 Maximizing host h2's terminal

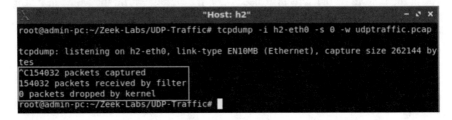

Fig. 123 Stopping the live traffic capture

Step 7. Use the $\boxed{\text{Ctrl+c}}$ key combination to stop live traffic capture. Statistics of the capture session will be displayed. 154,032 packets were recorded by the interface, which were then captured and stored in the new *udptraffic.pcap* file (Fig. 123).

While the UDP-based DoS attack did not generate as much network traffic as the TCP-based DoS attack, heavy amounts of traffic were generated by a single machine. Scaled to a large-scale attack, DoS attacks are extremely debilitating.

Fig. 124 Stopping the emulation

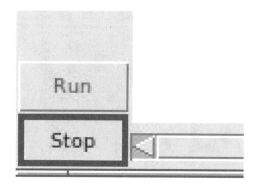

Step 8. Stop the current Mininet session by clicking the *Stop* button on the bottom left of the MiniEdit editor (Fig. 124) and close the MiniEdit editor by clicking the $\boxed{\text{x}}$ on the top right of the editor.

We will now return to the *Client* machine to process and analyze the newly generated network traffic.

16 Analyzing Collected Network Traffic

After successfully conducting both a TCP-based and UDP-based DoS attack, we can begin to analyze the collected network traffic using Zeek and the $\boxed{\text{zeek-cut}}$ utility commands to display the capture traffic.

16.1 Analyzing TCP-Based Traffic

Step 1. On the left side of the *Client* desktop, click on the Terminal icon as shown below (Fig. 125).

Step 2. Navigate to the *TCP-Traffic* directory to find the *tcptraffic.pcap* file (Fig. 126).

```
cd Zeek-Labs/TCP-Traffic/
```

Step 3. View the file contents of the *TCP-Traffic* directory to ensure that the *tcptraffic.pcap* file was successfully saved (Fig. 127).

```
ls
```

Fig. 125 Starting a terminal

Fig. 126 Navigating into *Zeek-Labs/TCP-Traffic/*'s directory

Fig. 127 Listing the contents of the current directory

Fig. 128 Processing the packet capture file using Zeek

Step 4. Use the following Zeek command to process the packet capture file (Fig. 128):

```
zeek -C -r tcptraffic.pcap
```

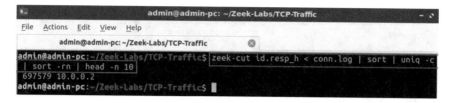

Fig. 129 Using the zeek-cut tool

Fig. 130 Using the zeek-cut tool

16.1.1 TCP Example Query 1

Example 8 Show the source IP addresses that generated the most network traffic, organized in descending order (Fig. 129).

```
zeek-cut id.resp_h <conn.log |sort |uniq -c |sort -rn
|head -n 10
```

The host *h2* received 657,579 TCP packets. This command, or a similar one, can be useful in real-world environments to detect vulnerable hosts within a network—allowing for the process of securing and mitigating possible threats.

16.1.2 TCP Example Query 2

Example 9 Show the destination ports that received the most traffic, organized in descending order (Fig. 130).

```
zeek-cut id.resp_p <conn.log |sort |uniq -c |sort -rn
|head -n 10
```

We can see that 697,579 packets were received by the host *h2* on port 80, which is the port we specified for the host *h1* to target. Additional ports may be discovered during processing, slightly variable due to LOIC attempting to establish connections; however, it is clear the most targeted port is the one we specified in the DoS attack.

Fig. 131 Navigating into *~/Zeek-Labs/UDP-Traffic/*'s directory

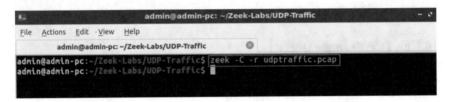

Fig. 132 Listing the contents of the current directory

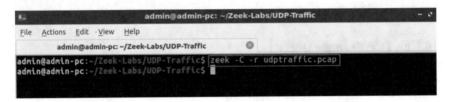

Fig. 133 Processing the packet capture file using Zeek

16.2 Analyzing UDP-Based Traffic

Step 1. Navigate to the *UDP-Traffic* directory to find the *udptraffic.pcap* file (Fig. 131).

```
cd ~/Zeek-Labs/UDP-Traffic/
```

Step 2. View the file contents of the *TCP-Traffic* directory to ensure that the *udptraffic.pcap* file was successfully saved (Fig. 132).

```
ls
```

Step 3. Use the following Zeek command to process the packet capture file (Fig. 133):

```
zeek -C -r udptraffic.pcap
```

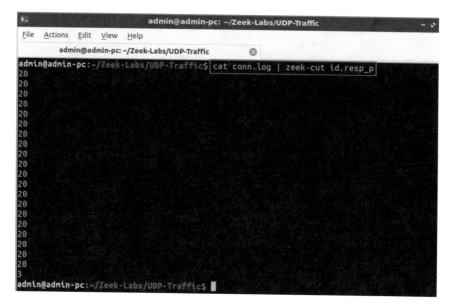

Fig. 134 Using the zeek-cut tool

Step 4. Show the list of ports that received network traffic (Fig. 134).

```
cat conn.log |zeek-cut id.resp_p
```

We can see that despite the large number of packets collected, very few were recorded by Zeek's event-based engine. We specified port 20 as the targeted port during our DoS attack; however, the number of identified packets is significantly lower than expected.

The primary cause of the decreased packet count is due to the number of UDP packets being dropped. Primarily due to firewalls, UDP packets may be traced on the interface, but may not reach the target destination. Furthermore, the default Zeek customization is primarily focused on TCP traffic, and is not designed to handle UDP traffic in such an in-depth manner, requiring additional scripts and policies that will be introduced in later labs.

16.3   Closing the Current Instance of Zeek

After you have finished the lab, it is necessary to terminate the currently active instance of Zeek. Shutting down a computer while an active instance persists will cause Zeek to shut down improperly and may cause errors in future instances.

Fig. 135 Stopping Zeek

Step 1. Stop Zeek by entering the following command on the terminal (Fig. 135). If required, type password as the password. If the Terminal session has not been terminated or closed, you may not be prompted to enter a password. To type capital letters, it is recommended to hold the Shift key while typing rather than using the Caps key.

```
cd $ZEEK_INSTALL/bin && sudo ./zeekctl stop
```

Chapter 6—Lab 18: Zeek Scripting

Overview
To conduct the experiment described in this section, please login into the Academic Cloud at http://highspeednetworks.net/ and reserve a pod for Lab 18.

This lab covers Zeek's scripting language. It introduces the major keywords and components required in a Zeek script. The lab then uses these scripts to analyze processed log files.

Objectives
By the end of this lab, students should be able to:

1. Develop scripts using Zeek's scripting language.
2. Analyze processed log files using Zeek scripts.
3. Modify log streams for creating additional events and notices.

Lab Topology
Figure 136 shows the lab topology. The topology uses 10.0.0.0/8, which is the default network assigned by Mininet. The *h1* and *h2* virtual machines will be used to generate and collect network traffic.

Lab Settings
The information (case-sensitive) in the table below provides the credentials necessary to access the machines used in this lab (Tables 11 and 12).

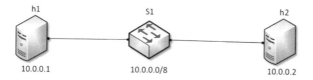

Fig. 136 Lab topology

Table 11 Credentials to access the Client machine

Device	Account	Password
Client	admin	password

Table 12 Shell variables and their corresponding absolute paths

Variable name	Absolute path
$ZEEK_INSTALL	/usr/local/zeek
$ZEEK_TESTING_TRACES	/home/zeek/admin/testing/btest/Traces
$ZEEK_PROTOCOLS_SCRIPT	/home/zeek/admin/scripts/policy/protocols

Lab Roadmap

This lab is organized as follows:

1. Section 17: Introduction to scripting with Zeek.
2. Section 18: Log file analysis using Zeek scripts.
3. Section 19: Modifying Zeek log streams.

17 Introduction to Scripting with Zeek

Zeek includes its own event-driven scripting language, which provides the primary means for an organization to extend and customize Zeek's functionality. By modifying Zeek's log streams, a more in-depth analysis can be performed on network events.

Since Zeek's scripting language is event-driven, we define which events we need Zeek to respond to when encountered during network traffic analysis.

17.1 Zeek Script Events

The script below shows events that will be explored during this lab. When developing a Zeek script, the script's functionalities are wrapped within respective events.

- `zeek_init` event: activated when Zeek is first initialized.
- `zeek_done` event: activated before Zeek is terminated.

- tcp_packet event: activated when a packet containing a TCP header is processed.
- udp_request event: activated when a packet containing a UDP request header is processed.
- udp_reply event: activated when a packet containing a UDP reply header is processed.

```
1 ▾ signature HTTP-sig {
2       ip-proto == tcp
3       dst-port == 80
4       payload /POST/
5       event "Found HTTP POST!"
6   }
```

Additional events and their required parameters are outlined and explained in Zeek's official documentation. To access the following link, users must have access to an external computer connected to the Internet, because the Zeek Lab topology does not have an active Internet connection.

https://docs.zeek.org/en/current/examples/scripting/

17.2 Zeek Module Workspace

The script below uses the module keyword that assigns the script to a *namespace*. Codes from other scripts can be accessed by including a matching module. The export keyword is used to export the code entered in its block with the module workspace.

```
1   @load-sigs
2
3   module ZeekScript;
4
5 ▾ export{
6       /* Append and define new log stream parameters */
7   }
```

- module ZeekScript : changes the module workspace to ZeekScript.
- export block : code entered here will be exported with the module workspace.

Exporting code with a module workspace allows more advanced scripts to be built on top of other scripts.

17.3   Zeek Log Streams

The script below shows the log stream functionality. When developing a Zeek script, all processed outputs will be sent to a specific log stream. These log streams will contain the format of the corresponding log file output. We can create new streams, modify original streams or append additional parameters to existing streams.

```
1   @load-sigs
2
3   module ZeekScript;
4
5   redef signature_files += "signature_file_path.sig"
```

- connection_established event: activated when a host makes a connection to a receiver.
- Log::create_stream: creates a new log stream, with a name, format structure, and path.
- Log::write: writes included data to the specified log stream.

Additional log stream commands are explained in detail in Zeek's official documentation.

18   Log File Analysis Using Zeek Scripts

With Zeek's event-driven scripting language, we can create specific event-based filters to be applied during packet capture analysis. This section shows example scripts for network analysis.

18.1   Starting a New Instance of Zeek

Step 1. From the top of the screen, click on the *Client* button as shown below to enter the *Client* machine (Fig. 137).

Step 2. The *Client* machine will now open, and the desktop will be displayed. On the left side of the screen, click on the Terminal icon as shown below (Fig. 138).

Step 3. Start Zeek by entering the following command on the terminal (Fig. 139). This command enters Zeek's default installation directory and invokes *Zeekctl* tool to start a new instance. To type capital letters, it is recommended to hold the *Shift*

Topology
Content
Status
Client

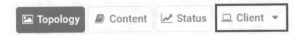

Fig. 137 Opening the *Client* machine

Terminal
Miniedit

Fig. 138 Opening the *Terminal*

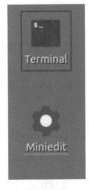

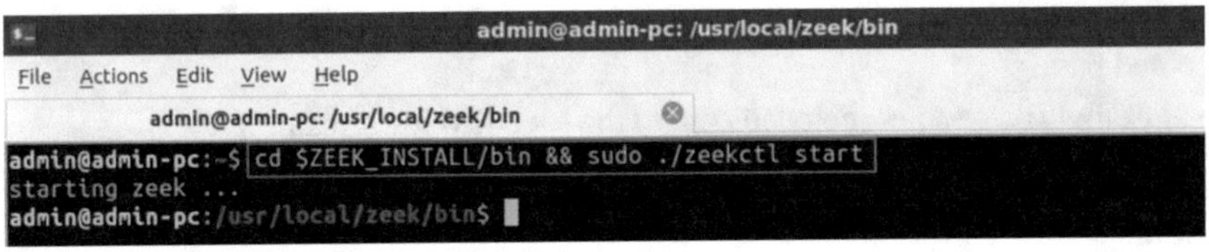
admin@admin-pc: /usr/local/zeek/bin
File Actions Edit View Help
admin@admin-pc: /usr/local/zeek/bin
admin@admin-pc:~$ cd $ZEEK_INSTALL/bin && sudo ./zeekctl start
starting zeek ...
admin@admin-pc:/usr/local/zeek/bin$

Fig. 139 Starting Zeek

key while typing rather than using the *Caps* key. When prompted for a password, type `password` and hit *Enter*.

```
cd $ZEEK_INSTALL/bin && sudo ./zeekctl start
```

A new instance of Zeek is now active, and we are ready to proceed to the next section of the lab.

18.2 Executing a UDP Zeek Script

This lab series includes a *Lab-Scripts* directory, containing all of the relevant Zeek scripts that will be used during the labs.

Step 1. Navigate to the *Lab-Scripts* directory (Fig. 140).

```
cd ~/Zeek-Labs/Lab-Scripts/
```

Within this directory, all lab scripts can be accessed, viewed, and modified.

Fig. 140 Navigating into *Zeek-Labs/TCP-Traffic/* directory

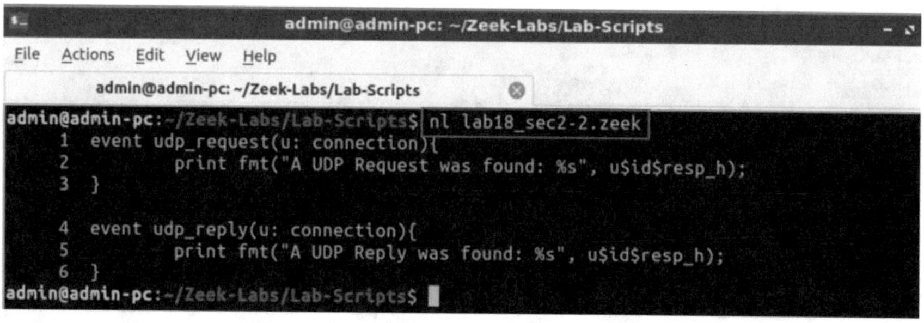

```
admin@admin-pc:~/Zeek-Labs/Lab-Scripts$ nl lab18_sec2-2.zeek
     1  event udp_request(u: connection){
     2          print fmt("A UDP Request was found: %s", u$id$resp_h);
     3  }

     4  event udp_reply(u: connection){
     5          print fmt("A UDP Reply was found: %s", u$id$resp_h);
     6  }
admin@admin-pc:~/Zeek-Labs/Lab-Scripts$
```

Fig. 141 Displaying the contents of the script *lab18_sec2-2.zeek*

Step 2. Display the content of the *lab18_sec2-2.zeek* Zeek script using nl
command. nl shows the line numbers in the file (Fig. 141).

```
nl lab18_sec2-2.zeek
```

The script is explained as follows. Each number represents the respective line number:

1. Event udp_request is activated when a packet containing a UDP Request header is processed. The related packet header information is stored in the connection data structure passed to the function through the u variable.
2. Prints the specified string. %s is a format specifier for strings with fmt . It indicates the position of the corresponding variable's information in the string. uidresp_h retrieves the destination IP address from the UDP packet.
3. End of the udp_request event.
4. Event udp_reply activated when a packet containing a UDP Reply header is processed. The related packet header information is stored in the connection data structure passed to the function through the u variable.
5. Prints the specified string. uidresp_h retrieves the destination IP address from the UDP packet.
6. End of the udp_reply event.

admin@admin-pc: ~/Zeek-Labs/UDP-Traffic
File Actions Edit View Help
admin@admin-pc: ~/Zeek-Labs/UDP-Traffic
admin@admin-pc:~/Zeek-Labs/Lab-Scripts$ cd ~/Zeek-Labs/UDP-Traffic/
admin@admin-pc:~/Zeek-Labs/UDP-Traffic$

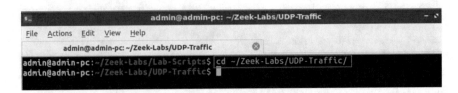

Fig. 142 Navigating into *Zeek-Labs/UDP-Traffic/* directory

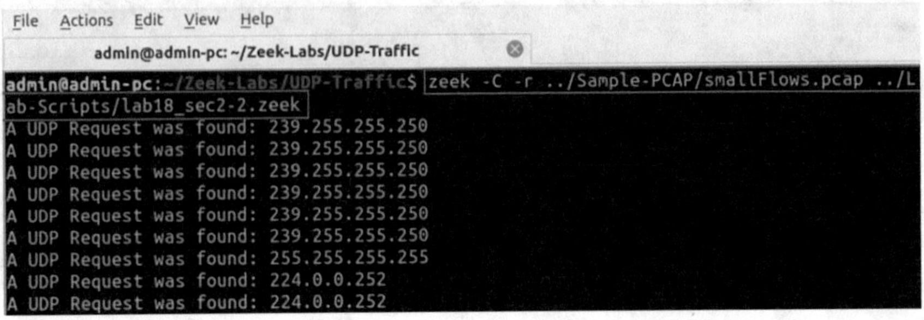
File Actions Edit View Help
admin@admin-pc: ~/Zeek-Labs/UDP-Traffic
admin@admin-pc:~/Zeek-Labs/UDP-Traffic$ zeek -C -r ../Sample-PCAP/smallFlows.pcap ../Lab-Scripts/lab18_sec2-2.zeek
A UDP Request was found: 239.255.255.250
A UDP Request was found: 239.255.255.250
A UDP Request was found: 239.255.255.250
A UDP Request was found: 239.255.255.250
A UDP Request was found: 239.255.255.250
A UDP Request was found: 239.255.255.250
A UDP Request was found: 255.255.255.255
A UDP Request was found: 224.0.0.252
A UDP Request was found: 224.0.0.252

Fig. 143 Processing a packet capture file using Zeek

Step 3. Navigate to the *UDP-Traffic* workspace directory (Fig. 142).

```
cd ~/Zeek-Labs/UDP-Traffic/
```

Step 4. Process a packet capture file using the Zeek script. It is possible to use the tab key to autocomplete the longer paths (Fig. 143).

```
zeek -C -r ../Sample-PCAP/smallFlows.pcap ../Lab-
Scripts/lab18_sec2-2.zeek
```

The packet capture file is processed into output log files. Since we did not create a new log stream, the script's output is displayed on the standard output (the screen). When udp_request or udp_reply events are triggered, the resulting packet information is displayed.

18.3 Executing a TCP Zeek Script

Step 1. Display the content of the *lab18_sec2-3.zeek* Zeek script using nl command (Fig. 144). nl shows the line numbers in the file. It is possible to use the tab key to autocomplete the longer paths.

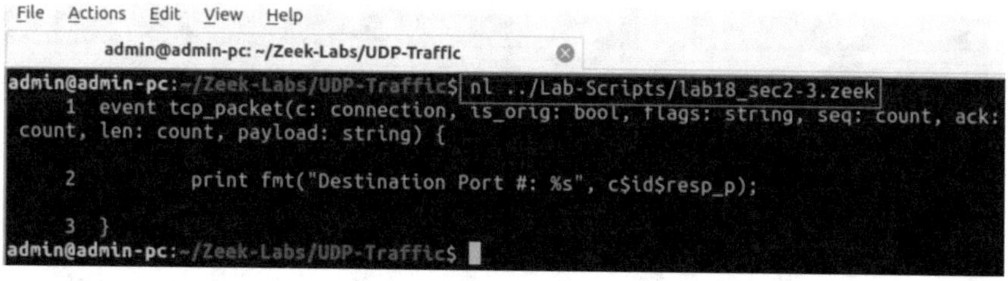

```
File  Actions  Edit  View  Help
        admin@admin-pc: ~/Zeek-Labs/UDP-Traffic         ⊗
admin@admin-pc:~/Zeek-Labs/UDP-Traffic$ nl ../Lab-Scripts/lab18_sec2-3.zeek
     1  event tcp_packet(c: connection, is_orig: bool, flags: string, seq: count, ack:
count, len: count, payload: string) {

     2          print fmt("Destination Port #: %s", c$id$resp_p);

     3  }
admin@admin-pc:~/Zeek-Labs/UDP-Traffic$ █
```

Fig. 144 Displaying the contents of the script *lab18_sec2-3.zeek*

Fig. 145 Processing a packet capture file using Zeek

```
nl ../Lab-Scripts/lab18_sec2-3.zeek
```

The script is explained as follows. Each number represents the respective line number:

1. Event tcp_packet is activated when a packet containing a TCP header is processed. The related packet header information is stored in the connection data structure passed to the function through the u variable. Additional TCP-related information is passed in a similar manner.
2. Prints the specified string. %s is a format specifier for strings with fmt . It indicates the position of the corresponding variable's information in the string. uidresp_h retrieves the destination IP address from the TCP packet.
3. End of the tcp_packet event.

Step 2. Process a packet capture file using the Zeek script (Fig. 145). It is possible to use the tab key to autocomplete the longer paths.

```
zeek -C -r ../Sample-PCAP/smallFlows.pcap ../Lab-
Scripts/lab18_sec2-3.zeek
```

The following output is produced (Fig. 146):

When the tcp_packet event is triggered, the resulting packet information is displayed. Highlighted is an example of Port 8443 and Port 80 traffic.

These examples highlight Zeek's capabilities of tracking specific traffic. For instance, a script can be designed to collect all Port 80 traffic daily and to export it to a log file. In the following section we introduce log streams.

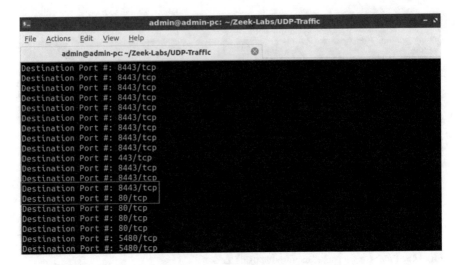

Fig. 146 Output of Zeek's processing

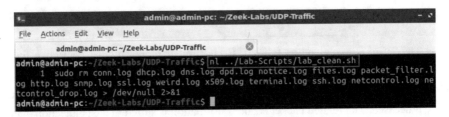

Fig. 147 Displaying the contents of the script *../Lab-Scripts/lab_clean.sh*

19 Modifying Zeek Log Streams

Zeek log streams determine where an event's output will be returned, as well as how it is formatted. It is possible to append new streams, modify default streams, or remove streams.

Before continuing, we must clear the lab workspace directory.

Step 1. Display the contents of the *lab_clean.sh* shell script using nl command (Fig. 147).

```
nl ../Lab-Scripts/lab_clean.sh
```

The shell script removes a list of files expected to be generated by Zeek's processing using default log streams. Executing this shell script will clear the directory of log files generated previously. Output messages from running this script are not displayed in the Terminal, instead the code >/dev/null 2>&1 will set errors and notices to be sent to a null folder, effectively eliminating them.

admin@admin-pc: ~/Zeek-Labs/UDP-Traffic
File Actions Edit View Help
admin@admin-pc: ~/Zeek-Labs/UDP-Traffic
admin@admin-pc:~/Zeek-Labs/UDP-Traffic$ ./../Lab-Scripts/lab_clean.sh
[sudo] password for admin:
admin@admin-pc:~/Zeek-Labs/UDP-Traffic$

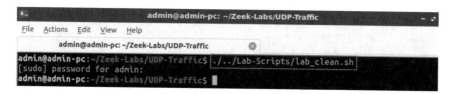

Fig. 148 Executing the script *../Lab-Scripts/lab_clean.sh*

admin@admin-pc: ~/Zeek-Labs/UDP-Traffic
File Actions Edit View Help
admin@admin-pc: ~/Zeek-Labs/UDP-Traffic
admin@admin-pc:~/Zeek-Labs/UDP-Traffic$ nl ../Lab-Scripts/lab18_sec3-1.zeek
1 event zeek_init(){
2
3 local update = Log::get_filter(Conn::LOG, "default");
4 update$path = "UpdatedConn";
5 Log::add_filter(Conn::LOG, update);
6 }
admin@admin-pc:~/Zeek-Labs/UDP-Traffic$

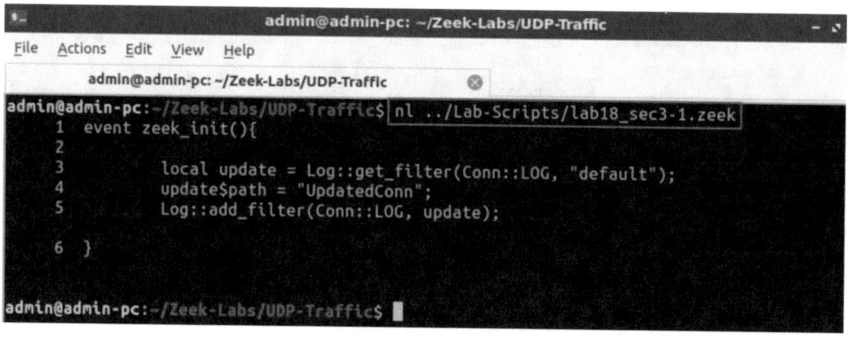

Fig. 149 Displaying the contents of the script *../Lab-Scripts/ lab18_ sec3-1.zeek*

Step 2. Execute the *lab_clean.sh* shell script (Fig. 148). It is possible to use the tab key to autocomplete the longer paths. If required, type password as the password.

```
./../Lab-Scripts/lab_clean.sh
```

With the workspace directory cleared, we can move to the next section.

19.1 Renaming the conn.log Stream

In this example, we will rename the *conn.log* file to be *UpdatedConn.log*. Renaming log streams can help with files organization, especially if a log file has been modified from its original functionality.

Step 1. Display the contents of the *lab18_sec3-1.zeek* Zeek script using the nl command (Fig. 149). It is possible to use the tab key to autocomplete the longer paths.

```
nl ../Lab-Scripts/lab18_sec3-1.zeek
```

The script is explained as follows. Each number represents the respective line number:

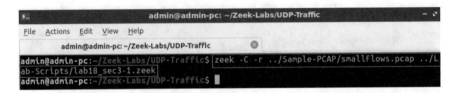

Fig. 150 Processing a packet capture file using Zeek

Fig. 151 Listing the contents of the current directory

1. Event zeek_init is activated when Zeek is first initialized.
3. Creates a local variable update initialized to the default Conn::LOG filter.
4. Sets the update variable's path to *UpdatedConn.log*.
5. Appends the new filter to the active log streams.
6. End of the zeek_init event.

Step 2. Process a packet capture file using the Zeek script (Fig. 150). It is possible to use the tab key to autocomplete the longer paths.

```
zeek -C -r ../Sample-PCAP/smallFlows.pcap ../Lab-
Scripts/lab18_sec3-1.zeek
```

Step 3. List the generated log files in the current directory (Fig. 151).

```
ls
```

Note the *UpdatedConn.log*, highlighted by the gray box. Since we did not change any formatting, it is an exact replica of the original *conn.log* file.

19.2   Updating the conn.log Stream

In this example, we modify the *conn.log* file to generate an additional *conn- http.log* file. This modification will split the *conn.log* contents between two log files, which is useful when organizing specific events—such as splitting UDP traffic from TCP traffic, or reply messages from requests.

admin@admin-pc: ~/Zeek-Labs/UDP-Traffic
File Actions Edit View Help
admin@admin-pc: ~/Zeek-Labs/UDP-Traffic
admin@admin-pc:~/Zeek-Labs/UDP-Traffic$ ./../Lab-Scripts/lab_clean.sh
[sudo] password for admin:
admin@admin-pc:~/Zeek-Labs/UDP-Traffic$

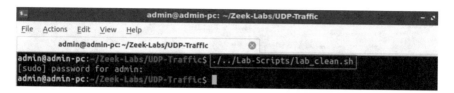

Fig. 152 Executing the script *../Lab-Scripts/lab_clean.sh*

admin@admin-pc: ~/Zeek-Labs/UDP-Traffic
File Actions Edit View Help
admin@admin-pc: ~/Zeek-Labs/UDP-Traffic
admin@admin-pc:~/Zeek-Labs/UDP-Traffic$ nl ../Lab-Scripts/lab18_sec3-2.zeek
1 function http_only(rec: Conn::Info) : bool {
2
3 return rec?$service && rec$service == "http";
4 }
5 event zeek_init(){
6 local filter: Log::Filter = [$name="http-only", $path="conn-http", $pred=http_only];
7 Log::add_filter(Conn::LOG, filter);
8 }
admin@admin-pc:~/Zeek-Labs/UDP-Traffic$

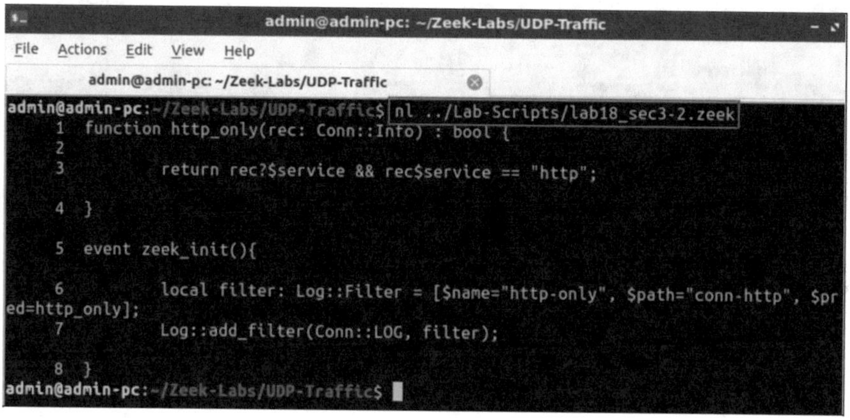

Fig. 153 Displaying the contents of the script *../Lab-Scripts/ lab18_ sec3-2.zeek*

Step 1. Execute the included *lab_clean.sh* shell script (Fig. 152). If required, type password as the password. It is possible to use the tab key to autocomplete the longer paths.

```
./../Lab-Scripts/lab_clean.sh
```

Step 2. Display the contents of *lab18_sec3-2.zeek* Zeek script using the nl command (Fig. 153).

```
nl ../Lab-Scripts/lab18_sec3-2.zeek
```

The script is explained as follows. Each number represents the respective line number:

1. Boolean function that has the parameter rec, an instance of Conn::Info.
3. Returns True if the service stored in rec is the HTTP protocol.
4. End of the function.
5. Event zeek_init is activated when Zeek is first initialized.
6. Creates a local filter with *http* related naming and pathing.
7. Appends the new filter to the active log streams.
8. End of the zeek_init event.

admin@admin-pc: ~/Zeek-Labs/UDP-Traffic
File Actions Edit View Help
admin@admin-pc: ~/Zeek-Labs/UDP-Traffic
admin@admin-pc:~/Zeek-Labs/UDP-Traffic$ zeek -C -r ../Sample-PCAP/smallFlows.pcap ../Lab-Scripts
/lab18_sec3-2.zeek
admin@admin-pc:~/Zeek-Labs/UDP-Traffic$

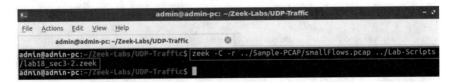

Fig. 154 Processing a packet capture file using Zeek

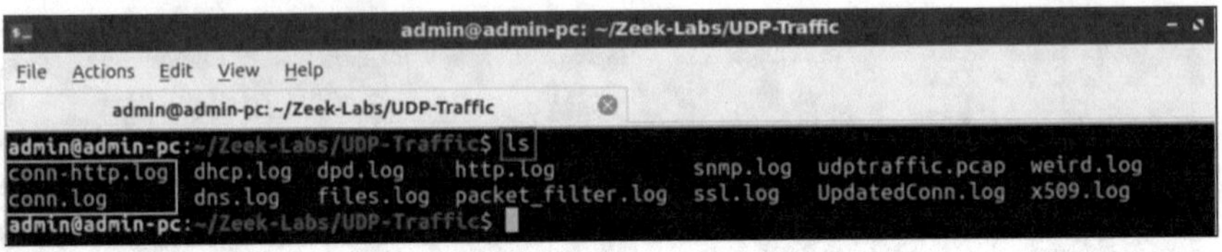

admin@admin-pc: ~/Zeek-Labs/UDP-Traffic
File Actions Edit View Help
admin@admin-pc: ~/Zeek-Labs/UDP-Traffic
admin@admin-pc:~/Zeek-Labs/UDP-Traffic$ ls
conn-http.log dhcp.log dpd.log http.log snmp.log udptraffic.pcap weird.log
conn.log dns.log files.log packet_filter.log ssl.log UpdatedConn.log x509.log
admin@admin-pc:~/Zeek-Labs/UDP-Traffic$

Fig. 155 Listing the contents of the current directory

Step 3. Process a packet capture file using the Zeek script (Fig. 154). It is possible to use the tab key to autocomplete the longer paths.

```
zeek -C -r ../Sample-PCAP/smallFlows.pcap ../Lab-
Scripts/lab18_sec3-2.zeek
```

Step 4. List the generated log files in the current directory (Fig. 155).

```
ls
```

Note the *conn-http.log* file in the first column. This file will have the same formatting as the *conn.log* file; however, it will only contain HTTP traffic. These files are highlighted by the orange box in the proceeding image.

19.3 Closing the Current Instance of Zeek

After you have finished the lab, it is necessary to terminate the currently active instance of Zeek. Shutting down a computer while an active instance persists will cause Zeek to shut down improperly and may cause errors in future instances.

Step 1. Stop Zeek by entering the following command on the terminal (Fig. 156). If required, type password as the password. If the Terminal session has not been terminated or closed, you may not be prompted to enter a password. To type capital letters, it is recommended to hold the Shift key while typing rather than using the Caps key.

```
cd $ZEEK_INSTALL/bin && sudo ./zeekctl stop
```

Fig. 156 Stopping Zeek

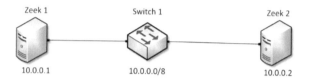

Fig. 157 Lab topology

Chapter 6—Lab 19: Zeek Signatures

Overview

To conduct the experiment described in this section, please login into the Academic Cloud at http://highspeednetworks.net/ and reserve a pod for Lab 19.

This lab covers Zeek's signature framework language. It introduces what network traffic signatures are and how they are matched to identify specific network events. This lab then reviews premade signature files and provides example usage for analysis.

Objectives

By the end of this lab, students should be able to:

1. Develop signatures using Zeek's signature framework.
2. Analyze processed log files using Zeek signatures.
3. Modify log streams for creating additional events and notices based on signatures.

Lab Topology

Figure 157 shows the lab topology. The topology uses 10.0.0.0/8, which is the default network assigned by Mininet. The hosts *h1* and *h2* will be used to generate and collect network traffic.

Lab Settings

The information (case-sensitive) in the table below provides the credentials necessary to access the machines used in this lab (Tables 13 and 14).

Lab Roadmap

This lab is organized as follows:

Table 13 Credentials to
access the Client machine

Device	Account	Password
Client	admin	password

Table 14 Shell variables and their corresponding absolute paths

Variable name	Absolute path
$ZEEK_INSTALL	/usr/local/zeek
$ZEEK_TESTING_TRACES	/home/zeek/admin/testing/btest/Traces
$ZEEK_PROTOCOLS_SCRIPT	/home/zeek/admin/scripts/policy/protocols

1. Section 20: Introduction to Zeek signatures.
2. Section 21: Log file analysis using Zeek signatures.
3. Section 22: Modifying Zeek signatures for advanced pattern matching.

20 Introduction to Zeek Signatures

Following the introduction of developing and implementing basic Zeek scripts, we can now begin generating Zeek signatures. Introduced in the beginning of this lab series, the Zeek event-based engine is the primary architecture for running Zeek as an efficient intrusion detection system. The Zeek event-based engine predominantly utilizes the extensive scripting language to develop policies in order to define the steps and notifications necessary to handle anomalies and exceptions.

However, oftentimes it is simpler to create a predetermined string, known as a signature, and parse packet capture files for the specific signature. Because signatures are used for low-level pattern matching, the Zeek signature framework does not provide the same in-depth functionality as the Zeek scripting language for its event-based engine. Zeek signatures are used to quickly aggregate related network packets through signature matching before analysts can perform further, in-depth analysis on such traffic.

It is important to understand and be familiar with signatures due to their widespread usage across many related Intrusion Detection Systems and application-level firewalls. Separate from Zeek, many alternative IDS, such as the popular *Snort*, rely on signature-based pattern matching for anomaly and malicious event detection. Therefore, in operational cybersecurity environments that analyze network traffic to mitigate and prevent malicious events, understanding Zeek's signature framework adds an additional tool for developing a comprehensive IDS.

This lab will begin by introducing Zeek signatures, detailing their unique file type, how to load them into the Zeek event-based engine, and include a number of examples of leveraging signature matching for log file analysis.

20.1 Zeek Signature Format

The signature below depicts a basic network traffic signature. Depending on their usage, signatures can either include stricter requirements, or be more lax to encompass a larger portion of the processed data.

```
 1 ▾ type Action: enum {
 2        ## Ignore this signature completely (even for scan detection).
 3        ## Don't write to the signatures logging stream.
 4        SIG_IGNORE,
 5        ## Process through the various aggregate techniques, but don't
 6        ## report individually and don't write to the signatures logging
 7        ## stream.
 8        SIG_QUIET,
 9        ## Generate a notice.
10        SIG_LOG,
11        ## The same as :zeek:enum:`Signatures::SIG_LOG`, but ignore for
12        ## aggregate/scan processing.
13        SIG_FILE_BUT_NO_SCAN,
14        ## Generate a notice and set it to be alarmed upon.
15        SIG_ALARM,
16        ## Alarm once per originator.
17        SIG_ALARM_PER_ORIG,
18        ## Alarm once and then never again.
19        SIG_ALARM_ONCE,
20        ## Count signatures per responder host and alarm with the
21        ## :zeek:enum:`Signatures::Count_Signature` notice if a threshold
22        ## defined by :zeek:id:`Signatures::count_thresholds` is reached.
23        SIG_COUNT_PER_RESP,
24        ## Don't alarm, but generate per-orig summary.
25        SIG_SUMMARY,
26    };
```

1. This line defines a new *signature* object, with the name *HTTP-sig*.
2. Defines the desired match's transport protocol to be TCP.
3. Defines the desired match's destination port to be 80.
4. Defines the desired match's payload to contain the regular expression equivalent to "POST."
5. Defines an event if the match is found. Currently, the event will post a "HTTP Packet Found!" message; however, these events can be developed with a more complex functionality if the need arises.

This signature can be loaded into the Zeek signature framework during network traffic analysis, in which Zeek will attempt to match packets with the signature's details. While each individual packet can only be matched one time, multiple signatures can be applied to any arbitrary data.

Additional signatures and their included variables are outlined and explained in Zeek's official documentation. To access the following link, users must have access to an external computer connected to the Internet, because the Zeek Lab topology does not have an active Internet connection.

```
https://docs.zeek.org/en/current/frameworks/signatures.html
```

20.2 Creating and Using Zeek Signatures

Similar to Zeek's policy scripting framework, Zeek signatures are saved in separate files denoted by the `.sig` file extension. There are three ways to initialize Zeek for network traffic analysis while leveraging the Zeek signature framework:

1. When initializing Zeek from the terminal, include the additional `-s` option:

```
zeek -r <pcap_file_location>-s <signature_file_location>
```

- `zeek`: command to invoke Zeek.
- `-r`: option signifies to Zeek that it will be reading from an offline file.
- `<pcap_file_location>`: indicates the pcap file location.
- `-s`: option signifies to Zeek that the next file contains signatures.
- `<script_location>`: indicates the script location.

2. When creating a Zeek policy script, include the `@load-sigs` directive:

3. When creating a Zeek policy script, extend the Zeek global `signature_files` variable by appending the $+ =$ operator followed by the signature file:

20.3 Zeek's Default Signature Framework

This section introduces the default Zeek signature file that is compiled and included after Zeek has been installed.

While this default Zeek script includes scan-based detection, it will not correctly identify every unique anomaly that may be encountered. However, it does provide a comprehensive starter code that can be reviewed and customized to understand the Zeek signature framework.

The default Zeek signature file is named *main.zeek*. More information on this script can be found in Zeek's documentation pages. To access the following link, users must have access to an external computer connected to the Internet, because the Zeek Lab topology does not have an active Internet connection.

```
https://docs.zeek.org/en/current/scripts/base/frameworks/si
gnatures/main.zeek.html
```

The file has been copied into the Zeek lab workspace directory and renamed to *ZeekSignatureFramework.zeek* for ease of access and name-reference clarity.

```
 1  type Info: record {
 2      ## The network time at which a signature matching type of event
 3      ## to be logged has occurred.
 4      ts:         time        &log;
 5      ## A unique identifier of the connection which triggered the
 6      ## signature match event.
 7      uid:        string      &log &optional;
 8      ## The host which triggered the signature match event.
 9      src_addr:   addr        &log &optional;
10      ## The host port on which the signature-matching activity
11      ## occurred.
12      src_port:   port        &log &optional;
13      ## The destination host which was sent the payload that
14      ## triggered the signature match.
15      dst_addr:   addr        &log &optional;
16      ## The destination host port which was sent the payload that
17      ## triggered the signature match.
18      dst_port:   port        &log &optional;
19      ## Notice associated with signature event.
20      note:       Notice::Type &log;
21      ## The name of the signature that matched.
22      sig_id:     string      &log &optional;
23      ## A more descriptive message of the signature-matching event.
24      event_msg:  string      &log &optional;
25      ## Extracted payload data or extra message.
26      sub_msg:    string      &log &optional;
27      ## Number of sigs, usually from summary count.
28      sig_count:  count       &log &optional;
29      ## Number of hosts, from a summary count.
30      host_count: count       &log &optional;
31  };
```

The figure above shows the options for signature match events within the *ZeekSignatureFramework.zeek* file. The options are explained as follows. Each number represents the respective line number:

4. SIG_IGNORE : if a signature is matched, do not write to the logging stream.

8. SIG_QUIET : if a signature is matched, process the included events but do not write to the logging stream.

10. SIG_LOG : if a signature is matched, generate a notice.

13. SIG_FILE_BUT_NO_SCAN : if a signature is matched and does not meet scan thresholds, write to the logging stream.

15. SIG_ALARM : if a signature is matched, generate a notice and set an alarm.

17. SIG_ALARM_PER_ORIG : if a signature is matched, generate a notice and set an alarm once per host that triggered the match.

19. SIG_ALARM_ONCE : if a signature is matched, generate a notice and set an alarm only one time, no matter the number of matches.

Fig. 158 Opening the *Client* machine

23. SIG_COUNT_PER_RESP : if a signature is matched, create a running count per responder host to compare against developed thresholds to identify and exclude scan traffic.

23. SIG_SUMMARY : generate a summary of all matched signatures based on the unique hosts that triggered a signature match.

Additional options and signature-specific events can be created using the Zeek scripting framework. Furthermore, Lab 8 of this series will enumerate upon the aforementioned scan thresholds and how Zeek determines if a host is probing a network.

The figure above shows the variables that store signature-specific packet information accessed in the *ZeekSignatureFramework.zeek* file. These variables can be accessed to extract the stored information for notifications and warnings. Furthermore, each variable can be printed to the logging stream, following the Zeek log file format reviewed in previous labs. Each variable is explained by its proceeding comments, denoted by the # character.

21 Log File Analysis Using Zeek Signatures

With Zeek's signature framework, we can create specific pattern-based signature filters to be applied during packet capture analysis. This section shows example signatures and their usage for network analysis.

21.1 Starting a New Instance of Zeek

Step 1. From the top of the screen, click on the *Client* button as shown below to enter the *Client* machine (Fig. 158).

Step 2. The *Client* machine will now open, and the desktop will be displayed (Fig. 159). On the left side of the screen, click on the Terminal icon as shown below.

Step 3. Start Zeek by entering the following command on the terminal (Fig. 160). This command enters Zeek's default installation directory and invokes Zeekctl tool to start a new instance. To type capital letters, it is recommended to hold the Shift

Fig. 159 Opening the
Terminal

Fig. 160 Starting Zeek

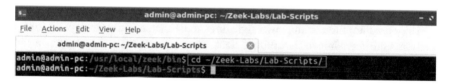

Fig. 161 Navigating into *Zeek-Labs/Lab- Scripts/* directory

key while typing rather than using the Caps key. When prompted for a password,
type password and hit Enter .

```
cd $ZEEK_INSTALL/bin && sudo ./zeekctl start
```

A new instance of Zeek is now active, and we are ready to proceed to the next
section of the lab.

21.2 Viewing a Premade Zeek Signature File

Step 1. Navigate to the *Lab-Scripts* directory (Fig. 161).

```
cd ~/Zeek-Labs/Lab-Scripts/
```

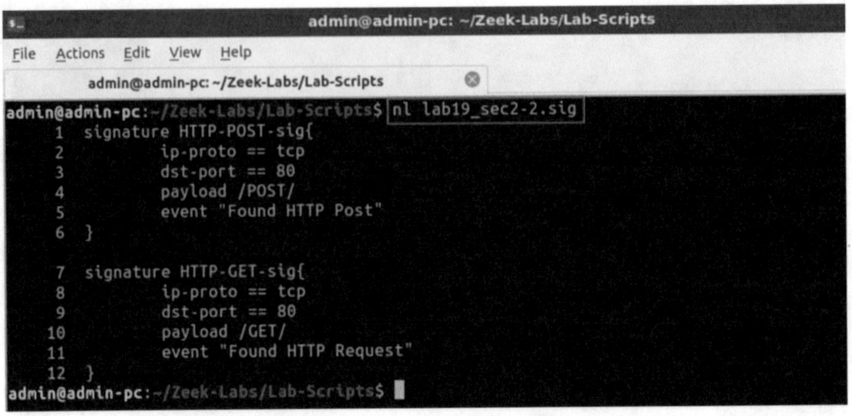

Fig. 162 Displaying the contents of the script *lab19_sec2-2.zeek*

Step 2. Display the contents of the *lab19_sec2-2.sig* file using $\boxed{\text{nl}}$ (Fig. 162).

```
nl lab19_sec2-2.sig
```

This signature file contains two signatures to be matched during network traffic analysis and is explained as follows. Each number represents the respective line number:

1. This line defines a new *signature* object, with the name *HTTP-POST-sig*.
2. Defines the desired match's transport protocol to be TCP.
3. Defines the desired match's destination port to be 80.
4. Defines the desired match's payload to contain the regular expression equivalent to "POST."
5. Defines an event if the match is found. Currently, the event will post a "Found HTTP Post" message.
7. This line defines a new *signature* object, with the name *HTTP-GET-sig*.
8. Defines the desired match's transport protocol to be TCP.
9. Defines the desired match's destination port to be 80.
10. Defines the desired match's payload to contain the regular expression equivalent to "GET."
11. Defines an event if the match is found. Currently, the event will post a "Found HTTP Request" message.

21.3   Executing the Premade Zeek Signature File

Step 1. Navigate to the *TCP-Traffic* directory (Fig. 163).

Fig. 163 Navigating into *../TCP-Traffic/* directory

Fig. 164 Processing a packet capture file using Zeek

Fig. 165 Listing the contents of the current directory

```
cd ../TCP-Traffic/
```

Step 2. Process the *smallFlows.pcap* packet capture file using the signature file *lab19_sec2-2.sig* (Fig. 164). It is possible to use the | tab | key to autocomplete the longer paths.

```
zeek -r ../Sample-PCAP/smallFlows.pcap -s ../Lab-
Scripts/lab19_sec2-2.sig
```

Step 3. List the generated log files in the current directory (Fig. 165).

```
ls
```

A new log file that has not been previously introduced is now displayed: *signatures.log*. This log file will contain all signature matches and their corresponding events and notices.

Step 4. View the contents of the *signatures.log* file using the | featherpad | text editor (Figs. 166 and 167).

Fig. 166 Opening the *signatures.log* file

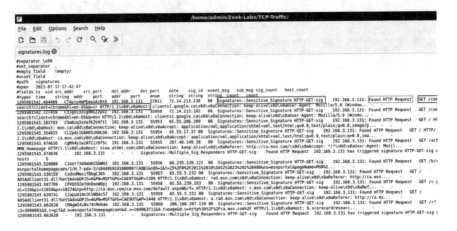

Fig. 167 The *signatures.log* file

```
featherpad signatures.log
```

The file is explained as follows:

- The red box indicates the name of the signature that was matched.
- The orange box indicates the event or message that was included when defining the signature.
- The blue box indicates the packet payload that was matched against the input signatures.

Step 5. Click the ☒ mark to close the │ featherpad │ window. Clear the contents of the TCP-Traffic directory (Fig. 168).

```
./../Lab-Scripts/lab_clean.sh
```

admin@admin-pc: ~/Zeek-Labs/TCP-Traffic
File Actions Edit View Help
admin@admin-pc: ~/Zeek-Labs/TCP-Traffic
admin@admin-pc:~/Zeek-Labs/TCP-Traffic$ ./../Lab-Scripts/lab_clean.sh
admin@admin-pc:~/Zeek-Labs/TCP-Traffic$

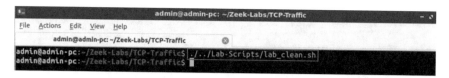

Fig. 168 Executing the script *./../Lab-Scripts/lab_clean.sh*

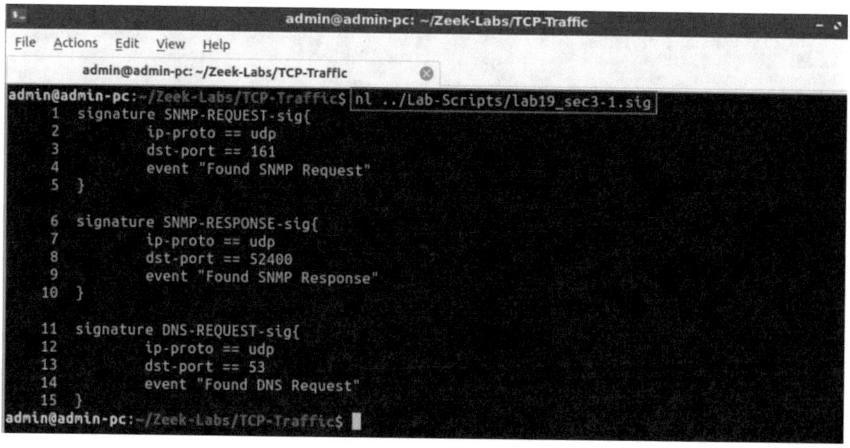

admin@admin-pc: ~/Zeek-Labs/TCP-Traffic
File Actions Edit View Help
admin@admin-pc: ~/Zeek-Labs/TCP-Traffic
admin@admin-pc:~/Zeek-Labs/TCP-Traffic$ nl ../Lab-Scripts/lab19_sec3-1.sig
1 signature SNMP-REQUEST-sig{
2 ip-proto == udp
3 dst-port == 161
4 event "Found SNMP Request"
5 }
6 signature SNMP-RESPONSE-sig{
7 ip-proto == udp
8 dst-port == 52400
9 event "Found SNMP Response"
10 }
11 signature DNS-REQUEST-sig{
12 ip-proto == udp
13 dst-port == 53
14 event "Found DNS Request"
15 }
admin@admin-pc:~/Zeek-Labs/TCP-Traffic$

Fig. 169 Displaying the contents of the script *../Lab-Scripts/ lab19_ sec3-1.zeek*

22 Executing Zeek Signature Matching for Network Traffic Analysis

This section modifies the existing signature file to generate additional signature events and notices. We will be modifying the previous signatures from TCP-based HTTP messages to UDP-based SNMP and DNS messages.

22.1 Modifying the Premade Zeek Signature File

Step 1. View the contents of the *lab19_sec3-1.sig* file using nl (Fig. 169).

```
nl ../Lab-Scripts/lab19_sec3-1.sig
```

Step 2. Open the *lab19_sec3-1.sig* file with the featherpad text editor (Fig. 170).

```
featherpad ../Lab-Scripts/lab19_sec3-1.sig
```

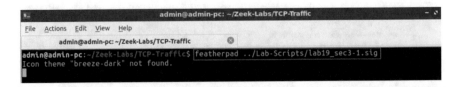

Fig. 170 Opening the *../Lab-Scripts/ lab19_sec3-1.sig* file

```
                                                           /home/admin/Zeek-Labs/Lab-Scripts/
 File   Edit   Options   Search   Help

 lab19_sec3-1.sig ⊗

signature SNMP-REQUEST-sig{
    ip-proto == udp
    dst-port == 161
    event "Found SNMP Request"
}

signature SNMP-RESPONSE-sig{
    ip-proto == udp
    dst-port == 52400
    event "Found SNMP Response"
}

signature DNS-REQUEST-sig{
    ip-proto == udp
    dst-port == 53
    event "Found DNS Request"
}
```

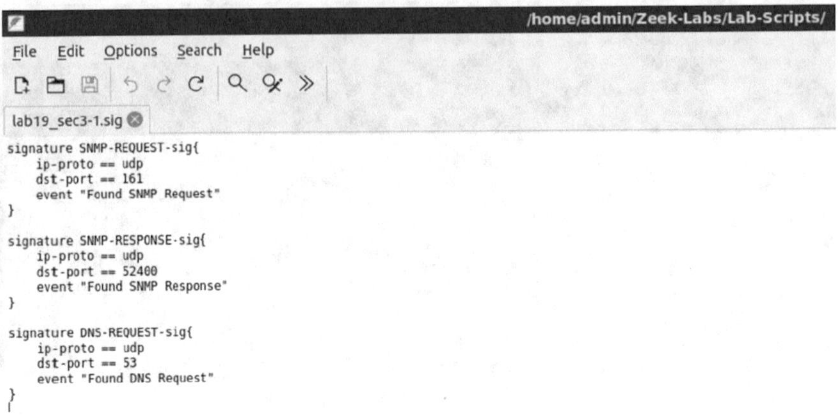

Fig. 171 The *../Lab-Scripts/ lab19_sec3-1.sig* file

Step 3. Update the *lab19_sec3-1.sig* file to include the following signatures. Then, close out the featherpad once finish editing (Fig. 171).

```
signature SNMP-REQUEST-sig{
        ip-proto == udp
        dst-port == 161
        event "Found SNMP Request"
}
signature SNMP-RESPONSE-sig{
        ip-proto == udp
        dst-port == 52400
        event "Found SNMP Response"
}
signature DNS-REQUEST-sig{
        ip-proto == udp
        dst-port == 53
        event "Found DNS Request"
}
```

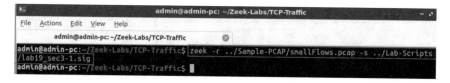

Fig. 172 Processing a packet capture file using Zeek

Fig. 173 Listing the contents of the current directory

22.2 Executing the Updated Zeek Signature File

Step 1. Process the *smallFlows.pcap* packet capture file using the signature file *lab19_sec3-1.sig* (Fig. 172). It is possible to use the tab key to autocomplete the longer paths.

```
zeek -r ../Sample-PCAP/smallFlows.pcap -s ../Lab-
Scripts/lab19_sec3-1.sig
```

Step 2. List the generated log files in the current directory (Fig. 173).

```
ls
```

The *signatures.log* file has been recreated and will contain the newly updated signature matches.

Step 3. View the contents of the *signatures.log* file using the featherpad text editor (Figs. 174 and 175). Then, close out the featherpad once finish examining the new file content.

```
featherpad signatures.log
```

The file is explained as follows:

- The red box indicates the DNS-REQUEST-sig signature match as well as the triggered IP address and event message.
- The orange box indicates the SNMP-REQUEST-sig signature match as well as the triggered IP address and event message.

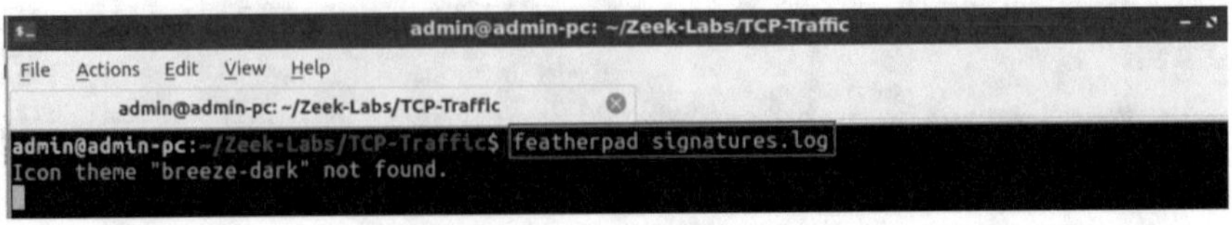

Fig. 174 Opening the *signatures.log* file

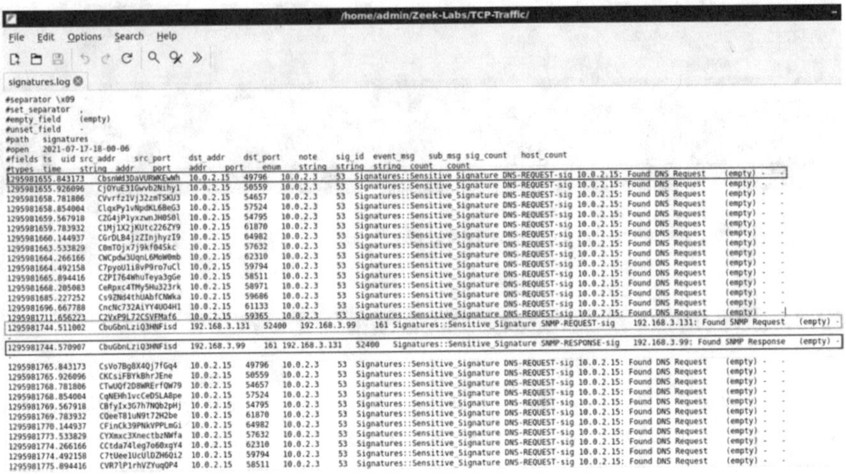

Fig. 175 The *signatures.log* file

Fig. 176 Stopping Zeek

- The blue box indicates the SNMP-RESPONSE-sig signature match as well as the triggered IP address and event message.

22.3   Closing the Current Instance of Zeek

After you have finished the lab, it is necessary to terminate the currently active instance of Zeek. Shutting down a computer while an active instance persists will cause Zeek to shut down improperly and may cause errors in future instances.

Step 1. Stop Zeek by entering the following command on the terminal (Fig. 176). If required, type ⎢password⎢ as the password. If the Terminal session has not been terminated or closed, you may not be prompted to enter a password. To type capital letters, it is recommended to hold the ⎢Shift⎢ key while typing rather than using the ⎢Caps⎢ key.

```
cd $ZEEK_INSTALL/bin && sudo ./zeekctl stop
```

References

1. K. Johnson, T. DeLaGrange, SANS survey on mobility/BYOD security policies and practices, SANS whitepaper, Oct. 2012. [Online]. Available: https://www.sans.org/reading-room/whitepapers/analyst/survey-mobility-byod-security-policies-practices-35175
2. Brown university (firewall) example. [Online]. Available: https://fasterdata.es.net/performance-testing/perfsonar/perfsonar-success-stories/brown-university-example/
3. Cisco nexus 3100 platform switch architecture, Cisco Systems White Paper, Oct. 2013. [Online]. Available: https://people.ucsc.edu/$\sim$warner/Bufs/cisco-3100-arch.pdf
4. E. Dart, L. Rotman, B. Tierney, M. Hester, J. Zurawski, The science DMZ: a network design pattern for data-intensive science, in *Proceedings of the International Conference on High Performance Computing, Networking, Storage and Analysis* (2013)
5. Cisco firepower NGIPS, Cisco Systems Data Sheet. [Online]. Available: https://www.cisco.com/c/en/us/products/collateral/security/ngips/datasheet-c78-738196.html
6. Snort open source intrusion prevention system. [Online]. Available: https://www.snort.org/
7. Processing of single stream large session (elephant flow) by the firepower services, Cisco Systems White Paper, Jan. 2017. [Online]. Available: https://www.cisco.com/c/en/us/support/docs/security/firepower-management-center/200420-Processing-of-Single-Stream-Large-Sessio.pdf
8. A. Sperotto, G. Schaffrath, R. Sadre, C. Morariu, A. Pras, B. Stiller, An overview of IP flow-based intrusion detection. IEEE Commun. Surv. Tutorials **12**(3), 343–356 (2010)
9. R. Hofstede, P. Celeda, B. Trammell, I. Drago, R. Sadre, A. Sperotto, A. Pras, Flow monitoring explained: from packet capture to data analysis with NetFlow and IPFIX. IEEE Commun. Tutorials **16**(4) (2014)
10. The bro network security monitor. [Online]. Available: http://www.broids.org
11. V. Stoffer, A. Sharma, J. Krous, 100G intrusion detection, Lawrence Berkeley National Laboratory Technical Report, Aug. 2015. [Online]. Available: https://www.cspi.com/wp-content/uploads/2016/09/Berkeley-100GIntrusionDetection.pdf
12. A. Gonzalez, J. Leigh, S. Peisert, B. Tierney, A. Lee, J. Schopf, Monitoring big data transfers over international research network connections, in *Proceedings of the IEEE International Congress on Big Data* (2017)
13. K. Miller, DDOS mitigation with sFlow. [Online]. Available: http://www.rn.psu.edu/2014/07/25/ddos-mitigation-with-sflow/
14. W. Kumari, D. McPherson, Remote triggered black hole filtering with unicast reverse path forwarding, Internet Request for Comments, RFC Editor, RFC 5635, Aug. 2009. [Online]. Available: https://tools.ietf.org/html/rfc5635
15. Y. Rekhter, T. Li, S. Hares, Border gateway protocol 4, Internet Request for Comments, RFC Editor, RFC 4271, Jan. 2006. [Online]. Available: https://tools.ietf.org/html/rfc4271
16. S. Peisert, W. Barnett, E. Dart, J. Cuff, R. Grossman, E. Balas, A. Berman, A. Shankar, B. Tierney, The medical science DMZ. J. Am. Med. Inform. Assoc. **23**(6), 1199–1201 (2016)

17. N. Pho, D. Magri, F. Redigolo, B. Kim, T. Feeney, H. Morgan, C. Patel, C. Botkaand T. Carvalho, Data transfer in a science DMZ using SDN with applications for precision medicine in cloud and high-performance computing, in *Proceedings of the International Conference for High Performance Computing, Networking, Storage and Analysis (SC15)* (2015)
18. K. Chard, S. Tuecke, I. Foster, Globus: recent enhancements and future plans, in *Proceedings of the XSEDE16 Conference on Diversity, Big Data, and Science at Scale* (2016)
19. W. Allcock, J. Bresnahan, R. Kettimuthu, M. Link, The globus striped GridFTP framework and server, in *Proceedings of the 2005 ACM/IEEE Conference on Supercomputing* (2005)
20. G. Vardoyan, R. Kettimuthu, M. Link, S. Tuecke, Characterizing throughput bottlenecks for secure GridFTP transfers, in *IEEE International Conference on Computing, Networking and Communications (ICNC)* (2013)
21. D. Hardt, The OAuth 2.0 authorization framework, Internet Request for Comments, RFC 6749, Oct. 2012. [Online]. Available: https://tools.ietf.org/html/rfc6749
22. S. Kent, S. Seo, Security architecture for the internet protocol, Internet Request for Comments, RFC Editor, RFC 4301, Dec. 2005. [Online]. Available: https://tools.ietf.org/html/rfc4301

Challenges and Open Research Issues

Owing to its proven efficiency to move large data sets, the number of deployed Science DMZs has been rapidly increasing in the last few years. However, there are still many challenges and open research issues that must be addressed.

1 Connectivity to the WAN

1.1 Cyberinfrastructure

The deployment cost of high-speed connections is still an unresolved problem in developing countries and many areas of developed countries. In the U.S., this is observed in areas such as remote Native lands, where there is a lack of cyberinfrastructure for WAN connectivity at Gbps rates. The deployment of fiber connections and access to POPs from such remote locations have prohibitive costs. As an example, in 2010, the U.S. Federal Communications Commission (FCC) released the National Broadband Plan, an effort to narrow the digital divide between urban and rural areas. Some key problems that must be addressed include:

- Terrain. Historically, service providers have dismissed the prospect of installing cables in areas located in remote, mountainous regions with extreme variations in elevation. The process of digging and laying underground fiber in these terrains is arduous, time-consuming, and expensive. Typically these areas are also far away from regional networks, exchange points, RENs and Internet2.
- Regulations pose a unique set of challenges. Many developing countries have only recently opened up to market forces, from a totally centralized scheme. Similarly, sovereign tribal nations in the U.S. require telecommunications providers to meet certain criteria to protect the land and culture, and most carriers are not interested in complying with additional rules on top of the regular bureaucracy. In developing countries and tribal nations, if a service provider is interested

J. Crichigno et al., *High-Speed Networks*, Practical Networking, https://doi.org/10.1007/978-3-030-88841-1_7

in laying fiber, it could see significant hurdles and consultations with local government. The provider would also have to perform environmental protection and historic preservation studies [1, 2].

- Cost. In rural areas with a limited potential subscriber base, RENs and service providers in particular see no possibility for any return on investment.

1.2   Connection-Oriented Networks

Since the early days of the Internet, there have been two camps regarding the service provided by the network layer: connectionless and connection-oriented. The first one has adopted the end-to-end argument [3] that shaped the Internet. However, many parts of the Internet and RENs are evolving to connection-oriented services as QoS becomes more important. For large data transfers and as a means to connect Science DMZs, a connection-oriented service provides bandwidth guarantee, a key advantage. An example of a connection-oriented service is the On-demand Secure Circuits and Reservation System (OSCARS) [4, 5], which connects WAN layer-2 circuits directly to the DTNs and provides bandwidth reservation and traffic engineering capabilities. Similarly, MPLS is used by large service providers, RENs, and Internet2 to provide QoS and establish long-term connections. Other connection-oriented schemes for bandwidth and delay guarantees have been also proposed [6, 7]. The adoption of connection-oriented technology to connect Science DMZs is expected to continue. Moreover, the wider adoption of this paradigm may encourage further development of upper-layer protocols. For example, given a guaranteed bandwidth, congestion control schemes based on pacing would be simpler to implement than current TCP congestion control schemes.

2   Data-Link and Network Layer Devices

2.1   Features for Large Flows

Typically, datacenter devices are designed for low-latency networks. These devices have a small amount of memory for buffering and use cut-through and fabric designs that are only suitable for small flows. Additionally, even when a device has sufficient memory to accommodate large flows, default configurations result in buffer underutilization. Since the technical expertise of cyberinfrastructure engineers mostly focuses on enterprise networks, suboptimal configurations are not uncommon. Fortunately, the market has recently noticed the need for Science DMZ-capable devices. Hence, many manufacturers such as Cisco [8], Brocade [9], and Ciena [10] are now providing features amenable for large flows, such as adequate buffer allocation and application-programming interfaces to automate processes and enforce preset policies (e.g., bypassing a firewall according to traffic type or trust level).

2.2 *Maximum Transmission Unit*

The maximum segment size has notable performance impact in high-throughput, high-latency networks, in particular under random-loss regimes. Unfortunately, supporting end-to-end jumbo frames is still an open challenge. Foremost, all hosts in a single broadcast domain must use the same MTU, and this can be difficult and error-prone. Additionally, Ethernet has no mechanism of detecting an MTU mismatch. A device that receives a frame larger than its MTU simply drops it silently. Secondly, since different administrative domains (ISPs, RENs) are independently operated, packets are routed through devices that either do not support jumbo frames or at best have different MTUs. Hence, there is a need to establish a standard for jumbo frames, so there is a reasonable guarantee that if vendors comply with the specifications, then there would be no interoperability problems.

3 TCP Optimization

3.1 *Congestion Control*

$$\text{buffer size} = C \cdot RTT. \tag{1}$$

Most TCP algorithms for congestion control use packet loss as a signal of congestion. According to Eq. (1), in order to achieve a throughput of 10 Gbps, TCP can only tolerate one segment loss for every 6,944,000,000 segments, which is incredibly small. The use of alternative congestion control mechanisms where packet loss is not a signal of congestion is a promising direction. The recently proposed BBR algorithm [11] has shown preliminary throughput improvement in medium- and high-loss packet regimes. Note that using TCP pacing to adjust the bit rate at an estimated bottleneck bandwidth is a departure from the traditional window-based congestion control mechanism. Additionally, since rate-based congestion control does not require constant congestion window updates, this approach avoids the long delays inherent in the receiver sending the congestion window. Moreover, the promising performance results of BBR may lead to the development of other congestion control algorithms. The use of parameters for detecting congestion and random losses that have stronger correlation to congestion than packet losses also needs to be explored.

3.2 Pacing

TCP FQ pacing has shown promising results in long fat networks. However, the main concern with this technique is finding the bottleneck link along the path between the end devices. Once the bottleneck link is identified, pacing packets at the bottleneck link's capacity mitigates the TCP sawtooth behavior and produces stable throughput. Pacing can also be easier in connection-oriented networks, as packets can be paced at the guaranteed bandwidth allocated to the connection.

3.3 TCP Extensions

Many TCP extensions have been proposed over the years, including selective acknowledgement, timestamp, window scale, and RTT measurement [12]. As most of these extensions were targeted to mitigate issues observed in the Internet's best-effort service model, they may not be suitable for large data transfers over well-conditioned networks such as Internet2 and other RENs. Hence, investigating the use of TCP extensions in Science DMZ environments is required.

4 Optimization in the Protocol Stack

As routers and switches are optimized for Science DMZs [8, 10], the protocol stack at DTNs may become the bottleneck for many implementations. Reducing DTN processing overheads is desirable to increase throughput.

Software techniques can help optimize the TCP performance on 10 Gbps WANs and above. However, optimizing a DTN to operate at 100 Gbps is currently a persistent challenge. Most TCP implementations have a considerable overhead and produce a very high CPU utilization, which raises questions about the viability of TCP as the network bandwidth continues to grow [13]. UDP-based tools such as Aspera FAST [14] and UDT [15] may suffer a performance penalty due to context switching and the process of copying data to user-space buffers. Kissel et al. [13] have recently proposed a new protocol called wide-area Remote Direct Memory Access (RDMA). RDMA decreases TCP processing overheads by using optimization techniques such as zero-copy and splice. Zero-copy is a procedure that relieves the CPU of copying data from one memory area to another (e.g., from lower-level layers to the TCP buffer). This technique saves CPU cycles and memory bandwidth when transmitting a file over a WAN. Similarly, splice is a system call used to move data between two file descriptors. Splice minimizes the movement of data between kernel space and user space.

Overall, zero-copy and splice are two techniques that can minimize the movement of data within a DTN. Similar cross-layer optimization techniques can further reduce processing overheads. For example, TCP and IP are usually implemented together, so that there is no need to copy the layer-3 payload when moving it from the network process to the transport process. This idea can be extended to the upper layers, i.e., from transport to application.

5 Applications

5.1 Data Transfer Tools

As the main applications used in Science DMZs, data transfer tools must be designed for high-throughput, high-latency networks. Namely, these tools should implement features such as parallel streams, large buffer sizes, and partial and restartable file transfers. At present, engineers rely on rule of thumbs to configure many of these features. For example, there is no formal solution to the problem of selecting the number of parallel TCP streams that should be open for a data transfer. Globus suggests that the number of streams should be between 2 and 8. Moreover, the optimal value may depend on the RTT, bandwidth, congestion control algorithm, etc.

Data transfer tools should minimize the time spent in input/output operations (which are expensive) and exploit the multicore capability of modern DTNs. For example, FDT [16] uses independent threads to read and write on physical devices in parallel. Data transfer tools should also avoid copying data multiple times within the DTN. Improvements may involve several layers, including transport and application.

The adoption of UDP-based data transfer applications has been minimal. Tests conducted in 10 Gbps networks indicate that the throughput is limited by the high CPU utilization [17]. Also, current UDP-based applications do not use parallel streams. Instead, they only open one stream per data transfer. Typically, the stream's process is tied to one core while other cores are idle. With this approach, UDP-based applications may only achieve higher rates by increasing the CPU's clock rate. However, increasing the CPU rate is a challenge. Instead, during the last decade, the throughput has been increased by using multicore CPUs. Thus, an open research issue includes the use of UDP-based applications using multiple streams, in particular when parallelism opportunities exist [4, 6].

5.2 Monitoring Applications

The effectiveness of perfSONAR in measuring end-to-end metrics and in detecting soft failures relies on its deployment across multiple domains [18]. While perfSONAR has been extensively deployed on RENs (e.g., ESnet [19], Internet2

[20], GEANT [21], CESNET [22], etc.), its deployment by ISPs is still lacking. A contributing factor here is the lack of familiarity of engineers who are more familiar with single-domain tools used in enterprise networks, such as SNMP, Syslog, and Netflow. Thus, there is a need to outreach to the networking community to widen the adoption of collaborative multi-domain tools without compromising the privacy and commercial interests of ISPs.

Integrating and correlating data collected from different applications is an immediate research direction. For example, SNMP and perfSONAR complement each other. The former can detect intra-domain hard failures while the latter can detect inter-domain soft failures. In this context, Gonzales et al. [23] describe a monitoring application integrating perfSONAR, SNMP, and other tools. The proposed platform also integrates data visualization and analytics modules. With the advent of SDN, this type of integration and the addition of network programmability are expected to continue.

5.3 Virtualization

The research community has been reluctant to adopt virtual components into Science DMZs, mainly because of the performance degradation of virtual DTNs. However, in small institutions where resources are often limited, using virtual DTNs is a cost-efficient alternative. Preliminary results suggest that virtual DTNs may be adequate for 10 Gbps Science DMZs, provided the physical server they run on has sufficient CPU capacity and the workload is minimal. However, when packet losses occur and DTNs require more processing capability for handling retransmissions, the performance degradation can be significant. Additionally, virtual components are unable to perform at 40/100 Gbps. Thus, research on minimizing processing overheads on virtual devices (virtual switch, virtual NIC, hypervisor) is still required.

6 Security

In general, not having web, email, and other general-purpose applications running on DTNs mitigates the delivery of malicious payloads via XML, SQL, cross-site injection, and other methods. However, since transfer rates are high, the data inspection in Science DMZs may be minimal. For example, the typical inspection rate of a payload-based IDS protecting a 100 Gbps Science DMZ connected to ESnet is between 2 and 4 Gbps [24], which is less than 5% of the total network input. While the reported number of malware attacks in current Science DMZs has been minimal, there is a trade-off between performance and security that should be carefully analyzed when deploying this type of IDS, in particular for 40/100 Gbps Science DMZs. A specific approach that can be explored for high rates may combine both flow-based and payload-based IDSs. A first layer of detection may preselect suspicious flows using a flow-based IDS, while a second layer may scan packets of the preselected flows using a payload-based IDS.

Confidentiality, integrity, and authentication are usually implemented at the application layer. Although current encryption algorithms are capable of performing at or near 10 Gbps, Globus' file integrity checks may introduce a penalty of up to 10%. Encryption rates of 40 and 100 Gbps are still uncommon in DTN deployments. However, recent development of specialized hardware indicates that a rate of 100 Gbps is achievable for in-transit encryption [25]. The use of medical Science DMZs [26, 27] and the need to comply with regulations [28, 29] are expected to accelerate these developments. Finally, preventing DoS and scanning attacks is also an ongoing research direction, as these attack types are continuously evolving.

References

1. J. Tveten, On American Indian reservations, challenges perpetuate the digital divide, ARS Technica, Jan. (2016). [Online]. Available: https://arstechnica.com/information-technology/2016/01/on-american-indian-reservations-challenges-perpetuate-the-digital-divide/
2. N. Sambuli, Challenges and opportunities for advancing Internet access in developing countries while upholding net neutrality. J. Cyber Policy 1(1), 61 (2016)
3. J. Saltzer, D. Reed, D. Clark, End-to-end argument in system design. ACM Trans. Comput. Syst. 2(4), 277–288 (1984)
4. J. Plante, D. Davis, V. Vokkarane, Parallel circuit provisioning in esnet's OSCARS, in *IEEE International Conference on Advanced Networks and Telecommunications Systems (ANTS)* (2014)
5. I. Monga, C. Guok, W.E. Johnston, B. Tierney, Hybrid networks: lessons learned and future challenges based on ESnet4 experience. IEEE Commun. Mag. 49(5), 114 (2011)
6. T. Orawiwattanakul, H. Otsuki, E. Kawai, S. Shimojo, Multiple classes of service provisioning with bandwidth and delay guarantees in dynamic circuit network, in *IEEE International Symposium on Integrated Network Management, May* (2015)
7. A. Gumaste, T. Das, K. Khandwala, I. Monga, Network hardware virtualization for application provisioning in core networks. IEEE Commun. Mag. 55(2), 152 (2017)
8. Event-based software-defined networking: build a secure science DMZ, Cisco Systems White Paper, 2015. [Online]. Available: https://www.cisco.com/c/en/us/products/collateral/cloud-systems-management/open-sdn-controller/white-paper-c11-735868.html
9. Software-driven science DMZ networks, Brocade White Paper, 2016. [Online]. Available: https://www.brocade.com/content/dam/common/documents/content-types/solution-brief/brocade-software-driven-science-dmz-networks-sb.pdf
10. Transform large-scale science collaboration, Ciena White Paper. [Online]. Available: http://media.ciena.com/documents/Science+DMZ+AN.pdf
11. N. Cardwell, Y. Cheng, C. Gunn, S. Yeganeh, V. Jacobson, BBR: congestion-based congestion control. Commun. ACM 60(2), 58–66 (2017)
12. D. Borman, B. Braden, V. Jacobson, R. Scheffenegger, TCP extensions for high performance, Internet Request for Comments, RFC 7323, Sep. 2014. [Online]. Available https://tools.ietf.org/html/rfc7323#section-4.2
13. E. Kissel, M. Swany, B. Tierney, E. Pouyoul, Efficient wide area data transfer protocols for 100 Gbps networks and beyond, in *Proceedings of the Third International Workshop on Network-Aware Data Management* (2013)

14. Ultra high-speed transport technology, Aspera White Paper. [Online]. Available: http://asperasoft.com/resources/white-papers/ultra-high-speed-transport-technology/
15. Y. Gu, R. Grossman, Udt: UDP-based data transfer for high-speed wide area networks. Comput. Netw. **51**(7), 1777–1799 (2007)
16. Fast data transfer (FDT). [Online]. Available: http://monalisa.cern.ch/FDT
17. UDP tuning in science DMZs. [Online]. Available: https://fasterdata.es.net/network-tuning/udp-tuning/#toc-anchor-1
18. J. Zurawski, S. Balasubramanian, A. Brown, E. Kissel, A. Lake, M. Swany, B. Tierney, M. Zekauskas, perfSONAR: on-board diagnostics for big data, in *Workshop on Big Data and Science: Infrastructure and Services* (2013)
19. The energy science network. [Online]. Available: https://www.es.net
20. Internet2. [Online]. Available: https://www.internet2.edu/
21. F. Farina, P. Szegedi, J. Sobieski, GEANT world testbed facility: federated and distributed testbeds as a service facility of GEANT, in *International Tele-traffic Congress* (2014)
22. K. Slavicek, V. Novak, J. Ledvinka, CESNET fiber optics transport network, in *IEEE International Conference on Networks* (2009)
23. A. Gonzalez, J. Leigh, S. Peisert, B. Tierney, A. Lee, J. Schopf, Monitoring big data transfers over international research network connections, in *Proceedings of the IEEE International Congress on Big Data* (2017)
24. V. Stoffer, A. Sharma, J. Krous, 100G intrusion detection, Lawrence Berkeley National Laboratory Technical Report, Aug. 2015. [Online]. Available: https://www.cspi.com/wp-content/uploads/2016/09/Berkeley-100GIntrusionDetection.pdf
25. Transpacific encryption success at 100Gbps, Ericsson Press Releases, Sep. 2017. [Online]. Available: https://www.ericsson.com/en/press-releases/2017/9/transpacific-encryption-success-at-100gbps
26. S. Peisert, E. Dart, W. Barnett, J. Cuff, R. Grossman, E. Balas, A. Berman, A. Shankar, B. Tierney, The medical science DMZ: a network design pattern for data-intensive medical science. J. Am. Med. Inform. Assoc. (JAMIA), Oct. 2017. [Online]. Available: https://academic.oup.com/jamia/article/doi/10.1093/jamia/ocx104/4367749/The-medical-science-DMZ-a-network-design-pattern
27. S. Peisert, W. Barnett, E. Dart, J. Cuff, R. Grossman, E. Balas, A. Berman, A. Shankar, B. Tierney, The medical science DMZ. J. Am. Med. Inform. Assoc. **23**(6), 1199–1201 (2016)
28. W. Lee, C. Lee, A cryptographic key management solution for HIPAA privacy/security regulations. IEEE Trans. Inform. Technol. Biomed. **12**(1), 34–41 (2008)
29. M. Alyami, Y. Song, Removing barriers in using personal health record systems, in *IEEE International Conference on Computer and Information Science (ICIS)* (2016)